# Plant Tissue Culture Concepts *and* Laboratory Exercises

# Plant Tissue Culture Concepts *and* Laboratory Exercises

*Edited by*
Robert N. Trigiano
Dennis J. Gray

**CRC Press**
**Boca Raton    New York    London    Tokyo**

**COVER PHOTOGRAPH:** Somatic embryos of orchardgrass (see chapter seventeen) developing directly from a cultured leaf segment. Courtesy of D. J. Gray, B. V. Conger and R. N. Trigiano (unpublished).

**Library of Congress Cataloging-in-Publication Data**

Plant tissue culture concepts and laboratory exercises / edited by
   Robert N. Trigiano and Dennis J. Gray.
      p.   cm.
   Includes bibliographical references and index.
   ISBN 0-8493-9409-0 (alk. paper)
   1. Plant tissue culture--Laboratory manuals.   I. Trigiano, R. N.
(Robert Nicholas), 1953-   .   II. Gray, Dennis J. (John Gray), 1953-
 .
   QK725.P5795   1996
   581′.0724--dc20                                                                        95-47400
                                                                                              CIP

# Preface

*Plant Tissue Culture Concepts and Laboratory Exercises* furnishes instructors and students alike with a broad consideration of the field, providing historical perspectives, discussing state-of-the-art techniques and methodologies and looking forward to future advances and applications. It is meant to serve as a primary text for both introductory and advanced courses on the topic. It also presents many useful protocols and procedures and, thus, should serve as a valuable reference. The book is divided into five sections, beginning with a brief History of Plant Tissue Culture, followed by Supporting Methodologies, Propagation Techniques, Crop Improvement Techniques, and concluding with chapters on Special Applications. The format of the book is rather informal — it provides the reader with a minimum number of references, but does not sacrifice essential information or accuracy. Broad topic chapters are authored by specialists with considerable experience in the field and are supported by one or more laboratory exercises illustrating central concepts of the topic. Collectively, the laboratory exercises are exceptionally diverse in nature, providing something for everyone — from beginning to advanced students. Importantly, the authors have successfully completed the exercises many times, often with either tissue culture classes or in their own research laboratories.

The laboratory projects are executed with many different agronomic and horticulturally important plants, including chrysanthemum, orchardgrass, syngonium, ajuga, petunia, watermelon, peanut, lilac and many others, all of which are either easy to grow or can be ordered from common sources for nominal fees or be purchased at the local market. This range of plants is substantially broader than that used in any previous text on the subject and should stimulate interest in students from many botanical and agricultural disciplines. A unique feature of the exercises is that the authors have provided in general terms what results should be expected from the experiments. At the end of each exercise, there is a series of questions designed to provoke individual thought and critical examination of the exercise and results.

We hope *Plant Tissue Culture Concepts and Laboratory Exercises* will facilitate training of students in the current principles and methodologies of our rapidly evolving field. We welcome comments from colleagues as they put the textbook to use.

**Robert N. Trigiano**
**Dennis J. Gray**

# Overview and scheduling of laboratory exercises

The book is arranged such that the "typical" or "common" supplies and media formulations for all of the laboratory exercises are listed in chapter two, General laboratory requirements, media and sterilization methods. However, unique supplies and media are given in each chapter when necessary. In particular, a "laboratory kit" containing specific supplies, such as scalpels and blades, is described in chapter two; those items are not listed in individual chapters. We recommend that each student or team of students be provided with a kit. As always, reference to specific products or suppliers does not constitute endorsement or criticism of similar ones not mentioned.

Although arranged by topic, several laboratory exercises from different topic areas use the same plant species. Therefore, it might be most efficient to conduct laboratories on a range of topics concurrently in order to efficiently utilize certain plant material. To assist in planning, the plant species are referenced to laboratory chapters in Table 1.

We encourage instructors to take the time to peruse the laboratory exercises well in advance of the anticipated laboratory starting date. Some experiments will require starting stock plants. Other experiments will depend on cultures started or obtained from outside sources weeks or months ahead of time. For example, most of the genetic transformation and in vitro plant pathology exercises require plant pathogenic bacteria, e.g., *Agrobacterium tumefaciens* and/or *Pseudomonas syringae*. *Note:* Regardless of the source of the pathogens (the American Type Culture Collection, a colleague at your institution etc.), be sure to acquire the proper documentation and/or permits for the cultures. Preliminary planning will save time and disappointment later on and will make for a better experience for students and instructors.

*Table 1* Correlation of Plants to Laboratory Exercises for Scheduling Purposes

| Plant | Chapter | Plant | Chapter |
|---|---|---|---|
| Ajuga | 36 | Peanut | 19 |
| Cabbage | 32 | Petunia | 15 |
| Carnation | 34 | Spruce | 22 |
| Carrot | 30 | Potato | 9, 24, 34 |
| Chrysanthemum | 12, 25, 34, 31 | Soybean | 37 |
| Cineraria | 20 | Syngonium | 8 |
| Lilac | 10 | Tobacco | 3, 24, 28, 30, 37 |
| Melon | 18, 32 | Torenia | 14 |
| Mint | 34 | Watermelon | 13 |
| Orchardgrass | 17, 25 | Yellow poplar | 21, 26 |

# *Editors*

**Robert N. Trigiano, Ph.D.,** is an Associate Professor of plant biotechnology in the Institute of Agriculture, Department of Ornamental Horticulture and Landscape Design at the University of Tennessee at Knoxville.

Dr. Trigiano received his B.S. degree with an emphasis in Biology and Chemistry from Juniata College, Huntingdon, Pennsylvania in 1975 and an M.S. in Biology (Mycology) from the Pennsylvania State University in 1977. He was an Associate Research Agronomist, mushroom culture and plant pathology, for Green Giant Co., Le Sueur, Minnesota until 1979 and then a Mushroom Grower for Rol-Land Farms, Ltd., Blenheim, Ontario, Canada during 1980. He completed a Ph.D. in Botany and Plant Pathology (co-majors) at North Carolina State University at Raleigh in 1983. After concluding postdoctoral work in the Plant and Soil Science Department at the University of Tennessee, he was appointed an Assistant Professor in the Department of Ornamental Horticulture and Landscape Design at the same university in 1987 and promoted to Associate Professor in 1991.

Dr. Trigiano is a member of the American Society for Horticultural Science, International Society of Plant Morphologists and the Mycological Society of America, and the honorary societies of Gamma Sigma Delta, Sigma Xi and Phi Kappa Phi. He has been an Associate Editor for the journals of the American Society for Horticultural Science, *Plant Cell, Tissue and Organ Culture* and *Plant Cell Reports.* He has received the T.J. Whatley Distinguished Young Scientist Award (The University of Tennessee, Institute of Agriculture, 1991) and the Gamma Sigma Delta Research Award of Merit (The University of Tennessee, 1991).

Dr. Trigiano has been the recipient of several research grants from the United States Department of Agriculture (USDA), Horticultural Research Institute and from private industries. He has published more than 80 research papers, book chapters and popular press articles. He teaches undergraduate/graduate courses in Plant Tissue Culture, Plant Disease Fungi and Electrophoresis of DNA and Proteins. Current research interests include somatic embryogenesis in ornamental species, fungal physiology and population analysis.

**Dennis J. Gray, Ph.D.,** is Professor of Horticulture–Developmental Biologist and a faculty member of the University of Florida's Institute of Food and Agricultural Sciences. He directs the cucurbit and grape biotechnology laboratory at the Central Florida Research and Education Center, Leesburg.

Dr. Gray graduated with a B.A. degree in Biology from California State College, Stanislaus in 1976 and received an M.S. degree in Mycology, with a minor in Botany, from Auburn University in 1979. He completed a Ph.D. degree in Botany, with a minor in Plant Pathology, at North Carolina State University in 1982. After a postdoctoral fellowship at the University of Tennessee in 1984, he joined the faculty at the University of Florida, reaching current rank in 1993.

Dr. Gray is a member of the American Association for the Advancement of Science, the American Institute of Biological Sciences, the American Society for Horticultural Science, the Botanical Society of America, the Council for Agricultural Science and Technology, the Florida State Horticultural Society, the Society for In Vitro Biology and the honorary society, Sigma Xi. He serves as a U.S. representative on the Commission Biotechnology of the International Society for Horticultural Science and is a member of the U.S. Advisory Council of the International Association of Plant Tissue Culture. He was an Associate Editor, then Managing Editor, of the internationally recognized refereed journal *Plant Cell Tissue and Organ Culture* from 1988 through 1994.

Dr. Gray has received research grants from the Binational Research and Development Fund, the Florida High Technology and Industry Council, the International Board for Plant Genetic Research, the United States Agency for International Development, the United States Department of Agriculture, and private industry. He has published more than 100 papers and holds one U.S. patent. His current interests include the initiation and ontogeny of regenerative plant cells and the application of contemporary and newly-emerging technologies to crop improvement.

# Contributors

**Stephen M. Attree**
Department of Biology
University of Saskatchewan
Saskatoon, Saskatchewan, Canada

**C. Jacyn Baker**
Microbiology and Plant Pathology
  Laboratory
United States Department of Agriculture
Beltsville, Maryland

**Charleen Baker**
Plant Science Department
Northern Plains Biostress Laboratory
South Dakota State University
Brookings, South Dakota

**Robert M. Beaty**
Department of Botany
The University of Tennessee
Knoxville, Tennessee

**Mark P. Bridgen**
Department of Plant Science
University of Connecticut
Storrs, Connecticut

**James D. Caponetti**
Department of Botany
The University of Tennessee
Knoxville, Tennessee

**Michael E. Compton**
School of Agriculture
University of Wisconsin-Platteville
Platteville, Wisconsin

**Bob V. Conger**
Department of Plant and Soil Science
Institute of Agriculture
The University of Tennessee
Knoxville, Tennessee

**Andrew J. Daun**
Knight Hollow Nursery, Inc.
Middleton, Wisconsin

**Larry C. Fowke**
Department of Biology
University of Saskatchewan
Saskatoon, Saskatchewan, Canada

**Victor P. Gaba**
Department of Virology
Institute of Plant Protection
The Volcani Center
Bet Dagan, Israel

**Effin T. Graham**
Department of Ornamental Horticulture
  and Landscape Design
Institute of Agriculture
The University of Tennessee
Knoxville, Tennessee

**Dennis J. Gray**
Central Florida Research and Education
  Center
Institute of Food and Agricultural Sciences
The University of Florida
Leesburg, Florida

Ernest Hiebert
Plant Pathology Department
The University of Florida
Gainesville, Florida

Michael E. Kane
Environmental Horticulture Department
The University of Florida
Gainesville, Florida

Michael J. Klopmeyer
Ball FloraPlant
West Chicago, Illinois

Chia-Min Lin
Department of Food Science and Human
  Nutrition
The University of Florida
Gainesville, Florida

Melina López-Meyer
Department of Biology
Texas A & M University
College Station, Texas

Ignacio E. Maldonado-Mendoza
Department of Biology
Texas A & M University
College Station, Texas

Kathleen R. Malueg
Department of Botany
The University of Tennessee
Knoxville, Tennessee

Roger A. May
Department of Horticulture
Michigan State University
East Lansing, Michigan

Deborah D. McCown
Knight Hollow Nursery, Inc.
Middleton, Wisconsin

Scott A. Merkle
Daniel B. Warnell School of Forest
  Resources
University of Georgia
Athens, Georgia

Norton M. Mock
Microbiology and Plant Pathology
  Laboratory
United States Department of Agriculture
Beltsville, Maryland

Craig L. Nessler
Department of Biology
Texas A & M University
College Station, Texas

Elizabeth W. Orlandi
Department of Botany
University of Maryland
College Park, Maryland

John E. Preece
Department of Plant and Soil Science
Southern Illinois University
Carbondale, Illinois

Sandra M. Reed
USDA/ARS
Nursery Crop Research Station
Tennessee State University
McMinnville, Tennessee

Patricia J. Rennie
Department of Biology
University of Saskatchewan
Saskatoon, Saskatchewan, Canada

James A. Saunders
USDA/ARS
Climate Stress Laboratory
Natural Resources Institute
Beltsville, Maryland

Otto J. Schwarz
Department of Botany
The University of Tennessee
Knoxville, Tennessee

Mary Catherine Scott
Department of Ornamental Horticulture
  and Landscape Design
Institute of Agriculture
The University of Tennessee
Knoxville, Tennessee

**Mary Ann L. Smith**
Department of Natural Resources and
  Environmental Sciences
University of Illinois
Urbana, Illinois

**Ellen G. Sutter**
Department of Pomology
University of California
Davis, California

**Gayle R. L. Suttle**
Microplant Nurseries, Inc.
Gervais, Oregon

**Leigh E. Towill**
USDA/ARS
National Seed Storage Laboratory
Fort Collins, Colorado

**Robert N. Trigiano**
Department of Ornamental Horticulture
  and Landscape Design
Institute of Agriculture
The University of Tennessee
Knoxville, Tennessee

**Richard E. Veilleux**
Department of Horticulture
Virginia Polytechnic Institute and State
  University
Blacksburg, Virginia

**Hazel Y. Wetzstein**
Department of Horticulture
University of Georgia
Athens, Georgia

# Acknowledgments

We would like to thank the contributing authors for their efforts and patience throughout this lengthy endeavor, and The University of Tennessee's Institute of Agriculture and The University of Florida's Institute for Food and Agricultural Sciences for providing financial support. We very much appreciate the efforts of Joy E. Webb (JEW) who was indispensable in producing the excellent computer drawings. We thank our wives, Kay Trigiano and Amy Gray, for assistance with proofreading and production and, of course, for their undying support. Finally, we are indebted to Andrew N. Trigiano and Bob D. Gray for providing constant inspiration and being welcome sources of diversions during the project.

# Contents

*Part I*

*History of plant tissue culture*

## chapter one

# History of plant tissue and cell culture

*James D. Caponetti, Dennis J. Gray and Robert N. Trigiano*

The field of in vitro plant culture is based on the premise that plants can be separated into their component parts (organs, tissues or cells), which then can be manipulated in vitro and, often, be grown back to complete plants. This idea of handling higher plants with the ease and convenience of microorganisms has conjured up many wonderful possibilities for their study; these possibilities, in turn, have been the long-standing stimuli driving research and development in the field.

It is generally thought that the first successful plant tissue and cell culture was accomplished by Gottlieb Haberlandt near the turn of the 20th century when he reported the culture of leaf mesophyll tissue and hair cells (see Steward, 1968; Krikorian and Berquam, 1969). This was a remarkable accomplishment considering that little was known about plant physiology at the time. In retrospect, however, Haberlandt must have drawn on a body of previous knowledge in plant biology. We must assume that he was familiar with the writings of early philosophers such as Aristotle, Theophrastus, Pliny the Elder, Dioscorides, Avicenna, Magnus, Angelicus and Goethe and that his studies surely must have included the anatomical observations of Hooke, Malpighi, Grew, Nageli and Hanstein. His own research must have led him to the investigations by early plant physiologists such as van Helmont, Mariotte, Hales, Priestly, Ingenhousz and Senebier, and he must have had access to the morphological and physiological investigations of 19th century botanical researchers such as Schleiden, Schimper, Pringsheim, Unger, Hedwig, Hofmeister, Vochting, Sachs, Goebel, Bower and Farlow plus knowledge of the inorganic "salts" solutions formulated by Sachs and Knop in the 1860s (Hoagland and Arnon, 1938). The information available from these sources coupled with improved light microscopes must have given Haberlandt the insight to culture plant cells and to predict that they could not only grow, but divide and develop into embryos and thence to whole plants, a scenario referred to as totipotency by Steward (1968).

Unfortunately, the cells that Haberlandt cultured did not divide, and thus, his ideas of plant development and totipotency did not come to fruition. The probable reason for the failure was that the plant growth regulators (PGRs) needed for cell division, proliferation and embryo induction were not known at the time. Apparently, Haberlandt became discouraged and pursued other physiological investigations. However, his ideas of embryo induction in culture did not go unnoticed in the scientific world. For example, Hannig (1904) cultured nearly mature embryos excised from seeds of several species of crucifers. Moreover, Haberlandt's lack of success was not in vain because one of his students, Kotte (1922), reported the growth of isolated root tips on a medium consisting primarily of inorganic salts. At the same time, and quite independently, Robbins (1922) reported a similar success with root tips and stem tips and White (1934) reported that, not only could cultured tomato root tips grow, but they could be repeatedly subcultured to fresh medium of inorganic salts supplemented with yeast extract, unbeknown to him, a good source of B vitamins.

Innovative plant tissue culture techniques progressed rapidly during the 1930s because of the discovery that B vitamins and natural auxin were necessary for the growth of isolated tissues containing meristems. The growth-promoting effects of thiamine on isolated tomato root tips were reported by White (1937). A series of ingenious experiments with oat seedlings by Fritz Went in the 1920s, along with the work of other plant physiologists, including Kenneth Thimann, led to the discovery of indoleacetic acid (IAA), a natural auxin. The events leading to the discovery of IAA are well documented in a book by Went and Thimann (1937). Duhamet (1939) reported the stimulation of growth of excised roots by IAA.

The avenue was now open for rapid progress in the successful culture of plant tissues during the 1930s. The culture of meristems other than root and shoot tips was explored. With improved culture media, La Rue (1936) achieved better success at culturing plant embryos compared with the efforts of Hannig 32 years earlier. Gautheret (1934) reported the successful culture of the cambium of several species of trees to produce callus on medium containing B vitamins and IAA. Further research with other meristematic tissues led Nobecourt (1937) and Gautheret (1939) to obtain callus from carrot root cambium and White (1939) to obtain tobacco tumor (crown gall) tissue. The rapid progress of plant tissue culture, however, was soon abruptly curtailed by the start of World War II in 1939.

Since one theater of World War II was concentrated in Europe, further progress in plant tissue culture by the European originators was almost impossible in the midst of the disruption and destruction and the shortages of laboratory supplies and equipment. Scientists in North America and other locations also suffered shortages of laboratory supplies, equipment and personnel. During the war years (1939–1945), nevertheless, some progress was made in plant tissue culture. Johannes Van Overbeek and his associates (1942) reported they were able to obtain seedlings from heart-shaped embryos by enriching culture media with coconut milk in addition to the usual salts, vitamins and other nutrients. Also by this time, Panchanan Maheshwari, his associates and students in India were very active in angiosperm embryology research that began before the war and progressed into the 1940s as described in his book (1950). Armin Braun (1943) reported on tumor induction related to crown gall disease. Experiments on tobacco tissues by Folke Skoog (1944) demonstrated organ formation in cultured tissues and organs. In Western Europe, some progress occurred in that Guy Camus (1943) was the first to report grafting experiments in tissue cultures and Georges Morel (1944) reported on developing techniques to culture parasitized plant tissues.

With the end of the war in 1945, a resumption of prewar tissue culture activity could not occur immediately. It took several years to reach a measure of recovery from a holocaust of such magnitude. Even North America required some time to overcome the shortages of equipment and supplies. Many adjustments had to be made in employment and educational opportunities for returning military personnel and others who worked for the war effort. But again, some progress in plant tissue culture technology occurred. Ernest Ball (1946) reported on greatly improving the potential for culturing shoot tips, beginning with those of nasturtium and lupine. Albert Hildebrandt and co-workers (1946) made improvements on the medium for the growth of tobacco and sunflower tissues, and Morel (1948) was well into applying tissue culture techniques to the study of parasites associated with plant tissues. Herbert Street and his colleagues (1949, 1950) began a series of extensive studies on the nutrition of excised tomato root tips.

After 1950, rapid progress was made in plant tissue culture techniques. Also, many advancements in the knowledge of plant development were accomplished, especially in the area of the effects of PGRs. While intensive studies on the in vivo and in vitro effects of auxins continued, other classes of PGRs were now recognized. Skoog and his associates (Skoog and Tsui, 1951) continued investigations on the nutritional requirements of tobacco callus and extended their research to include the induction of bud formation on tobacco stem segments by adenine sulfate. These initial and subsequent investigations led to the discovery of the cell division promoter, kinetin, by Skoog and Carlos Miller (1957), which then progressively led to the discovery of other cytokinins during the latter half of the 1950s and into the 1960s. Intensive research then began with adenine, kinetin and several other newly discovered cytokinins on their in vitro shoot-promoting effects for the rapid propagation of plants, especially the economically important horticultural and agronomic cultivars.

Meanwhile, according to Stowe and Yamaki (1957), the "western world" became familiar with the gibberellins through the efforts of a group at Imperial Chemical Industries in Britain and a group at the USDA. These groups "discovered" research on gibberellins that had been conducted by many Japanese plant physiologists beginning with Kurosawa (1926) in what was then Formosa. After the "western world" became aware of the gibberellins in the early 1950s, research on the in vivo effects of gibberellins accelerated at a rapid pace worldwide. In vitro research with gibberellins was slow to begin in the late 1950s and increased substantially in the 1960s and 1970s.

A similar situation occurred with the PGRs ethylene and abscisic acid. Crocker et al. (1935) were the first to propose that ethylene is involved in fruit ripening. In vivo investigations in the 1960s confirmed the role of ethylene in fruit ripening, as first proposed in the 1930s, and also in early seedling growth, leaf abscission and epinasty, senescence, and other growth-promoting and inhibitory effects. In vitro studies with ethylene did not begin until the 1970s. The in vivo effects of abscisic acid on plant development were not recognized until the 1960s. Studies into the 1970s were directed toward leaf and fruit abscission and bud dormancy. Other plant growth-promoting and growth-inhibiting studies soon followed. In vitro investigations with abscisic acid began in the 1970s with experiments involving zygotic and somatic embryogenesis.

By the middle 1950s, plant tissue culture methods had progressed to the point of making a major impact on research in plant developmental biology. Those investigators who had studied plant development over the previous 50 years using the conventional techniques of morphology, anatomy, physiology, biochemistry, cytology and genetics progressively incorporated the "tool" of plant tissue culture into their research. Major breakthroughs of knowledge in plant development came sooner than would have been possible without the techniques of sterile culture. The progress also was hastened by innovations in laboratory equipment and supplies and by improved methods of worldwide transportation and communication between developmental botanists around the world. In particular, the development of HEPA filters, which screen fungal spores and bacteria from air, made laminar flow transfer hoods possible. Laminar flow hoods became commonly available in the late 1960s to early 1970s and, finally, made sterile culture a routine task. Some of the European and American originators of the basic techniques of plant tissue culture from the 1930s and 1940s traveled around the world after the 1950s to many research laboratories as Visiting Botanists or to attend meetings, and became the "teachers" of the basic methods to numerous colleagues and students.

Concomitantly, another aspect of plant tissue culture greatly aided the increased research activity in plant development. The methods of obtaining plant callus tissue from several sources were well developed by the middle 1950s. The study of the causes of mammalian cancer (especially human cancers) became very popular during this time period. It was logical, then, that plant callus development was equated to mammalian tumor development. Thus, research on plant callus and cell suspension cultures intensified in many research laboratories because several agencies that were interested in mammalian cancer research awarded generous grants to plant developmental botanists to study "plant cancers". Among these agencies were the American Cancer Society, the National Cancer Institute, the National Institutes of Health and the U.S. Department of Health, Education and Welfare. A few notable pioneer investigators, among many, in the study of plant callus and cell suspensions in the early 1950s to the middle 1960s included Morel, Wetmore,

Henderson, Bonner, Steward, Muir, Braun, Nickell, Partanen, Das, Nitsch, Torrey, Reinert, Klein, Bergmann, Murashige and Skoog.

In the 1950s, the early prediction that somatic plant cells could undergo embryogenesis finally was validated by Steward et al. (1958) and Reinert (1959) who showed in many subsequent publications that somatic cells of carrot would differentiate into embryos when cultured within a proper nutrient–PGR regime (see Steward, 1968 and chapter sixteen). The ability to regenerate plants from single somatic cells through such a "normal" developmental process was envisioned to have great applications in propagation and genetic engineering (see below). Nonzygotic embryogenesis now has been demonstrated in most species of higher plants that have been tested.

In the early 1960s, Murashige completed a study while working in Skoog's laboratory leading to a commercial application for tissue culture. A culture medium developed by Murashige was originally devised for the rapid growth and bioassays with tobacco callus (Murashige and Skoog, 1962). However, research by several nurseries in California and elsewhere, through the advice of Murashige and knowledge of the work of Morel (1965) on orchid propagation, showed that practical shoot tip propagation of several ornamental plants could be accomplished on Murashige and Skoog medium (1962). Today, the ornamental plant industry depends heavily on tissue and organ culture micropropagation to supply high-quality, low-cost stocks of many species (see chapters seven and thirty-eight).

Reports concerning the recovery of plants from haploid cells began to appear in the 1960s (see chapter twenty-seven). The first successes were obtained with *Datura* (Guha and Maheshwari, 1966) and tobacco (Bourgin and Nitsch, 1967). This discovery received significant attention from plant breeders, since plants recovered from doubled haploid cells are homozygous and express all recessive genes, making them ideal pure breeding lines. Haploid-based breeding programs now are in place for several major agronomic crops.

Plant protoplasts were isolated and began to be cultured in the 1960s (Cocking, 1960). The removal of the plant cell wall allowed many novel experiments on membrane transport to be undertaken and, with totipotent protoplasts, allowed the somatic hybridization of sexually incompatible species to be accomplished. Somatic hybridization now has been used successfully in a number of cultivar development programs (see chapter twenty-three). In addition, the uptake of DNA into the plant cell was facilitated by removal of the wall, leading to the first reports of plant genetic transformation (see chapter twenty-nine and below).

Perhaps the greatest stimulus and change in the direction of plant cell and tissue culture research occurred after the discovery of restriction endonuclease enzymes in the early 1970s. These enzymes cleave DNA molecules at predictable sites and allow specific genes to be removed, modified and inserted into other DNA strands. The watershed of technological development that occurred after this breakthrough, and still is occurring today, has provided the basic tools needed for genetic engineering (see chapter twenty-nine). The term "biotechnology" was coined to denote this new research field. The availability of totipotent plant cells that, conceivably, could be altered by insertion of specific genes caused a revolution in plant research, because the obvious implications to agriculture of such genetically altered plants were so great. The potential commercial value of genetically engineered plants attracted an unprecedented amount of industrial and investor interest such that, in the late 1970s and 1980s, a number of new plant biotechnology-based companies sprang up around the world, but primarily in the United States. Many of these companies exist today or have been absorbed by larger companies. The result of this intense research activity, utilizing plant cell and tissue culture, has been the recent release or near release (as of 1995) of crop plants that exhibit quality and insect and disease resistance levels that were not obtainable by conventional breeding.

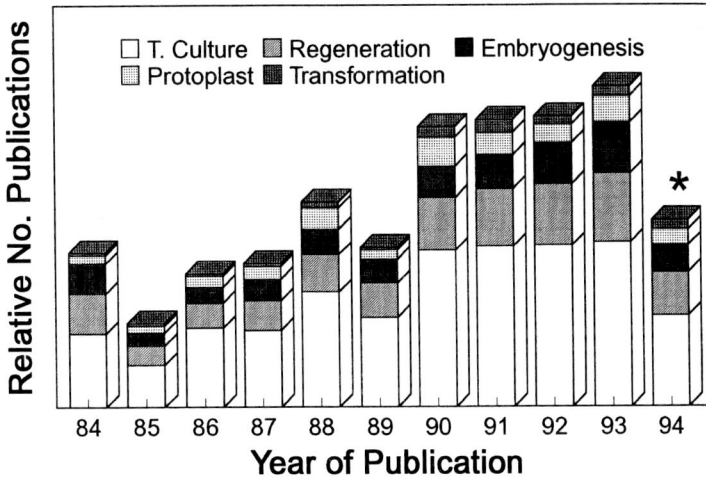

**Figure 1.1** Change in the relative number of tissue culture-related publications from 1984 to 1994. Data were generated by querying the CAB International Silverplatter 3.1 data base with the following key phrases: (1) "plant tissue culture", (2) "plant tissue culture" and "regeneration", (3) "plant tissue culture" and "embryogenesis", (4) "plant tissue culture" and "protoplast", and (5) "plant tissue culture" and "genetic transformation". Thus, all publications had the phrase "plant tissue culture" either in their title or listed as a key word, with or without the other words or phrases. *Only partial data were available for 1994.

Figure 1.1 details the recent growth and changes in the field of plant tissue and cell culture that have occurred since 1984 — for a listing of tissue culture reports prior to 1985, see Bhojwani et al. (1986). The data were obtained by literature search using the primary search phrase "plant tissue culture". The relative change from year to year in the number of publications that listed this phrase as a key word is reflected. The other phrases "regeneration", "embryogenesis", "protoplast" and "genetic transformation" were searched only when linked to "plant tissue culture". Hence, only publications that listed one of these four phrases along with "plant tissue culture" were counted. Note that there has been a steady increase of publication in all topic areas (Figure 1.1), while only a small part of the total number of plant tissue culture publications, those dealing with genetic transformation, has been steadily increasing. Although data for 1994 were incomplete, the trends suggest that relevant publications for that year have equaled or exceeded those for 1993.

Tissue and cell culture increasingly is used as a tool by plant scientists and biotechnologists. However, as will become evident throughout this book, much remains to be discovered in terms of the methods and procedures needed for efficient growth of plants in vitro.

## Literature cited

Ball, E. 1946. Development in sterile culture of stem tips and subadjacent regions of *Tropaeolum majus* L. and of *Lupinus albus* L. *Am. J. Bot.* 33:301–318.

Bhojwani, S.S., V. Dhawan and E.C. Cocking. 1986. *Plant Tissue Culture: A Classified Bibliography.* 789 pp. Elsevier Science Publishers, Inc., New York.

Bourgin, J.P. and J.P. Nitsch. 1967. Obtention de *Nicotiana* haploides a partir d'etamines cultivees in vitro. *Ann. Physiol. Veg.* 9:377–382.

Braun, A.C. 1943. Studies on tumor inception in the crown gall disease. *Am. J. Bot.* 30:674–677.

Camus, G. 1943. Sur le greffage des bourgeons d'endive sur des fragments de tissues cultives in vitro. *C. R. Acad. Sci.* 137:184–185.

Cocking, E.C. 1960. A method for the isolation of plant protoplasts and vacuoles. *Nature.* 187:962–963.

Crocker, W., A.E. Hitchcock and P.W. Zimmerman. 1935. Similarities in the effects of ethylene and the plant auxins. *Contrib. Boyce Thompson Inst.* 7:231–248.

Dormer, K.J. and H.E. Street. 1949. The carbohydrate nutrition of tomato roots. *Ann. Bot. N. S.* 13:199–217.

Duhamet, L. 1939. Action de l'heteroauxine sur la croissance de racines isolees de *Lupinus albus*. *C. R. Acad. Sci. Paris.* 208:1838–1840.

Gautheret, R.J. 1934. Culture de tissu cambial. *C. R. Acad. Sci. Paris.* 198:2195–2196.

Gautheret, R.J. 1939. Sur la possibilite de realiser la culture indefinie des tissus de tubercules de carotte. *C. R. Acad. Sci. Paris.* 208:118–120.

Guha, S. and S.C. Maheshwari. 1966. Cell division and differentiation in the pollen grains of *Datura* in vitro. *Nature.* 212:97–98.

Hannig, E. 1904. Zur physiologie pflanzlicher Embryonen. I. Uber die Kultur von cruciferen-embryonen Ausserhalb des Embryosacks. *Bot. Ztg.* 62:45–80.

Hildebrandt, A.C., A.J. Riker and B.M. Duggar. 1946. The influence of the composition of the medium on growth in vitro of excised tobacco and sunflower tissue culture. *Am. J. Bot.* 33:591–597.

Hoagland, D.R. and D.I. Arnon. 1938. The water culture method for growing plants without soil. *Calif. Agric. Expt. Sta. Circ.* 347, Berkeley, Ca.

Kotte, W. 1922. Kulturversuche mit isolierten Wurzelspitzen. *Beitr. Allg. Bot.* 2:413–434.

Krikorian, A.D. and D.L. Berquam. 1969. Plant cell and tissue culture — the role of Haberlandt. *Bot. Rev.* 35:58–88.

Kurosawa, E. 1926. Experimental studies on the secretions of *Fusarium heterosporum* on rice plants. *Trans. Nat. Hist. Soc. Eormosa.* 16:213–127.

La Rue, C.D. 1936. The growth of plant embryos in culture. *Bull. Torrey Bot. Club.* 63:365–382.

Maheshwari, P. 1950. *An Introduction to the Embryology of Angiosperms.* p. 453. Mcgraw-Hill Book Co., Inc., New York.

Morel. G. 1944. Le developpement du mildiou sur des tissues de vigne cultives in vitro. *C. R. Acad. Sci.* 218:50–52.

Morel, G. 1948. Recherches sur la culture associee de parasites obligatoires et de tissus vegetaux. *Ann. Epiphyt. N. S.* 14:123–234.

Morel, G. 1965. Clonal propagation of orchids by meristem culture. *Cymbidium Soc. News* 20:3–11.

Murashige, T. and F. Skoog. 1962. A revised medium for rapid growth and bioassays with tobacco tissue cultures. *Physiol. Plant.* 15:473–497.

Nobecourt, P. 1937. Cultures en serie de tissus vegetaux sur milieu artificiel. *C. R. Seanc. Soc. Biol.* 205:521–523.

Robbins, W.J. 1922. Cultivation of excised root tips and stem tips under sterile conditions. *Bot. Gaz.* 73:376–390.

Skoog, F. 1944. Growth and organ formation in tobacco tissue cultures. *Am. J. Bot.* 31:19–24.

Skoog, F. and C.O. Miller. 1957. Chemical regulation of growth and organ formation in plant tissues cultured in vitro. *Symp. Soc. Exp. Biol.* 11:118–131.

Skoog, F. and C. Tsui. 1951. Growth substances and the formation of buds in plant tissues. In: *Plant Growth Substances.* F. Skoog., Ed., pp. 263–285. University of Wisconsin Press, Madison.

Steward, F.C. 1968. *Growth and Organization in Plants.* p. 564. Addison-Wesley Publishing Co., Reading, MA.

Steward, F.C., M.O. Mapes and M.J. Smith. 1958. Growth and organized development of cultured cells. I. Growth and division of freely suspended cells. *Am. J. Bot.* 45:693–703.

Stowe, B.B. and T. Yamaki. 1957. The history and physiological action of the gibberellins. *Annu. Rev. Plant Physiol.* 8:181–216.

Street, H.E. and J.S. Lowe. 1950. The carbohydrate nutrition of tomato roots. II. The mechanism of sucrose absorption by excised roots. *Ann. Bot.* 14:307–329.

Van Overbeek, J., M.E. Conklin and A.F. Blakeslee. 1942. Cultivation in vitro of small *Datura* embryos. *Am. J. Bot.* 29:472–477.

Went, F.W. and K.V. Thimann. 1937. *Phytohormones.* p. 294. The Macmillan Co., New York.

White, P. R. 1934. Potentially unlimited growth of excised tomato root tips in liquid medium. *Plant Physiol.* 9:585–600.

White, P.R. 1937. Vitamin $B_1$ in the nutrition of excised tomato roots. *Plant Physiol.* 12:803–811.

White, P.R. 1939. Potentially unlimited growth of excised plant callus in an artificial medium. *Am. J. Bot.* 26:59–64.

# Part II

## Supporting methodologies

*chapter two*

# General laboratory requirements, media and sterilization methods

*Ellen G. Sutter*

Teaching plant tissue culture so that students develop an understanding of underlying principles, as well as the ability to apply them, necessarily involves the use of laboratory exercises. Tissue culture courses generally are heavily dependent on laboratory experience and, because of the relative slow growth of plant cultures, exercises initiated during one period might not reach a termination stage until several laboratory periods have passed. For this reason, it is important for procedures to be well organized and streamlined so that several exercises can be initiated and managed at one time.

This chapter serves as the source of general procedures, supplies and basic formulae used in succeeding laboratory exercises; it is referenced at appropriate points throughout the text. In particular, formulations for the most commonly used media are given and a brief discussion of the role of various nutrients is provided. A "kit" containing basic instruments and supplies that should be available at each workstation or provided to each student or team is described below. The familiarity of students in use of kit components is presumed and will not be repeated in the laboratory exercises. Addresses of some major suppliers are listed in Appendix I. The staging of plants for experiments is discussed in the "Overview and scheduling of laboratory exercises" found at the beginning of this book. In addition to this chapter, specialized procedures and media formulae can be found in individual laboratory exercises, as required. Interested readers also should consult chapter thirty-eight for additional information on laboratory organization.

## Setting up a laboratory for teaching plant tissue culture

Plant tissue culture laboratories need not be very expensive to set up. The most expensive items will be instrumentation. Below is a list of instruments required for a plant tissue culture laboratory and the number of items needed for teaching a class of 20.

* Autoclave that is larger than a desktop model
* Two or more balances having the capacity of 100 g with precision of 0.01 g (plant growth regulators [PGRs] must be weighed to the nearest milligram and thus could be done by the instructor with an analytical balance)
* Two or three pH meters with 4.0 and 7.0 standards for calibration — see operating manual for instructions
* Five or more hot/stir plates
* Water purifier/deionizer or glass still
* One refrigerator to store media, stocks, etc.

0-8493-9409-0/96/$0.00+$.50
© 1996 by CRC Press, Inc.

* Five or more stereomicroscopes with light sources (fiber optic illuminators are preferred)
* Four or more multistation laminar flow hoods
* One or more double-sized incubators; light shelves may be substituted
* One or more rotary shakers for suspension cultures, etc.
* One or more compound microscopes and, if possible, equipped for epifluorescence (see chapter twenty-five for filter requirements)

General glassware supplies include assorted sizes (50, 100, 250, 500, 1000 ml and larger) of Erlenmeyer flasks and beakers, graduated cylinders (10, 25, 100, 500 and 1000 ml) and culture tubes. Much of the glassware may actually be plasticware. Polymethylpentene is clear and enables one to measure accurately in graduated cylinders. Plasticware cannot be placed on hot plates, but much of it can be sterilized by autoclaving (check catalog). Plastic sterile petri dishes, culture tubes and/or larger culture vessels, such as baby food jars and Magenta Corp. GA-7 vessels, are used for culturing explants. Media may be dispensed using plastic syringes, graduated cylinders or, in the case of petri dishes, simply poured. Other general laboratory supplies include a supply of 70 and 95% ethanol, sterile magnetic stir bars, scissors, a supply of undiluted commercial bleach, various size core borers, aluminum foil, plastic wrap, cotton batting, cheesecloth, sterile plastic petri dishes (100 and/or 60 mm diam.), Parafilm™, sterile glass and/or plastic pipettes (1, 5, 10 and 25 ml), glass microscope slides and coverslips, spatulas, weighing paper or boats, heat-resistant gloves and vinyl gloves.

## Kit of basic supplies

Throughout the text, laboratory exercises assume that each work station, student or team is equipped with a kit of basic supplies. The kit includes a spray bottle with 70% ethanol for washing the hood, forceps (short, long and fine tipped), scalpel handle (#3), disposable scalpel blades (#10 and #11), safety razor blades, large nonsterile waste container, pipette bulb or pump, sterile water, bunsen burner or alcohol lamp, and a sterile surface for cutting explants.

Sterile plastic petri dishes (100 × 15 mm) often are used for cutting explants, but the cost of using these can become prohibitive. Other ideas are metal commercial ashtrays that can be washed and sterilized, glass plates, sterile paper towels and sterile paper plates. Paper goods can be autoclaved in plastic autoclaving bags. A rack for keeping the tools off the surface of the work area (similar to a pipette holder) can be fashioned from a clothes hanger. A vessel for holding alcohol for flaming instruments may be prepared from a 25-mm-diam. culture tube placed in an Erlenmeyer flask or beaker for stability. Be sure to place a wad of cheesecloth on the bottom of the vessel to avoid breaking the glass with instruments. A marking pen and marking tape are also necessary.

## Culture media

Nutrient media for plant tissue culture are designed to enable explants to grow in a totally artificial environment. In order for plants to grow in vitro, scientists have devised growth media that provide the nutrients that are usually available in soil. In addition to mineral elements that make up the macro- and micronutrients present in soils and fertilizers, nutrient media also contain organic compounds such as vitamins, PGRs and a carbon source. Chapter three illustrates the effects of nutrition and media composition on growth of callus from tobacco and other plants.

## Mineral elements

One of the most successful media, commonly known as MS medium, was devised by Murashige and Skoog (Murashige and Skoog, 1962). MS medium was formulated by analyzing the inorganic components in tobacco plants and then experimentally adding them to media in amounts similar to those found in the plants. Not only was it found that the ions themselves were important, but the form in which they were supplied was critical as well. The following sections discuss commonly used components of culture media. In addition, a laboratory exercise concerning nutrition in callus cultures is presented in chapter three.

### Macroelements

Macroelements consist of calcium (Ca), magnesium (Mg), nitrogen (N), phosphorus (P), potassium (K) and sulfur (S).

*Calcium (Ca).* Calcium functions as a cofactor with many enzymes and is particularly important in cell wall synthesis. In plant cultures, calcium deficiency may result in shoot tip necrosis. Calcium (1–3 mM) is supplied mostly as calcium chloride or calcium nitrate.

*Magnesium (Mg).* Magnesium is critical for the functioning of enzymes and is an integral component of the chlorophyll molecule. In plants, magnesium is a cation that balances negative ions. Magnesium sulfate is the inorganic compound most used to supply the magnesium ion. It is present in concentrations similar to that of calcium.

*Nitrogen (N).* Nitrogen is required for general growth and is essential to plant life. Most inorganic nitrogen is converted to amino acids and then to proteins. The two most widely used forms of inorganic nitrogen used in plant nutrient media are the nitrate ion ($NO_3^-$ oxidized) and the ammonium ion ($NH_{4+}$ reduced) that are added as inorganic salts. Nitrate is usually added at 25–40 mM and ammonium at 2–20 mM. In poorly buffered media, use of both forms helps maintain a stable pH. Many plants appear to grow best if given both forms, although the reason for this is not known. In devising media, both the total amount of nitrogen as well as the relative amounts of $NO_3^-$ and $NH_{4+}$ are important. When the ammonium ion is used alone, it may be toxic. Inorganic nitrogen generally ranges from 25–60 mM in media. Nitrogen may also be added in an organic form as amino acids, hydrolysates (such as casein hydrolysate) and organic acids. The organic forms are often useful when added to media that do not contain ammonium. One advantage of using organic nitrogen is that it is already reduced to the form in which most nitrogen exists in the plant, and thus may be taken up more readily than inorganic nitrogen. Organic forms of nitrogen, however, cannot totally replace inorganic forms.

*Phosphorus (P).* In plants, phosphorus is required because it is an integral part of nucleic acids and other structural compounds. Phosphorus is supplied to plant cultures as phosphate ($PO_4^-$) in sodium or potassium hydrogen phosphates and ranges from 1–3 mM.

*Potassium (K).* Potassium serves as the major positive ion in plants that balances negative ions. The amount of potassium required in plant medium varies widely, depending on the plant species. Generally, the amount of potassium in media is closely correlated with that of nitrate and ranges between 20 and 30 mM.

*Sulfur (S).*   Sulfur is critical in the structure of proteins by means of its presence in several important sulfur-containing amino acids. Sulfur is supplied as the $SO_4^-$ ion generally with magnesium (rarely with ammonium) as the cation in concentrations ranging from 1–3 mM.

### Micronutrients

Micronutrients used in plant tissue culture consist of boron (B), cobalt (Co), copper (Cu), iodine (I), iron (Fe), manganese (Mn), molybdenum (Mo) and zinc (Zn).

*Boron (B).*   Boron is an essential element involved in enzyme activity involved with lignin biosynthesis and metabolism of phenolic acids. Boron is supplied as boric acid in culture medium. Boron deficiency results in the death of shoot tip meristems.

*Cobalt (Co).*   Cobalt is not considered to be an essential mineral by plant physiologists, but is included in many of the most widely used media formulations. Cobalt is supplied in concentrations similar to that of copper, 0.1 µM. It may be toxic at higher concentrations.

*Copper (Cu).*   Although copper is required in plant culture media in extremely low concentrations (0.1 µM), it is critical in many enzyme reactions, including the cytochrome oxidase system. One must be careful in adding copper because high amounts can be toxic. Copper is added as cupric sulfate.

*Iodine (I).*   The effect of iodine varies depending on the plant species. It is not considered an essential element, but is added to plant culture media because it has been found to improve growth of roots and callus in culture.

*Iron (Fe).*   Iron is required for chlorophyll synthesis as well as in many oxidation/ reduction reactions. It is critical that the medium be at the proper pH for iron to be taken up by plant tissues. The use of chelating agents, compounds that bind metal ions, makes iron more stable and available to plant tissues over wider pH ranges. The major problem with supplying iron is that it forms insoluble compounds in alkaline pH, which is particularly evident in liquid cultures. Although there are several chelating agents, the sodium or potassium form of ethylenediaminetetraacetic acid (EDTA), is used most often in plant tissue culture media. EDTA has the advantage of not being as toxic as other chelating agents and enables iron to be available to cultures over a wider pH range than other agents. Fe-EDTA may be purchased as a salt, or it may be prepared from ferric sulfate and EDTA (see Table 2.3). Iron is present in media at approximately 1 µM.

*Manganese (Mn).*   Manganese is required for enzyme reactions, particularly in respiratory and photosynthetic processes. It is most often supplied as manganese sulfate ranging from 5–30 µM.

*Molybdenum (Mo).*   Molybdenum is a cofactor with two enzymes involved in transforming nitrate to ammonium. It is supplied as sodium molybdate and is present in very low concentrations (1 µM) in culture medium.

*Zinc (Zn).*   Zinc is also required in many enzyme activities and appears to be important for chloroplast development. The amount of zinc added to plant media ranges widely (5–30 µM) and does not appear to be toxic in higher concentrations. The most common form in which zinc is added is as the sulfate salt.

*Other elements.*    Aluminum (Al), nickel (Ni) and silicon (Si) are present in a few media formulations. These elements have not been found to be necessary for most plant species in vitro.

## Organic compounds

A variety of organic compounds often are added to plant culture media. Some, such as sugars, are required for growth, whereas others, such as vitamins, undefined compounds and organic acids, may be useful because they enhance growth significantly.

### Sugars

Sugars serve as the energy source for plant cultures. Most cultures are not autotrophic, that is, capable of fixing carbon through photosynthesis. Because of limitations in $CO_2$ availability, the efficiency of photosynthesis is low, and plant tissues require that sugar be added to media for adequate growth. The most commonly used sugar in plant tissue culture is sucrose, although glucose, fructose and sorbitol have also been effective under certain situations. During metabolism in plant cells, sucrose is broken down into glucose and fructose. Sucrose is also the major sugar that is transported in most plant species. It is partially hydrolyzed into glucose and fructose during autoclave sterilization, especially when present with other compounds. Glucose can substitute for sucrose under some conditions. Fructose has been found to be more effective than sucrose for some orchid cultures, and sorbitol has been used in apple cultures as a partial or total replacement for sucrose. The concentration of sugars in nutrient media generally ranges from 20–40 g/L.

Sugars also act as an osmoticum in the medium. Nutrient salts contribute from 20–50 % of the osmotic potential of media, with sucrose making up the rest. When sucrose is hydrolyzed, as during autoclave sterilization, the contribution of sugar to osmotic potential is further increased. Osmotic effects are particularly important during protoplast isolation and culture. Osmotic potential has also been found to affect growth and morphogenesis in embryo culture, somatic embryogenesis, anther culture and morphogenesis of callus cultures.

### Vitamins

Of all vitamins, only thiamine ($B_1$) has been shown to be essential for many species of plants in culture. Thiamine is required for carbohydrate metabolism and the biosynthesis of some amino acids. Murashige and Skoog medium (1962), as well as many others, also include nicotinic acid (niacin) and pyridoxine ($B_6$). Other vitamins such as biotin, folic acid, ascorbic acid (vitamin C) and vitamin E (tocopherol) are used in some media formulations. Most vitamins have been found to improve, but are not absolutely necessary for, plant growth. Concentrations of vitamins are generally very low.

### Myo-inositol

Characterized as a vitamin, but actually a sugar alcohol, myo-inositol is added to most plant culture media. It is thought to be important in membrane and cell wall development. Although myo-inositol is not absolutely essential for growth to occur, its effect on the growth of many plant species is significant. It is considered to be necessary for monocots as well as some dicots and gymnosperms.

### Complex organics

Complex organics include undefined supplements such as coconut milk and water, yeast extract, fruit juices and fruit pulps. Complex organics were used early in the history of plant tissue culture. Today they are used when no other combination of known components supports growth of tissues. The disadvantage of using complex organics is that the exact composition of the supplements is unknown and may vary among batches. In addition, it is not known exactly what component in the organic supplement is responsible for the improved growth of the tissues. Complex organics supply amino acids, vitamins, PGRs and probably secondary plant metabolites that have not yet been identified. Coconut milk, liquid endosperm of *Cocos nucifera* L. fruits, was used first in the 1940s for *Datura* cultures (Van Overbeek et al., 1941) and later became an important staple of carrot embryogenic culture media. Of the many undefined organic supplements available, coconut water (from mature coconut fruits) is used most commonly today. Although available commercially, fresh coconut milk is difficult to obtain because it is extracted from immature, green coconuts.

## Activated charcoal

Activated charcoal is added to many media in order to adsorb toxic compounds released by the plant tissues. It is particularly useful in cultures to manage the production of brown exudates and oxidized phenolics. Activated charcoal also aids in root induction. Activated charcoal adsorbs not only toxic compounds, but PGRs and other compounds that are added to the medium (Nissen and Sutter, 1990). Thus, one must be aware that the effects of PGRs may be significantly limited or even negated when activated charcoal is used.

# Support systems

Support systems are required to keep plant cultures from being submerged in the medium. Liquid medium is used for many plants, but it must often be agitated to provide sufficient oxygen. Among support systems are gelling agents, such as agar, Gelrite™, agarose and gelatin, and mechanical supports, such as filter paper bridges and a variety of commercially available polypropylene membrane raft systems (see Sigma Catalog, Sigma Chemical Co., St. Louis, MO).

## Agar

Agar, the most common gelling agent, is a mixture of polysaccharides derived from red algae. Agar has several advantages that make it more desirable than other gelling agents. It is convenient to work with since it melts at about 100 C and solidifies at about 45 C. In addition, agar does not react with media components nor is it digested by plant enzymes. Concentrations generally range between 0.5 and 1.0%. Agar does not gel well under acidic conditions and thus may not be appropriate for situations in which a low pH (<4.5) is required. The inclusion of activated charcoal in media may also inhibit gelling of agar at normal concentrations.

All agar contains impurities, the amounts of which differ depending on the manufacturer and purity of the product. Among impurities are inorganic salts, organic compounds, phenolics and long chain fatty acids. These compounds often do not interfere with the growth of cultures. If necessary, agar may be washed to remove inhibitory impurities. The concentration of agar may be critical to plant growth in culture. Media that is too soft may produce hyperhydric plants, whereas media that is too hard may cause reduced plant growth (Singha, 1982).

## Agarose

For greater purity, agarose may be used. Agarose is extracted from agar leaving behind agaropectin and its sulfate groups. Because of the additional purification, agarose is considerably more expensive than agar. Agarose also has higher gel strength than agar, and, thus, less is required for solidification of media. Agarose is used in situations where the impurities of agar are a major disadvantage, such as in protoplast and anther culture.

## Gellan gum

Gellan gums, such as Gelrite and Phytagel™, consist of a polysaccharide produced by the bacterium *Pseudomonas elodea*. It has the advantage of being clear, rather than translucent as agar is. Consequently, contamination is more easily detected at an early stage. Impurities in gellan gum contain inorganic ions, but no organic compounds. Gellan gum requires more stirring than does agar when being added to media. Unlike agar, gellan gum cannot be reheated and gelled successfully. One limitation of gellan gum is that the concentration of divalent cations such as calcium and magnesium ions must be within the range of 4–8 mM. Concentrations of these two ions either less than or greater than this range result in the media not gelling. Gellan gum may also produce hyperhydric plants when used at low concentrations.

The selection of gelling agent for specific plants is often empirical. One major consideration is the degree of hyperhydricity induced by the different gelling agents. Also, for unknown reasons, some species grow more vigorously on one gelling agent than on another. For example, apple shoots become hyperhydric on Gelrite (Pasqualetto et al., 1986), whereas walnut shoots grow best on Gelrite and poorly on agar (McGranahan et al., 1988). One solution to overcome the negative effects of different gelling agents on hyperhydricity is to combine both agar and either Gelrite or Phytagel™ in the medium.

## Mechanical supports

Mechanical supports, such as filter paper folded into wicks and polyethylene rafts, are often used with liquid media to ensure an adequate supply of oxygen. Many other materials have been used as well, including rock wool, cheesecloth and pieces of foam. Explants can be grown in rotating vessels that enable the tissue alternately to be bathed in liquid and exposed to air.

# Plant growth regulators

The major classes of PGRs used in plant tissue culture are auxins, cytokinins, gibberellins and abscisic acid. In most cultures only auxins and cytokinins are used. Skoog and Miller (1957) first demonstrated that the ratio between the auxin and cytokinin added to medium determines their morphogenetic effects on cultured tissues. It was later shown that the absolute amount of PGRs is also an important factor in morphogenesis. The concentrations of PGRs range from 0.1–100 µM depending on the specific PGR and the situation in which it is used.

## Auxins

Auxins are known to affect a great many processes in plants including cell elongation, apical dominance and adventitious root formation. The auxins used in tissue culture are both natural and synthetic. Naturally occurring auxins include indole-3-acetic acid (IAA) and indole-3-butyric acid (IBA). Until recently, IBA was considered to be synthetic, but has been found to occur naturally in many plants including olive and tobacco (Epstein et al., 1989). Synthetic auxins most often used in plant tissue culture are 1-naphthalene acetic acid (NAA) and 2,4-dichlorophenoxyacetic acid (2,4-D). Both IAA and IBA have been shown to decompose in light over time, with IAA being considerably less stable than IBA (Nissen and Sutter, 1990). Auxins may be used as mixtures as well as alone.

Auxins are used for induction of callus from explants (see chapter three), root and shoot morphogenesis (see chapters seven and eleven) and somatic embryogenesis (see chapter sixteen). Auxins are often most effective when combined with cytokinins. For organogenic culture systems, a general rule is that an auxin:cytokinin ratio greater 1 leads to root formation and an auxin:cytokinin ratio less than 1 results in shoot formation. The use of 2,4-D is more limited than other auxins. It is used mostly in somatic embryogenesis and callus induction and at much lower concentrations than other auxins. The use of 2,4-D is more limited than that of IAA and IBA for morphogenesis because it inhibits organogenesis in many plant species; it is used almost exclusively in the induction and maintenance of embryogenic cultures.

## Cytokinins

Cytokinins act as antagonists to auxin in apical dominance, as antisenescent compounds, and as effectors of cell division in the whole plant. Both naturally occurring and synthetic cytokinins are used in plant culture media. Naturally occurring cytokinins used most in plant tissue culture include zeatin, 6-(γ,γ-dimethylallylamino)purine (2-iP) and adenine. Although the cost of 2-iP and adenine is reasonable, that of zeatin is relatively high, and, thus, zeatin is usually used only in research. Other naturally occurring cytokinins such as dihydrozeatin and zeatin riboside are rarely used. Synthetic cytokinins are used more often than natural forms. Synthetic cytokinins consist of either substituted purines such benzyladenine (BA) and 6-furfurylaminopurine (kinetin) or as phenylureas such as thidiazuron. In culture, cytokinins are used mostly to induce shoot bud induction, development and multiplication. Cytokinins stimulate cell division and both formation and growth of axillary and adventitious shoots in plant tissue culture. Cytokinins often inhibit embryogenesis and root induction.

## Gibberellins

The main effect of gibberellins in plants is to cause stem elongation and flowering. There exist over 80 different compounds in plants classified as gibberellins. Usually only two forms are used in plant tissue culture, $GA_3$, gibberellic acid, and $GA_{4+7}$. In tissue culture, gibberellins often inhibit the induction of organogenesis, particularly adventitious root formation. Gibberellins have been used for stem elongation in cases where the shoots do not elongate naturally. In most plants, however, gibberellins produce abnormally long shoots with excessively narrow leaves.

## Abscisic acid

In the whole plant, abscisic acid (ABA) is involved in regulation of stomatal closure, as well as bud and seed dormancy. In tissue culture its greatest use is during somatic embryogenesis (see chapter sixteen). ABA has been found to be important during maturation and germination of normal somatic embryos (Ammirato, 1974; Roberts et al., 1990).

## Ethylene

Ethylene, a gas, is the main PGR responsible for abscission and flower senescence. Its role in adventitious root formation remains contradictory and in tissue cultures is unclear. It is produced by cultures and accumulates within well-sealed containers. It is rarely added to

**Table 2.1** Ionic Composition of Six Commonly Used Tissue Culture Media

| Ions | Murashige Skoog | Linsmaier Skoog | White | Gamborg B5 | Schenk & Hildebrandt | Nitsch & Nitsch |
|------|------|------|------|------|------|------|
| NO$_3$ | 40 | 40 | 3.8 | 25 | 25 | 18.5 |
| NH$_4$ | 20 | 20 | — | 2 | 2.5 | 9 |
| Total N | 60 | 60 | 3.8 | 27 | 27.5 | 27.5 |
| P | 1.5 | 1.5 | 0.15 | 1 | 2.5 | 0.5 |
| K | 21.5 | 21.5 | 1.65 | 25 | 25 | 10 |
| Ca | 3 | 3 | 1.5 | 1 | 1.5 | 1.5 |
| Mg | 1.5 | 1.5 | 3 | 1 | 1.5 | 0.75 |
| Cl | 6 | 6 | 0.85 | 2 | 3 | 3 |
| Fe | 100 | 100 | 15 | 50 | 55 | 100 |
| S | 1730 | 1730 | 4545 | 2112 | 1619 | 985 |
| Na | 202 | 202 | 3180 | 1102 | 111 | 202 |
| B | 100 | 100 | 25 | 50 | 80 | 150 |
| Mn | 100 | 100 | 20 | 60 | 60 | 100 |
| Zn | 30 | 30 | 10 | 7 | 3.5 | 35 |
| Cu | 0.1 | 0.1 | 0.04 | 0.1 | 0.8 | 0.1 |
| Mo | 1 | 1 | 0.007 | 1 | 0.4 | 1 |
| Co | 0.1 | 0.1 | — | 0.1 | 0.4 | — |
| I | 5 | 5 | 4.5 | 4.5 | — | — |
| Inositol | 550 | 550 | — | 550 | 550 | 550 |
| Nicotinic acid | 4 | — | 0.8 | 8 | 40 | 40 |
| Pyroxidine HCl | 2.5 | — | 0.5 | 5 | 2.5 | 2.5 |
| Thiamine HCl | 0.3 | 1.2 | 1.5 | 30 | 15 | 1.5 |
| Glycine | 25 | — | 40 | — | — | 25 |
| Cysteine | — | — | — | — | — | — |
| Casein hydrolysate | 1 | — | — | — | — | — |

cultures, but its inhibition by addition of ethylene inhibitors to medium has resulted in better initiation of embryogenic cultures of corn (*Zea mays* L.).

## Media formulations

A comparison of the ions in several different media is given in Table 2.1. Composition of several media by salts and organic compounds is given in milligrams per liter and molar units in Table 2.2. Although the most common medium formulation is MS (Murashige and Skoog, 1962), several others are used regularly. As discussed above, MS medium was developed for culture of tobacco explants and formulated based on an analysis of the mineral compounds present in the tobacco tissue itself. It is noted for its high salt levels, particularly K and N salts. Linsmaier-Skoog medium (Linsmaier and Skoog, 1965) is a revision of the MS medium, in which the organic constituents have been modified and only thiamine and myo-inositol retained. White's medium (White, 1963) was devised early in the history of plant tissue culture for the culture of tomato (*Lycopersicon esculentum* L.) roots and has lower concentrations of salts than MS medium. Gamborg's B5 medium (Gamborg et al., 1968) was devised for soybean callus culture and contains a much greater proportion of nitrate compared to the ammonium ion. Schenk and Hildebrandt (1972) developed their SH medium for the culture of callus of both monocotyledons and dicotyledons. Nitsch and

*Table 2.2*   Composition of Five Commonly Used Tissue Culture Media in Milligram per Liter and Molar Concentration

| Compounds | White | Schenk Hildebrandt | Murashige Skoog | Gamborg B-5 | WPM | Nitsch & Nitsch |
|---|---|---|---|---|---|---|
| Macronutrients, mg/L(mM) | | | | | | |
| $KNO_3$ | 80 (0.8) | 2500 (24.8) | 1900 (18.8) | 2500 (24.8) | — | 950 (9.4) |
| $NH_4NO_3$ | — | — | 1650 (20.6) | — | 400 (5.0) | — |
| $NH_4SO_4$ | — | — | — | 134 (1.0) | — | — |
| $NH_4H_2PO_4$ | — | 300 (2.6) | — | — | — | — |
| $Ca(NO_3)_2 \cdot 4H_2O$ | 285 (1.2) | — | — | — | 556 (2.4) | — |
| $CaCl_2 \cdot 2H_2O$ | — | 151 (1.0) | 332.2 (2.3) | 150 (1.0) | 96 (0.7) | 166 (1.1) |
| $MgSO_4 \cdot 7H_2O$ | 740 (3.0) | 400 (1.6) | 370 (1.5) | 250 (1.0) | 370 (1.5) | 185 (0.75) |
| $KCl$ | 65 (0.9) | — | — | — | — | — |
| $KH_2PO_4$ | — | — | 170 (1.3) | — | 170 (1.3) | 68 (0.5) |
| $Na_2SO_4$ | 200 (1.4) | — | — | — | — | — |
| $NaH_2PO_4$ | 16.5 (0.12) | — | — | 130.5 (0.9) | — | — |
| Micronutrients, μM (approx.) | | | | | | |
| $MnSO_4 \cdot H_2O$ | 7.0 (41) | 13.2 (78) | 16.9 (100) | 10.0 (59) | 29.4 (174) | 25 (148) |
| $ZnSO_4 \cdot 7H_2O$ | 2.67 (9) | 1 (3) | 8.6 (30) | 2.0 (7.0) | 8.6 (30) | 10 (35) |
| $H_3BO_3$ | 1.5 (25) | 5 (80) | 6.2 (100) | 3.0 (49) | 6.2 (100) | 10 (162) |
| $KI$ | 0.75 (5) | 0.1 (0.6) | 0.83 (5) | 0.75 (5) | — | — |
| $CuSO_4 \cdot 5H_2O$ | 0.01 (0.04) | 0.2 (0.08) | 0.025 (0.1) | 0.025 (0.1) | 0.25 (1) | 0.025 (.1) |
| $NaMoO_3$ | 0.001 (0.001) | — | — | — | — | — |
| $Na_2MoO_4 \cdot 2H_2O$ | — | 0.1 (0.4) | 0.25 (1) | 0.25 (1) | 0.25 (1) | 0.25 (1) |
| $CoCl_2 \cdot 6H_2O$ | — | 0.1 (0.4) | 0.025 (0.1) | 0.025 (0.1) | — | — |
| $FeSO_4 \cdot 7H_2O$ | — | 15 (54) | 27.8 (100) | 27.8 (100) | 27.8 (100) | 27.8 (100) |
| $Fe_2(SO_4)_3$ | 2.5 (6.2) | — | — | — | — | — |
| $Na_2EDTA$ | — | 20.1 (54) | 37.3 (100) | 37.3 (100) | 37.3 (100) | 37.3 (100) |
| Organics, μM | | | | | | |
| Myo-inositol | — | 1000 (5500) | 100 (550) | 100 (555) | 100 (555) | 100 (555) |
| Glycine | 3.0 (40) | — | 2.0 (26.6) | — | 2.0 (26.6) | 2.0 (26.6) |
| Nicotinic acid | 0.5 (4.1) | 5.0 (41) | 0.5 (4.1) | 1.0 (8.2) | 0.5 (4.1) | 5 (40.6) |
| Pyridoxine HCl | 0.1 (0.45) | 0.5 (2.4) | 0.5 (2.4) | 1.0 (4.8) | 0.5 (2.4) | 0.5 (2.4) |
| Thiamine HCl | 0.1 (0.3) | 5.0 (14.8) | 0.1 (0.3) | 10.0 (30) | 0.1 (0.3) | 0.5 (1.5) |
| Biotin | — | — | — | — | — | 0.2 (0.05) |

*Table 2.3*   Macro- and Micronutrient Stock Solutions for Murashige–Skoog (MS) Medium (1962) (Each of the following stocks is prepared using 1000 ml of tap water and should be added at the rate of 20 ml/L of medium)[a]

| Stock (50×) | Component | Amount |
|---|---|---|
| *Nitrate* | $NH_4NO_3$ | 82.5 g |
| | $KNO_3$ | 95.0 g |
| *Sulfate* | $MgSO_4 \cdot 7H_2O$ | 18.5 g |
| | $MnSO_4 \cdot H_2O$ | 0.85 g |
| | $ZnSO_4 \cdot 7H_2O$ | 0.43 g |
| | $CuSO_4 \cdot 5H_2O$ | 1.25 mg[b] |
| *Halide* | $CaCl_2 \cdot 2H_2O$ | 22.0 g |
| | KI | 41.5 mg |
| | $CoCl_2 \cdot 6H_2O$ | 2.5 mg[c] |
| *P B Mo* | $KH_2PO_4$ | 8.5 g |
| | $H_3BO_3$ | 310 mg |
| | $Na_2MoO_4$ | 12.5 mg |
| *NaFeEDTA* | $FeSO_4 \cdot 7H_2O$ | 1.39 g[d] |
| | NaEDTA | 1.87 g[d] |

[a] To prepare stocks for any other media listed in Table 2.2, multiply the amount of mg of a chemical component shown by 50 and dissolve in 1 L of distilled water. Be sure to group individual chemicals as above to avoid reactions and possible precipitation.

[b] Dissolve 125 mg of $CuSO_4 \cdot 5H_2O$ in 100 ml of water. Add 1 ml to the sulfate stock solution.

[c] Dissolve 125 mg of $CoCl_2 \cdot 6H_2O$ in 100 ml of water. Add 1 ml to the halide stock.

[d] Mix together and heat, while stirring, until orange and store in an amber bottle.

Nitsch (1969) developed Nitsch's medium for anther culture; it contains lower salt concentrations than that of MS, but not as low as that of White's medium. Other media that are used commonly include Lloyd and McCown's Woody Plant Medium (WPM) (Lloyd and McCown, 1980) that was developed for *Kalmia latifolia* L., an ericacious plant and has been used successfully for a great many tree species, and Knudson's medium used to culture a great many orchid species from seed (Knudson, 1946).

## Preparation of media

### Stock solutions

Mineral salts and vitamin mixtures may be prepared as stock solutions ranging from 10–1000 times the final concentrations. Stock solutions may be prepared as two solutions that contain all macronutrients and all micronutrients. These solutions must be kept fairly dilute (10–20×) in order to avoid precipitation of calcium and magnesium phosphates and sulfates. A more common method is to arrange stock solutions according to the ions they contain. A series of solutions is prepared as follows: nitrate, sulfate, halide, PBMo (phosphorus, boron, molybdenum) and iron. Vitamins are prepared separately. Using this method, salt stocks can be prepared in 100× concentrations and vitamins in 1000× concentrations. Iron-EDTA chelate can be purchased as K or Na-ferric–EDTA. It is easily prepared from iron sulfate and Na-EDTA by mixing the proper amounts of the two compounds with water and then heating slowly, stirring until the yellow solution becomes orange. It then must be stored in a dark container. Compositions of stock solutions for Murashige-Skoog medium macro and minor salts are given in Table 2.3; stocks solutions for other media may be prepared in a similar fashion. Although the initial outlay for chemicals may be substantial, the advantage of using stock solutions is that the ongoing cost is less than using

***Table 2.4***  Stock Preparation of Commonly Used PGRs

| PGR | mg/10 μM | Solvents (10 ml) | Comments |
|---|---|---|---|
| IAA | 17.5 | 95% ethanol, 100% propanol or KOH | To use KOH, IAA is first dissolved in small amount (1 ml) of 1N KOH and then water is added to final volume. |
| IBA | 20.3 | As above | As above |
| NAA | 18.6 | As above | As above |
| 2,4-D | 22.1 | As above | As above |
| Kinetin | 21.5 | KOH | As above. Does not dissolve in alcohol. |
| BA | 22.5 | 95% ethanol, 100% propanol, KOH | To use KOH, see comment concerning IAA |
| 2-iP | 20.3 | As above | As above |
| Zeatin | 21.9 | As above | As above |
| TDZ | 22.0 | DMSO | Does not dissolve in alcohol or base |
| ABA | 26.4 | KOH | To use KOH, see comment concerning IAA |

prepared mixes. In addition, changes can be made in individual components. Stock solutions, however, require storage in a refrigerator and must be replaced regularly if not used.

## Prepared mixes

Several companies sell prepared salt and vitamin mixtures as powders. These are easily handled by adding the proper amount of powder to water. Prepared mixtures avoid the necessity of storing many different stocks. The mixtures may be purchased as complete media or as salts with or without PGRs, agar and sucrose. A list of some suppliers is given in Appendix 1.

A small amount of precipitate may be seen using either of the two methods mentioned above. This precipitate results from iron being displaced from the chelate over time as media is stored. Although unsightly, the precipitate has not been shown to be a liability.

## Vitamins, PGRs and other addenda

Most organic addenda are added in relatively low concentrations, too low to be weighed out accurately. Consequently, stock solutions ranging from 100–1000× are generally prepared. Vitamin solutions can be divided into aliquots and stored in the freezer at –20 C. If stored in the refrigerator, they must be checked regularly for fungal contamination. PGRs are stable for up to a month in stock solutions. The solvents used for dissolving PGRs vary (Table 2.4). When working in molar units it is often convenient to prepare PGRs as 10 ml of a 10-μM solution. This is easily remembered as 0.1 × MW (in milligrams) per 10 ml solvent. Thus, to prepare 10 ml of a 10-μM solution of IAA, one weighs out 17.5 mg. When working in milligrams per liter a convenient concentration is 10 mg/10 ml. It is very difficult to weigh out less than 10 mg accurately. The advantage of using stock concentrations in the 10-μM range is that one needs to add only small volumes. Thus, when necessary, alcohol can be used as a solvent because it is generally added at less than 1%, an amount that is not toxic.

## Calculation of PGR concentrations

Literature reports provide PGR concentrations as either mg/L or μM, and it is important to be able to convert between the two sets of units. In some respects, molar units are more

*Table 2.5*  Conversion of PGRs from mg/L to µM

Plant Growth Regulators[a] (µM)

| mg/L | ABA | BA | 2-iP | 2,4-D | GA$_3$ | IAA | TDZ | Zea | IBA | K-IBA | Kin | NAA |
|---|---|---|---|---|---|---|---|---|---|---|---|---|
| 0.1 | 0.38 | 0.44 | 0.49 | 0.45 | 0.29 | 0.57 | 0.45 | 0.46 | 0.49 | 0.41 | 0.46 | 0.54 |
| 0.2 | 0.76 | 0.88 | 0.98 | 0.90 | 0.58 | 1.14 | 0.91 | 0.91 | 0.98 | 0.83 | 0.93 | 1.07 |
| 0.5 | 1.89 | 2.22 | 2.46 | 2.26 | 1.44 | 2.85 | 2.27 | 2.28 | 2.46 | 2.07 | 2.32 | 2.69 |
| 0.8 | 3.03 | 3.55 | 3.94 | 3.62 | 2.31 | 4.57 | 3.63 | 3.65 | 3.94 | 3.32 | 3.72 | 4.3 |
| 1.0 | 3.78 | 4.44 | 4.92 | 4.52 | 2.89 | 5.71 | 4.54 | 4.56 | 4.90 | 4.14 | 4.65 | 5.37 |
| 10 | 37.8 | 44.4 | 49.2 | 45.2 | 28.9 | 57.1 | 45.4 | 45.6 | 49.0 | 41.4 | 46.7 | 53.7 |

[a] Abbreviations: ABA, abscisic acid; BA, benzyladenine; 2-iP, 2-isopentenyl adenine; 2,4-D, 2,4-dichloroacetic acid; GA$_3$, gibberellic acid; IAA, indolacetic acid; TDZ, thidiazuron; Zea, zeatin; IBA, indolbutric acid; K-IBA, the potassium salt of IBA; Kin, kinetin; NAA, naphthaleneacetic acid.

accurate since they reflect the actual number of molecules present, regardless of their different molecular weights. A table of conversions for several PGRs is presented in Table 2.5. A simple formula, however, makes it possible to convert any concentration, but first it is necessary to know the definitions of some terms:

> Gram molecular weight (also known as mole or mol): The formula weight of a compound expressed in grams
> Molarity: One mole of a compound in a liter of solution

In order to convert from mg/L to µM, divide by the gram molecular weight. For example, the gram molecular weight of IAA is 175.2 g or 175,200 mg. Thus, a 2 mg/L solution of IAA = 2/175,200 = 0.0000114 M or if expressed as µM, move decimal 6 places to the right or 11.4 µM.

## Water

The quality of water used in media can make a significant difference in success or failure of culture procedures. It is essential that an adequate source of high quality water is available for use in the laboratory exercises. Type I water is most desirable because it has very little solids and is produced through distillation, ion exchange, reverse osmosis or a combination of methods if the water is particularly hard. Under no circumstances should tap water be used.

## Adjustment of pH

The pH of most plant tissue culture media is adjusted to 5.5–5.8 with KOH prior to autoclaving. If the pH is much lower than 4.5, agar will be acid hydrolyzed during autoclave sterilization and will not gel. Although pH drops somewhat during autoclave sterilization, it has not been shown to be necessary to readjust the pH of the medium after autoclaving.

## Procedure

1. Place a volume of glass-distilled or deionized water, equal to approximately half the total volume of media to be prepared, in a beaker or Erlenmeyer flask.
2. Add stock solutions or the proper amount of salt mixture.
3. Add vitamins and myo-inositol.
4. Add appropriate amount of crystalline sucrose, usually 20–30 g/L.

5. Add PGRs and any other compounds if required and bring to about 90% of the final volume.

6. Adjust pH.

7. Adjust volume to the final desired amount with glass-distilled or deionized water.

8. For liquid media, dispense into flasks or appropriate culture vessels.

9. For solidified media, add agar or Gelrite. Gelrite must be added while the liquid medium is rapidly stirred to assure its subsequent even distribution. If media is to be poured into autoclavable vessels, it must be melted first, dispensed into the vessels, and then autoclaved. The gelling agent can be melted using an autoclave, heating plate or a microwave oven. When media is to be poured into sterile petri dishes, it does not have to be melted prior to autoclave sterilization.

10. Seal flasks of media to be poured after sterilization with aluminum foil, cotton or foam plugs or place prepoured and capped culture vessels into autoclave without additional covers.

11. Autoclave for 15 min at 121 C.

12. It is preferable to pour media into petri dishes as soon as the flasks are cool enough to handle (about 60 C). It is often desirable to stack the petri dishes containing molten media about 5–10 high. By doing so, excessive condensation of water on the inner surface of the dish lid may be avoided. If there is not enough time to pour dishes, the medium can be allowed to solidify in the flask. It then is remelted by heating it in the autoclave for 3–5 min. Since reheating media tends to degrade certain compounds and cause the agar to become semiliquid, it is not advised. Note that Gelrite and Phytagel™ cannot be remelted.

## Sterilization of media

### Autoclave sterilization

The standard conditions for autoclave sterilization are 121 C for 15 min under 1.05 kg/cm$^2$. Longer times are necessary for larger volumes (Burger, 1988). For example, as much as 30 min might be required for 1 L of medium, but often the standard times will suffice. Most inorganic compounds are heat stable and thus may be autoclave sterilized safely. Sucrose may break down into fructose and glucose during autoclave sterilization. Organic compounds, such as some PGRs, amino acids and vitamins, may be degraded during autoclave sterilization and thus should be filter sterilized. In spite of heat lability and subsequent degradation, however, some compounds such as thiamine and IAA are sterilized by autoclaving if results are found to be reliable and reproducible. Compounds that are considered heat labile and, thus, should be not be autoclave sterilized include ABA, ascorbic acid, cystine, citric acid, folic acid, gibberellic acid, IAA, IBA, kinetin, nicotinamide, pantothenic acid, pyridoxine, 2-iP, thiamine and zeatin. However, often ABA, IAA, IBA, kinetin, pyridoxine, 2-iP and thiamine are nevertheless autoclave sterilized, since good results are obtained.

### Filter sterilization

Compounds that are degraded by autoclaving can be sterilized by filtration through a 0.22-μm membrane. Several manufacturers produce membranes made of nitrocellulose that can be sterilized by autoclaving. They are placed between sections of a filter unit and sterilized as one piece. Other filters come presterilized and fit on the end of a syringe. The size of the filter one selects depends on the volume of the solution to be sterilized and the components of the solution. The compounds may be filtered directly into the autoclaved, cooled (less than 60 C) medium. Swirl the contents vigorously to mix, and pour into petri dishes immediately.

## Storage of media

Media may be stored for up to 1 month if well sealed and kept in the dark. As noted above, light destroys several PGRs, and thus media stored in the light would contain reduced amounts when the explants are incubated. Although it is possible to store media for longer periods at 4 C, condensation usually forms and accelerates contamination.

## Selection of proper medium

When culturing a new plant species, one may start with several established media (see also chapter thirty-eight). The formulations are selected based on the species being cultured. Changes in macro- or micronutrients, organics or growth regulators may be required to obtain optimum growth and morphogenic response.

Adherence to strict quality standards for water, chemicals and gelling agents is essential to insure that success is obtained in the exercises that follow in this text.

## Literature cited

Ammirato, P.V. 1974. The effects of abscisic acid on the development of somatic embryos from cells of caraway (*Carum carvi* L.) *Bot. Gaz.* 135:328–337.

Burger, D.W. 1988. Guidelines for autoclaving media used in plant tissue culture. *HortScience.* 23:1066–1068.

de Fossard, R. 1976. *Tissue Culture for Plant Propagators.* University of New England, Armidale, Australia.

Epstein, E., K.-H. Chen and J.D. Cohen. 1989. Identification of indole-3-butyric acid as an endogenous constituent of maize kernels and leaves. *Plant Growth Regul.* 8:215–223.

Gamborg, O.L., R.A. Miller and K. Ojima. 1968. Nutrient requirements of suspension cultures of soybean root cells. *Exp. Cell Res.* 50:151–158.

Knudsen, L. 1946. A new nutrient solution for the germination of orchid seed. *Am. Orchid Soc. Bull.* 15:214–217.

Linsmaier, E. and F. Skoog. 1965. Organic growth factor requirements of tobacco tissue cultures. *Physiol. Plant.* 18:100–127.

Lloyd, G. and B. McCown. 1980. Commercially feasible micropropagation of mountain laurel, *Kalmia latifolia*, by use of shoot tip culture. *Intl. Plant Prop. Soc. Proc.* 30:421–427.

McGranahan, G., C.A. Leslie and J.A. Driver. 1988. In vitro propagation of mature Persian walnut cultivars. *HortScience.* 23:220.

Murashige, T. and F. Skoog. 1962. A revised medium for rapid growth and bio-assays with tobacco tissue cultures. *Physiol. Plant.* 15:473–497.

Nissen, S.J. and E.G. Sutter. 1990. Stability of IAA and IBA in nutrient medium to several tissue culture procedures. *HortScience.* 25:800–802.

Nitsch, J.P. and C. Nitsch. 1969. Haploid plants from pollen grains. *Science.* 163:85–87.

Pasqualetto, P.L., R.H. Zimmerman and I. Fordham. 1986. Gelling agent and growth regulator effects on shoot vitrification of 'Gala' apple in vitro. *J. Am. Soc. Hort. Sci.* 111:976–980.

Roberts, D.R., B.S. Flinn, D.T. Webb, F.B. Webster and B.C. Sutton. 1990. Abscisic acid and indole-3-butyric acid regulation of maturation and accumulation of storage proteins in somatic embryos of interior spruce. *Physiol. Plant.* 78:355–360.

Schenk, R.V. and A.C. Hildebrandt. 1972. Medium and techniques for induction and growth of monocotyledonous and dicotyledonous plant cell cultures. *Can. J. Bot.* 50:199–204.

Singha, S. 1982. Influence of agar concentration on in vitro shoot proliferation of *Malus* sp. 'Almey' and *Pyrus communis* 'Seckel'. *J. Am. Soc. Hort. Sci.* 107:657–660.

Skoog, F. and C.O. Miller. 1957. Chemical regulation of growth and organ formation in plant tissues cultured in vitro. *Symp. Soc. Exp. Biol.* 11:118–131.

Van Overbeek, J., M.E. Conklin and A.F. Blakeslee. 1941. Factors in coconut milk essential for growth and development of very young *Datura* embryos. *Science.* 94:350–351.

White, P.R. 1963. *The Cultivation of Plant and Animal Cells.* 2nd Ed. Ronald Press Co., New York.

*chapter three*

# Nutrition of callus cultures

*James D. Caponetti*

Callus is a relatively undifferentiated tissue consisting primarily of parenchymatous cells. Callus tissue can serve as an experimental system to investigate and solve a broad range of basic research problems in plant cytology, physiology, morphology, anatomy, biochemistry, pathology and genetics. It can also be used to resolve applied research problems in organogenesis and embryogenesis related to the propagation of horticultural and agronomic plants.

Tissues from various organs from many species of plants can be induced to form callus. However, many seemingly unrelated factors may determine the ability of a specific tissue to form callus. Among these are chemical factors such as mineral nutrition and plant growth regulators (PGRs); environmental factors such as light, temperature and humidity; and the genetic constitution or the genotype of the plant. A medium that induces good callus growth in one genotype or species may fail to do so in another closely related plant. The same holds true for a particular combination and concentration of PGRs added to the medium. Also, tissues from different plant species and genotypes can respond differently from one another under various conditions of light, temperature and humidity. Therefore, extensive empirical research is required to determine which combination and concentration of medium nutritional ingredients, PGRs and environmental factors are best for maximizing callus production for each species or genotype. Examples of such studies are Murashige and Skoog (1962) and Linsmaier and Skoog (1965).

The following laboratory experiments illustrate the technique of obtaining callus from mature tissues of several angiosperm stems and roots and compare callus growth on a standard medium formulation. Also, we will contrast callus growth of one species on several media to which ingredients are added in a serial sequence. Specifically, the laboratory experiences will familiarize students with the following concepts: (1) that one medium may be sufficient to support callus growth from some species, but is not able to support growth from others and (2) that some ingredients in a medium are more vital for callus tissue growth when provided in combination rather than separately. Each of the above mentioned experiments will be detailed below, including procedures and anticipated results. In all cases, growth will be measured by fresh and dry weight of developed callus.

## General considerations

### Source of plant material

The tobacco plants (*Nicotiana tabacum* L.) needed for these experiments can be cultivated in a greenhouse from seed obtained from a farmer's co-op store or from a biological supply company (see chapters twenty-eight and thirty-seven for details). Alternatively, tobacco plants of the appropriate size may be available from other laboratories/green-houses on campus. Plants that are about 1.5 m tall would be the appropriate size. The tobacco varieties that produce good results include WI 38, Xanthi, Burley 14, Burley 21, KY 9, KY 17, VA 509, SpG-28 and Hicks. The other plant material needed for these experiments can be purchased from a grocery store produce counter. Depending on what is available at the time of year that these experiments are conducted, good results can be obtained with Irish potato tuber (*Solanum tuberosum* L.), turnip tuber (*Brassica rapa* L.), Jerusalem artichoke rhizome (*Helianthus tuberosus* L.), sweet potato root (*Ipomoea batatas* L.), carrot root (*Daucus carota* L.) and parsnip root (*Pastinaca sativa* L.).

## Exercises

### Experiment 1. Growth potential of callus from several species of plants on a single complete medium

An important basic concept in plant tissue culture is that tissue from different species will not grow equally on one specific medium. This experiment was designed to illustrate this concept by comparing the growth of tobacco stem pith on a medium designed for it with the growth of stem and root tissues of several other plants on the same medium. The experiment can best be performed by teams consisting of two students each. Refer to Figure 3.1 for a diagrammatic representation of this exercise. The experiment requires 4 to 8 weeks in order to obtain measurable results.

#### Materials

The following laboratory items are needed for each team of students:

* Several tobacco stem sections
* One or more of the plant tissues listed above (see Figure 3.1)
* Coconut milk
* Several 150 × 25 mm culture tubes with plastic or metallic caps
* Supply of wire and wooden racks
* Number 3 cork borer and metal punch
* 10% solution of commercial bleach to which several drops of dishwashing liquid have been added
* Drying ovens set for 60 C
* Aluminum foil weighing cups

The culture medium is modified from Murashige and Skoog (1962) and Linsmaier and Skoog (1965) and was designed specifically for the culture of tobacco stem pith tissue. Each liter of medium is composed of Murashige and Skoog (1962) major and minor salts (see chapter two for composition), 100 mg of myo-inositol, 0.4 mg of thiamine HCl, 2.0 mg (10.7 µM) of naphthaleneacetic acid (NAA), 0.5 mg (2.3 µM) of kinetin, 30 g (88 mM) of sucrose and 10 g of agar. Before adding the agar, adjust the pH of the medium to 5.7. Heat to dissolve the agar and distribute about 20 ml of medium to each of 50–150 × 25 mm culture tubes. Cover each tube with a plastic or metallic cap, place the tubes in a metal rack and autoclave for 20 min. For each tissue type used in the experiment, 20–30 tubes should be prepared by each team. If facilities are limited, each team could process one tissue.

#### Procedure

Refer to Figure 3.1 for a diagrammatic representation of procedures of plant organ preparation and tissue explant culture. All plant tissues should be surface disinfested for 10 min

**Figure 3.1** Diagrammatic representation for obtaining explants from various plants. A. Tobacco plant. B. Tobacco internode segment. $C_1$ to $C_6$. All tissue removed using a number 3 cork borer after surface disinfestation. $C_1$. Tobacco pith; $C_2$. Jerusalem artichoke pith; $C_3$. Irish potato pith; $C_4$. Turnip pith; $C_5$. Carrot and parsnip cortex and phloem tissues; $C_6$. Sweet potato cortex and phloem tissues. D. Cylinder of tissue removed from the cork borer and placed into a sterile, plastic petri dish. E. Tissue cut into equal-sized disks for culture and determining fresh and dry weights. F. Tissue placed in culture tubes.

in 10% bleach solution in 1500 ml beakers; sufficient bleach should be used to completely submerse the tissue. The bleach is removed by rinsing several times with sterile distilled water. Sterilize forceps, scalpels and cork borers by dipping them in the 95% ethanol and burning off in the flame. Allow instruments to cool for 5–10 sec before using. Force the cork borer through the tissue and withdraw — the tissue should be contained within the cylinder. Express the tissue from the barrel of the cork borer with a sterile metallic punch into a sterile plastic petri dish and, with a scalpel, cut into 50–60 equally thick sections. Typically, one piece of tissue is placed on the medium surface in the middle in such a way that it does not dislodge, but be careful not to submerge the explant below the medium surface. Culture tubes are placed in a randomized complete block design in plastic or wooden racks designed to hold about 30 culture tubes.

For each tissue type, weigh 30 disks collected in a tared aluminum foil cup and record the fresh weight. Divide the weight by 30 to obtain the average fresh weight of one disk. Place all cups into the drying oven set at 60 C for several days to obtain dry weights. Again, divide the weight by 30 to obtain the dry weight of one disk. The fresh and dry weights of the explant tissues will be subtracted later from the callus weights at the termination of the experiment in order to determine net gain in weight of tissues.

Place the racks of prepared cultures in a growth chamber or incubator where cultures may receive white light of about 25 $\mu mol \cdot m^{-2} \cdot sec^{-1}$ at 25 C and relative humidity of 60–70%. If a growth chamber or incubator is not available, place cultures in the laboratory on a table near a north-facing window, i.e., not in direct sunlight.

When the tobacco explant disks have each grown into callus that almost fills the diameter of the tube (about 4–8 weeks), it is time to terminate the experiment. Pool uncontaminated calluses from tissue of the different species into tared aluminum foil weighing cups, record the average fresh weight and divide by the number of calluses to obtain the fresh weight of one callus. Then place all cups in the drying oven for several days to obtain dry weights and calculate the dry weight of one callus as before. Don't forget to subtract fresh and dry weights of explant disks in order to obtain accurate weight gains of calluses. Student teams should share and analyze the data by one of the procedures described in chapter six.

### Anticipated results
Since the culture medium used was designed for tobacco, the tobacco explant disks will produce the largest and heaviest calluses when the experiment is terminated. The other plant tissues will produce callus, but not to the extent that tobacco disks do.

### Questions

* What is the function of each ingredient of the medium in the growth of callus?
* Why are fresh and dry weights recorded for calluses of each plant tissue?
* Why is it important that all explant disks be of uniform size?


## Experiment 2. Initiation and growth of tobacco callus on serially composed media

Another important basic concept in plant tissue culture is that most tissues require a combination of specific nutrients and PGRs to produce the appropriate growth response of the tissues. This experiment was designed to illustrate this concept by starting with a medium without nutrients, and then preparing eight other media by adding one category of nutrients at a time to each medium, ending with a medium that contains all the essential nutrients. Two additional media test the effects of coconut milk, which supposedly contains complete nutrition and PGRs for growth of callus from explant tissues of tobacco.

All 11 media are different, but are based on the MS medium described previously and should be prepared as directed in experiment 1. The 11 media all include distilled water and 1% agar and are as follows (per liter):

1. Water only;
2. MS salts;
3. MS salts, 3% sucrose;
4. MS salts, 3% sucrose, 0.4 mg thiamine HCl;
5. MS salts, 3% sucrose, 100 mg myo-inositol;
6. MS salts, 3% sucrose, 0.4 mg thiamine HCl, 100 mg myo-inositol;
7. MS salts, 3% sucrose, 0.4 mg thiamine HCl, 100 mg myo-inositol, 0.5 mg (2.3 μM) kinetin;
8. MS salts, 3% sucrose, 0.4 thiamine HCl, 100 mg myo-inositol, 2.0 mg (10.7 μM) NAA;
9. MS salts, 3% sucrose, 0.4 mg thiamine HCl, 100 mg myo-inositol, 0.5 mg kinetin, 2.0 mg NAA;
10. 20% coconut milk (200 ml of coconut milk and 800 ml of water);
11. However, the amount of nutrients in medium 10 may not be sufficient to produce the full potential of callus growth. Therefore, it is necessary to test the performance of tobacco pith on a medium 11 that consists of all the ingredients in medium 9 and

20% coconut milk. Coconut milk can be obtained from fresh nuts, as a canned concentrate from an ethnic Asian grocery store or from Sigma (St. Louis, MO).

### Procedure

This experiment is best performed by teams of two students each. Refer to Figure 3.1 for a diagrammatic representation for obtaining tobacco pith disks for this experiment. Use tobacco plants, laboratory items and procedures outlined in experiment 1 including procedures for obtaining fresh and dry weight of tobacco callus from each of the 11 media. Don't forget to subtract fresh and dry weights of explant disks in order to obtain accurate weight gains of tissues at the conclusion of the experiment. The experiment usually requires 4–8 weeks for the pith disks on medium 11 to produced callus that almost fills the diameter of the tube. Student teams should share and analyze the data by one of the methods described in chapter six.

### Anticipated results

Growth of callus from the disks on media 1 through 8 will be relatively poor. Growth from disks on medium 9 will be good because all essential nutrients are present. Callus growth on medium 11 will be best because it contains all the nutrients of medium 9 plus the nutrients from coconut milk. Growth of callus on medium 10 will occur, but will be less than that on media 9 or 11.

### Questions

* What concept of tissue growth does this experiment demonstrate?
* Why does the experiment work best with tobacco pith tissue?
* Would the experiment work with other explant tissues? Explain.
* Would raising the temperature of the growth chamber to 35 C improve the growth of the tobacco tissues? If so, why? If not, why not?

## Literature cited

Linsmaier, E.M. and F. Skoog. 1965. Organic growth factor requirements of tobacco tissue cultures. *Physiol. Plant.* 18:100–127.

Murashige, T. and F. Skoog. 1962. A revised medium for rapid growth and bioassays with tobacco tissue cultures. *Physiol. Plant.* 15:473–497.

*chapter four*

# Histological techniques

*Robert N. Trigiano, Kathleen R. Malueg and Effin T. Graham*

Plant histology can be defined simply as the study of the microscopic structures or characteristics of cells and their assembly and/or arrangement into tissues and organs. Several histological techniques are commonly used for examining plant tissues, each providing somewhat similar gross information, but differing in resolution of details and the medium in which the samples are prepared. These techniques include those for brightfield and fluorescent microscopy, in which specimens can be prepared for cutting into thick sections (15–40 μm) either without a stabilizing medium (fresh sections), in cryofluids (frozen sections), or embedded in paraffin-like materials or in various formulations of plastic. Other techniques that employ electron microscopy either do not require specialized embedding medium for specimen preparation (scanning electron microscopy) or have samples embedded in plastics (transmission electron microscopy) that can be cut into ultrathin sections (65–100 nm). Explanation of the details involved in all of these techniques is beyond the scope of this chapter.

One might reasonably ask why include a chapter on histology in a plant tissue culture book? In many situations, histological techniques provide essential information that may not be evident upon only visual inspection. Much of our understanding of the in vitro developmental processes that are presented throughout this book largely result from detailed histological research. For example, somatic embryos (chapter sixteen) can be produced on the surface of a leaf explant, but may be so morphologically aberrant that they are unrecognizable. By using histological techniques and scrutinizing anatomical features, the characteristics of somatic embryos can be more readily seen. Another example for using histological techniques is to investigate the origin of specific structures, i.e., adventitious shoots and roots (chapter eleven), embryos, etc., that develop in culture. Histological development can be studied over time by periodically sampling tissues and/or the result can be examined in the mature structure. One must always be mindful that growth of tissues is dynamic and changing from moment to moment, whereas histological sections, for example, are static and fixed in time, and only present a very narrow glimpse of the developmental process. Nevertheless, the origin of tissues and organs may be convincingly inferred from serial observation of many samples and sections.

The field of plant histology and microtechnique is quite broad and cannot be conveyed in one simple chapter. Therefore, the intent of this chapter is to briefly summarize the essentials and provide a "primer" of sorts for preparing specimens from plant tissue cultures for histological examination using elementary paraffin and scanning electron microscopy techniques. The techniques presented herein are considerably easier and avoid many of the toxic materials found in more traditional protocols. For a broader view of general histological methodologies, students should consult one or more of the following references: Berlyn and Miksche, 1976; Sass, 1968; Jensen, 1962; Johansen, 1940.

# General considerations

## Equipment and supplies

The equipment and materials needed to complete a paraffin histological study are listed below:

* Indelible marker (Scientific Products [SP], Atlanta, GA)
* 500- or 1000-ml plastic beaker
* Aluminum pan
* Rotary microtome with disposable blade holder (SP)
* Warming oven (58–60 C)
* Two slide warming trays (40 and 50 C)
* Water bath (50 C)
* Hot/stir plate
* 1% agar or agarose in a 150-ml beaker
* Disposable microtome blades (SP)
* Paraplast® embedding medium (SP)
* Disposable plastic base molds and embedding rings (SP)
* 5-ml microbeakers (SP)
* Alcohol lamp and metal spatula
* Screw-capped Coplin jars
* Histochoice® fixative (Amresco, Inc., Solon, OH) plus 20% ethanol
* Disposable snap cap plastic 10-ml specimen vials
* Wooden applicator sticks
* Clean glass slides and cover glasses
* Camel or sable hair artist brush
* Test tube racks
* Eukitt cover glass resin (EMS, Fort Washington, PA)
* Microclear® (Micron Environmental Industries, Fairfax, VA)
* 30, 50, 75, 95 (all aqueous) and 100% isopropanol
* 0.2% aqueous solution of alcian blue 8GX (Sigma) (200 mg/100 ml of water)
* Quick-mixed hematoxylin (Graham, 1991) — Stock A: 750 ml distilled water, 210 ml propylene glycol, 20 ml glacial acetic acid, 17.6 g aluminum sulfate and 0.2 g sodium iodate; Stock B: 100 ml propylene glycol and 10 g certified hematoxylin (Sigma, St Louis, MO). Stain is prepared by adding 2 ml of Stock B to 98 ml of Stock A.

## General protocol for paraffin studies

The procedure described below will work well with most tissues, including those from stock plants as well as those cultured in vitro. Moreover, this procedure has several important advantages over more traditional paraffin protocols. First, it circumvents the use of toxic aldehydes (formaldehyde) and heavy metals (chromium and mercury) as fixatives. Second, it avoids the use of specimen/slide adhesive agents, such as Haupt's, that require the use of formaldehyde. Third, it does not use toxic substances, such as xylene, for deparaffinizing sections. Fourth, sections are stained directly through the paraffin, greatly reducing the number of operational steps in the process. The advantages listed above combine to make preparing tissue for histology and slides a safe and less intimidating process for both students and instructors alike.

### Fixation

The first step in the protocol is to identify typical specimens (i.e., embryos on leaf sections), remove them from the culture dishes, and place in a small petri dish containing water. Most explants are too large to be adequately fixed and usually the subject (embryos) occupies a very small area. The specimen then should be carefully trimmed using a razor blade or a scalpel to include the subject and a small area of surrounding tissue. Five or fewer specimens are placed in about 5 ml of Histochoice®, a nonaldehyde, nontoxic fixative amended with 20% ethanol, contained in each specimen vial. If possible, the open vials should be aspirated (–1 atm or less) for about 30 min to remove air from the samples and promote infiltration of the fixative. Replace the original fluid with fresh fixative. The vials are closed and the tissue allowed to remain in the fixative at room temperature for a minimum of 24 h; samples may be stored at room temperature for weeks or months without degradation of structures. If long-term storage is needed, we suggest using screw-capped rather than snap-capped vials.

### Dehydration and infiltration

After fixation, the samples must be dehydrated and placed in a solvent compatible with paraffin. Traditionally, this has been accomplished with graded series of ethanol and transitioning to tertiary butyl alcohol. This process can be considerably shortened by using isopropanol to both dehydrate the tissue and dissolve the paraffin. Histochoice® fixative can be removed from the vials with a long, thin pipette and poured down the drain with plenty of water. Specimens are dehydrated using 30, 50, 75, 95 and three changes of 100% isopropanol — each for about 30 min. After the last change of pure isopropanol, fill the vials about one quarter full with fresh 100% isopropanol, add a few pellets of Paraplast® embedding medium, loosely recap the vials and incubate at 60 C. Periodically over the next several hours, swirl the contents and add a few more pellets of Paraplast® to each of the vials. At the end of the day, remove the caps from the vials to allow the isopropanol to evaporate. At this time, fill a 500- or 1000-ml plastic beaker with Paraplast® pellets, place several pellets into a number of base molds contained in an aluminum pan and put in the 60 C oven. The next morning all of the isopropanol should be dissipated, the specimens completely infiltrated with Paraplast®, and the pellets in the beaker and base molds melted.

### Casting specimens into blocks

Using a sharp razor blade, whittle a flattened paddle-shaped end on several wooden applicator sticks and store in a vial with the flattened end immersed in molten paraffin in the oven. Transfer a specimen into the molten paraffin contained in a base mold and position it so that the sectioning plane of the specimen is parallel to the bottom of the base mold. For example, if a transverse section of a stem is required, the stem should be placed perpendicular to the base mold; the long axis of the stem should be "sticking up" straight in the molten paraffin. If the stem is positioned flatly in the base mold, longitudinal sections will result. With more complex specimens, like single somatic embryos or shoots growing from basal tissue, this orientation step becomes more critical. It is important to visualize the exact desired section plane at this step. Once the sample is oriented correctly, quickly transfer the base mold to a cool surface — an inverted aluminum pan containing ice works well. The paraffin surrounding in the bottom of the base mold will quickly become cloudy (congealed), and the specimen will be immobilized. Rapidly attach an embedding ring to the base mold and fill the apparatus with molten paraffin from the plastic beaker. This is a very vulnerable point in the process. If the immobilizing paraffin is allowed to congeal too far, the added molten paraffin will not bond with it and the block will split apart. At this time, an identifying label consisting of white paper and text written in pencil (do not use ink) may be inserted partially into the molten paraffin. Place the casted blocks in the refrigerator or they may be floated in a large beaker containing ice water. The base mold may be removed after the block is hardened.

### Microtomy and mounting sections on slides

A word of caution before beginning to section. Regardless of whether a stainless steel knife or a disposable razor blade is used to cut sections, they are both extremely sharp and cut flesh very easily. Exercise care while working with the microtome, and if the microtome is left unattended, place a note on the knife holder warning in bold print that a knife or blade has been installed. The rotary action should be kept locked at all times except when sectioning.

One of the advantages of using embedding molds is that the cutting face of the block is square. If the block were sectioned as is, the resulting "ribbon of very large sections" would be straight. However, seldom does the specimen occupy the entire block face. Therefore, it is often desirable to trim away some of the excess paraffin from around the sample. The horizontal edges of the block must be parallel with each other or, another way of visualizing it is, the sides of the block must be of equal length. If the ribbon of sections curves to the right or left, the side of the block opposite the direction of curvation is too long. Remove the block from the chuck and retrim so that the sides are parallel and of equal length. Do not trim the block while mounted in the chuck over the knife!

Mount a new disposable blade in the blade holder — only a small portion of the blade should be clear of the carriage. Typically, new blades are coated with oil and may be cleaned with a small wad of tissue moistened with a little Microclear® — carefully wipe tissue perpendicular to the blade surface and allow to completely dry. Set the microtome to cut 10-µm sections. Now, mount the specimen block in the chuck and orient it so the bottom of the cutting face is parallel to the edge of the blade. Position the specimen at the level of the knife and carefully and gently slide the knife holder toward the block until they are very close. With the utmost care, turn the wheel clockwise causing the specimen to advance 10 µm per revolution. The first sections may not include the entire block face — continue advancing the block until a complete section is obtained. Lock the advancing wheel into position, carefully remove the partial sections and clean the knife as before. After the Microclear® has evaporated, continue to operate the microtome and after 5 or more sections are produced, place a camel hair brush underneath the ribbon for support while it is being held away from the knife holder. A 30–50 section ribbon can be produced and easily manipulated. Lay the ribbon out on a clean piece of dark paper and, using a razor blade, divide it into lengths that do not exceed about 75% the length of a glass microscope slide.

Prewash glass microscope slides in a mild dishwashing detergent solution, rinse well with tap water and rinse with distilled water. Allow the slides to dry in test tube racks with an aluminum foil base. Washed and dried slides can be stored indefinitely in slide boxes. Label the frosted portion of the slide with appropriate information using an indelible marker or lead pencil and cover the specimen surface of the slide with distilled water. Moisten the tip of a brush, touch it to the surface of the first section to the left, and lift the entire ribbon off of the paper. Now, touch the first section to the right onto the water and lay the rest of the ribbon down so that air is not trapped underneath the sections. Place the slide on the 50 C warming table to expand the sections. After about 10 min, set the slide on a sponge and slightly tilt it to drain off most of the water. Affix the sections to the glass by placing the slide on the 40 C warming table for 12–16 h or overnight. The sections are now ready to be stained.

### Staining sections

The following staining procedure is adapted from Graham and Joshi (1995). Preheat the alcian blue solution in a closed Coplin jar. Place the slides in the Coplin jar and stain for

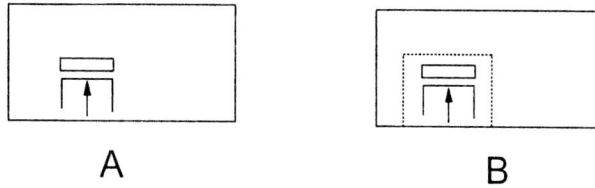

**Figure 4.1** Immobilization. A. A specimen (stem) is placed on an agar-coated slide, then covered with molten agar. A white slip of paper with an arrow drawn in pencil is also affixed with agar. B. After the agar has hardened, the preparation may be trimmed and removed (dotted lines) from the glass slide.

1 h at 50 C in a water bath. Gently rinse the slides in a stream of cold tap water and carefully blot dry the remaining water droplets. Transfer the slides to Coplin jars containing hematoxylin stain for 15 min at room temperature, rinse and dry as described before, and place on the 40 C slide table for a minimum of 30 min. The paraffin in the sections is removed with three 5-min soaks in Microclear® followed by three 5-min changes of 100% isopropanol. After the slides have air-dried, pool some resin (Eukitt) in the center of the slide and gently lower a cover glass so that the resin spreads evenly over the surface without air bubbles. The resin may be cured overnight on a 40 C warming table.

Do not become discouraged if your results are less than perfect the first time — don't give up — try again! Like any other technique, the operator becomes more skilled after some experience.

### Immobilization of specimens for paraffin sectioning

This sample preparation technique involves immobilization and precise orientation of specimens between two layers or a "sandwich" of agar. It offers several advantages over more conventional methods of preparation, especially when working with small and/or delicate tissues. Small samples, e.g., embryos or callus, are often lost or damaged during the dehydration sequences when fluids are removed and replaced. By immobilizing the specimens in agar, it is nearly impossible to lose samples and the agar minimizes damage from handling. However, the real advantage to the technique is that the sample can be oriented in the agar before fixation eliminating the need to position it while casting the block.

*Procedure.* This procedure is adapted primarily from Hock (1974). Prepare 1% water agar (1 g agar per 100 ml of water in a tall 150-ml beaker) on a hot/stir plate and after the agar is melted store in a 50 C water bath. Dip a clean glass microscope slide into the molten agar and then place it in a 100 × 15 mm petri dish containing moistened filter paper. The remainder of the immobilization procedure is diagrammatically represented in Figure 4.1. Select a specimen and place it on top of the cool and hardened agar. With a pasteur pipette, cover the sample with a small amount of molten agar. The position of the sample is now secure. Draw an arrow with pencil on a piece of white paper about the width of the sample. Using a stereomicroscope scope, if necessary, and fine forceps, place the arrow directly opposite or behind the desired cutting face of the tissue and tack it into position using molten agar. Cover the entire preparation with a thin coating of molten agar. Once the agar has hardened, trim any excess agar away using a razor blade. The specimen can be removed from the slide by "slipping" the razor blade under the bottom of the agar sandwich and placing it in fixative. Follow the previously described protocols for fixation of tissue and infiltration with paraffin.

Instead of using an embedding mold, use a 5-ml plastic microbeaker to cast the specimen into a paraffin block. Draw a straight line on the bottom of the beaker with an

**Figure 4.2**   Top view of positioned specimen
cast in a plastic microbeaker.

indelible pen. Fill the container with molten paraffin and transfer the sample to the beaker so that the sample is parallel, yet slightly behind, and the arrow is positioned orthogonally (at a right angle) to the line on the bottom (Figure 4.2). Place the microbeaker on a cold surface and label with a paper slip as described before.

### Dissection and mounting of specimens

The block of paraffin in which the specimen was cast cannot be mounted on the microtome. Therefore, it is necessary to dissect the specimen and a small amount of the surrounding block and mount it on a block formed by an embedding mold. Usually there are some old blocks around from previous projects or newly cast ones may be used. Before the microbeaker is removed, take a sharp razor blade and cut through the beaker and score the paraffin in the block underlying the line on the microbeaker. This will ensure that the cutting face is identified even if the specimen and arrow are not clearly visible in the block. Dissect the specimen from the block. Create a flat field on the face of the embedding mold block. Melt a thin layer of the embedding block using the flat side of a spatula heated with an alcohol burner. Quickly attach the side opposite of the cutting face of the specimen block to the embedding mold block. Reinforce the attachment by melting small chips of paraffin with a hot spatula tip along all four sides of the specimen block (Figure 4.3). The block may be sectioned and slides prepared as previously described.

## Preparation of specimens for scanning electron microscopy

The sections prepared for paraffin histology present a two dimensional (flat) interior view of a small area of a much larger three dimensional object. In contrast, scanning electron microscopy (SEM) allows for three-dimensional topical or internal views of an entire specimen. Samples for SEM do not need to be embedded in paraffin nor do they need to be stained. Moreover, the resolving power of a scanning electron microscope is much greater than a compound light microscope. However, in many ways, preparation of samples for SEM is technically easier than for paraffin histology. Note: methodologies for

**Figure 4.3**   Mount the dissected specimen (A) to a flat block cast using a embedding mold (B). Note the paraffin reinforcement along the sides of the specimen.

critical point drying and sputter coating are not presented and should be performed by a competent SEM technician or other person familiar with these techniques.

The following materials will be needed to prepare samples for SEM:

* 3% glutaraldehyde in 0.05 M potassium phosphate buffer, pH 6.8 — store in the refrigerator
* 1% osmium tetroxide in the same buffer (optional) — store in a shatterproof bottle in the freezer
* Aluminum mounting stubs

All of the above may be purchased from Electron Microscopy Sciences, Fort Washington, PA.

* Double-sided sticky tape
* 0.05 M potassium phosphate buffer, pH 6.8 — prepare by dissolving 0.87 g $K_2HPO_4$ and 0.68 g $KH_2PO_4$ in 200 ml of water.
* Graded series of ethanol or acetone (10, 30, 50, 75, 95 and 100%)

*Note:* Glutaraldehyde and osmium tetroxide are extremely toxic and should only be handled wearing gloves and in a certified fume hood. For additional information, consult material safety data sheets.

Fix and aspirate tissue in a cold (4 C) buffered solution of glutaraldehyde for at least 24 h. Dispose of the fixing solution in a waste bottle and rinse specimens at least three times with cold phosphate buffer. With great care and working in a fume hood, dispense enough of the cold, buffered osmium tetroxide (osmium) solution to cover the samples and recap the vials. After 2 h, dispose of the osmium in a properly labeled hazardous waste bottle and rinse the tissues with three changes of cold buffer. In many instances, the tissue will be blackened after postfixing with osmium. Dehydrate the tissue for 30 min in each member of the graded ethanol or acetone series except 100%. The final steps of dehydration are three changes of room temperature absolute ethanol or acetone over 30 min. The samples may be stored in absolute ethanol for a few days until they can be critical point dried by the SEM technician. Note: we suggest that if the specimens cannot be dried within a few days, that they be stored in 75% ethanol or acetone.

Place a piece of double-sided clear tape onto a stub and mount the dried sample on the tape. The samples are ready to be gold–palladium coated and viewed with the aid of a technician.

## Literature cited

Berlyn, G.P. and J.P. Miksche. 1976. *Botanical Microtechnique and Cytochemistry.* Iowa State University Press, Ames, IA.

Graham, E.T. 1991. A quick-mixed aluminum hematoxylin stain. *Biotechnic Histochem.* 66:279–281.

Graham, E.T. and P.A. Joshi. 1995. Novel fixation of plant tissue, staining through paraffin with alcian blue and hematoxylin, and improved slide assembly. *Biotechnic Histochem.* 70:263–266.

Hock, H.S. 1974. Preparation of fungal hyphae grown on agar coated microscope slides for electron microscopy. *Stain Technol.* 49:318–320.

Jensen, W.A. 1962. *Botanical Histochemistry.* W.H. Freeman Co., San Francisco.

Johansen, D.A. 1940. *Plant Microtechnique.* McGraw-Hill, New York.

Sass, J. E. 1968. *Botanical Microtechnique.* Third ed. Iowa State University Press, Ames, IA.

*chapter five*

# Photographic methods for plant cell and tissue culture

*Dennis J. Gray*

Plant cell and tissue culture inherently is based on direct visual observation, unlike other disciplines, such as molecular genetics, virology and physiology, where results from experiments often are abstract. With in vitro culture, the confirmation that the basic experimental response has occurred is based primarily on visual proof. For example, whether or not cultured tissue undergoes embryogenesis or organogenesis is confirmed by visualizing the presence of somatic embryos or organs. Thus, the fundamental documentation of plant cell and tissue culture research often is a photograph.

As an essential component of cell and tissue culture research, accurate photography is needed in order to satisfy the doctrine of scientific research, which demands that experimental methods and results be documented and repeatable so that they can be confirmed and built upon by others. Unfortunately, many research reports are plagued by poor or inadequate photographic technique. Because of the pivotal nature of photography in plant cell and tissue culture research, poor photographic documentation severely detracts from or, in many cases, jeopardizes the credibility of a research report. This is unfortunate, since high-quality photography is not difficult given proper equipment and supplies.

In-depth instruction in photographic technique is beyond the scope of this book and will not be attempted. Instead, we will assume that a general knowledge of photography and microscopy preexists or is readily attainable. In this chapter, the use of equipment, supplies and techniques that are required for basic photographic documentation will be described. Although advanced techniques that utilize photography, such as confocal microscopy and video imaging are becoming more common, they remain highly specialized and will not be included in this chapter. Similarly, scanning and transmission electron photomicroscopy routinely are used in plant cell and tissue culture research, but generally are taught as separate courses and will not be discussed here.

## Types of photography used in plant cell and tissue culture research

Several different photographic techniques are required in order to document various aspects of plant cell and tissue culture. Because objects of interest vary greatly in size and type (e.g., from a cell to a culture vessel), but often are very small, equipment to accomplish both macro- and microphotography is required.

**Figure 5.1**   Typical setup of specimen in petri dish on stereomicroscope. Note fiber optic illuminator bundles positioned on each side of the specimen.

## Macrophotography

When cultures contained in whole petri dishes or groups of petri dishes must be photographed, a simple 35-mm SLR camera equipped with a close-up lens and mounted on a copy stand is preferable. In this instance the photographic field might range from 2 cm (a large callus culture) to 10 cm (a petri dish) to 40 cm diam. (a group of petri dishes).

## Microphotography using stereomicroscopes

For characterization of callus cultures, small shoots, nonzygotic embryos and other objects that range in size from 0.5 mm to 1.5 cm, stereomicrography is most appropriate. A stereomicroscope or "dissecting scope" allows the surface morphology of a specimen to be examined. Because specimens often possess much surface relief, the ability to photograph a wide depth of field is essential in stereomicrography. Depth of field can be adjusted on those stereomicroscopes that incorporate an iris diaphragm between the objective lens and the eyepieces. Depth of field increases as the iris diaphragm is closed (see the example below). Stereomicroscopes often use a separate light source. The ideal type of light source for stereomicrography consists of a fiber optic illuminator with a bifurcated fiber bundle. This arrangement allows for great flexibility in lighting, allowing an optimum view of the specimen to be obtained (Figure 5.1).

## Microphotography using compound microscopes

A compound microscope is used when smaller objects such as early-stage embryos, or cells, or the internal structure of cells and tissues must be examined. A compound microscope can resolve objects from the 10 µm to 1.5 mm size range, depending on optical quality. However, a compound microscope differs from a stereomicroscope in that it utilizes light that is transmitted through the specimen. Therefore, the specimen must be thin and lucent enough to allow for light to pass. Typically, histological processing of tissue is employed to produce lucent sections (see chapter four), or else cells and tissues that can be pressed thin enough for light to pass are utilized. Compound microscopes are configured as either "upright", in which the specimen is below the objective lens, or "inverted", in which the specimen is placed above the objective lens. Upright microscopes are most common, typically have the highest resolving power, and usually are used to view specimens mounted on microscope slides with coverslips. Inverted microscopes are used to examine cells and tissues in liquid, such as suspension cultures or protoplasts, since they can view specimens through the bottom of the culture vessel. Inverted microscopes are indispensable in protoplast research, where living cultures must be monitored in situ.

## Focusing and cropping the image

Methods of adjusting the camera to obtain sharply focused photographs differ from camera to camera and between macrophotography and types of microphotography, with stereomicrography generally being the most difficult to accurately focus.

Focusing the subject on a copy stand usually is not difficult and amounts to following the directions for a given SLR camera. One note of caution, however, is to make sure that the camera angle is the same as that of the copy stand bed in order to maintain a consistent focal plane across the field of view. This parameter frequently is out of alignment, but can be measured and adjusted by using a small bubble level (obtained from a hardware store) to compare and adjust the angle of camera vs. that of the copy stand bed.

Focusing the subject through microscopes is more difficult, since most photomicrography systems must be adjusted to synchronize the optical focal plane (i.e., the image that the eye sees) with the film focal plane. Since everyone's eyes focus differently, users must learn to adjust the camera to their own preference. Focal systems vary between photomicroscopes (individual owner's manuals should be consulted), but often consist of a photoreticle (an inscribed pattern of black lines visible through one eyepiece) that must be brought into sharp focus along with the specimen by a combination of adjusting both the microscope focus and the focusing ring on the individual eyepiece. Once this is accomplished, the optical and film focal planes are synchronized so that focusing the specimen through the eyepiece results in the camera also being in focus. It is recommended that those who normally wear corrective lenses use them during photomicrography in order to obtain a consistent sharp focus.

Generally, the specimen should be magnified to fill most of the viewing area. A common mistake is to produce a photograph of a large empty field with only a small specimen in the center. This is not only a waste of film, but does little to illustrate the specimen.

## Film types

Film is the recording medium of choice for photography. Typically, both color positive (slide) and black and white negative (B & W) film types in 35-mm format are used. Color slide film is used to provide lifelike, detailed images. 35-mm slides are conveniently stored and are the desired medium for oral presentation of results. B & W film provides a negative that can be used to produce B & W prints, which is the most commonly used method of presenting results in a publication. Use of the correct film type often determines whether or not an acceptable photograph can be obtained.

### Color slide film

Color film is highly sensitive to color balance of the light used for illumination. Since most light used to illuminate copy stands, stereomicroscopes and compound microscopes is "tungsten balanced", a compatible color film type must be used. Tungsten balanced film usually has a *T* or the word "tungsten" printed after the film type. Excellent film for color photography is Kodak Ektachrome ASA 64 T (EPY 135), for use with copy stands or microscopy where abundant light is present, and Kodak Ektachrome ASA 160 T (EPT 135), for use in low-light situations, such as high magnification microscopy or fluorescence microscopy. In all instances, the light source, be it on a copy stand, fiber optic illuminator or built into the microscope, must be adjusted to the maximum or near maximum level in order to achieve the proper color balance to be compatible with tungsten film.

### B & W film

While B & W film is not subject to color balance problems, quality of prints produced from color negatives varies tremendously with the film type. For very low-light conditions, any fast (high ASA number) film can be used. However, for high-quality images, Kodak Technical Pan (TP 135) is recommended. This film may be used at several different film speeds (ASA levels) and with several different developers. However, the highest-quality negatives result from utilizing a film speed of ASA 25 and development in Kodak Technidol developer. The resulting negatives may be sent out to have prints produced or, preferably, be printed by the investigator, which allows the ideal enlargement and exposure to be obtained and generally results in the best composition.

## Composition

While proper equipment and materials are essential in order to accomplish acceptable photography, the composition of the subject in the viewing area often makes the difference between a poor photograph and an excellent photograph. As mentioned above, the specimen should be carefully cropped so as to fill most of the viewing area. Also, the orientation and lighting of the specimen is critical. The following example illustrates the importance of depth of field adjustment and specimen orientation to obtain an optimized micrograph with a stereomicroscope.

Figure 5.1 shows the typical setup of a stereomicroscope and fiber optic illuminator with a bifurcated fiber bundle. A petri dish containing cotyledon bases of *Cucumis melo* L. on agar, which were cultured to produce somatic embryos (see chapter eighteen), is on the microscope stage. The petri dish lid is removed briefly during each exposure. The microscope stage has a black surface, which generally provides better contrast. The fiber bundles are positioned to provide optimum lighting of the specimen, as determined by viewing the specimen through the microscope while maneuvering the bundles. The illuminator is turned to the maximum power setting if color film is to be used. The specimen within the dish is adjusted on the agar so as to show the desired view (in this case, embryos) and the magnification is adjusted to fill the viewing area to the greatest extent possible.

### Depth of field

With stereomicrography, depth of field, which can be adjusted on those microscopes that possess an iris diaphragm between the objective and eyepiece lenses, is of critical importance. Figure 5.2A shows a typical photomicrograph obtained with the iris diaphragm set to full open. The focus was set on the apical dome and foreground of the cotyledon (large arrow). Note that the background area of the cotyledon (smaller arrow) is out of focus as is the globular stage embryo in the foreground (smaller arrow). This demonstrates the limited focal plane obtained under these conditions due to reduced depth of field. An improved view of the specimen is shown in Figure 5.2B, which was obtained by adjusting the iris diaphragm on the stereomicroscope so that it was nearly closed, thus increasing the depth of field — no other adjustments were made. Note now that both the fore- and backgrounds are in focus, resulting in a superior micrograph.

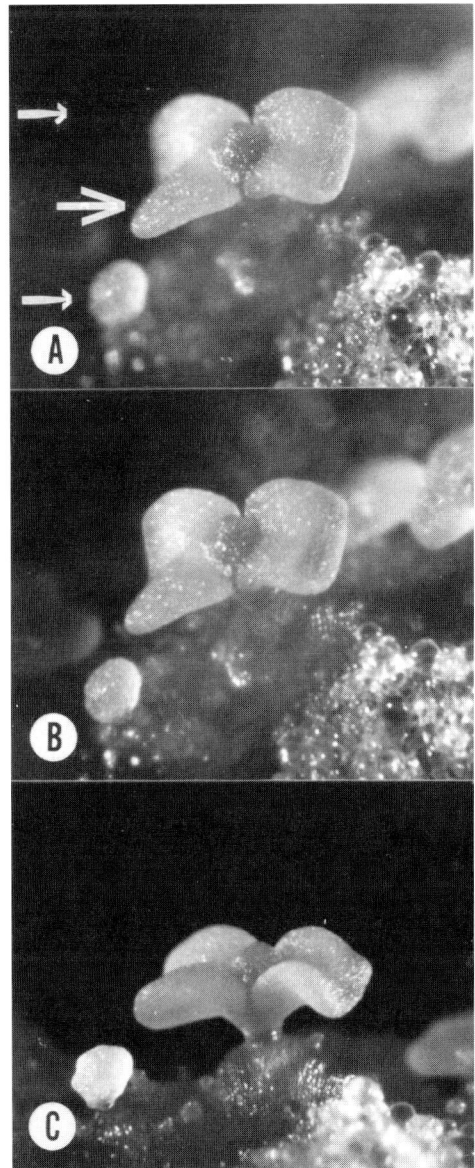

**Figure 5.2** Optimization of depth of field and orientation of *Cucumis melo* somatic embryos for photography with a stereomicroscope. A. Specimen photographed with poor depth of field. Note area that is in focus (large arrow) vs. out-of-focus areas in the fore- and background (small arrows). B. Specimen photographed with increased depth of field obtained by closing the stereomicroscope iris diaphragm. Note that fore- and background now are in focus. C. Same embryos as in A and B demonstrating improved photograph obtained by careful orientation of the specimen.

## Orientation

Specimen orientation, particularly when accomplishing stereomicroscopy, makes a dramatic difference in the clarity of documentation. Figure 5.2C shows the same embryos as in Figures. 5. 2A and B, except that the specimen was positioned to show the embryos from the side. In this orientation, much more of the important morphological attributes of the larger embryo can be seen, including the attachment of the embryo to cotyledon tissue. Note also that the morphology of the subtending globular stage embryo can be better determined.

## Conclusion

Photography is an important tool in tissue and cell culture research. Due to the ranges in size and detail of tissue and cell culture specimens, several different types of photography usually are needed to provide adequate documentation. While a photograph is relatively simple to obtain, proper photographic and microscopic equipment and supplies are required to accomplish high-quality photography. With equipment in place, photographic technique then becomes of paramount importance. Details of composition often determine the difference between acceptable and unacceptable photography.

## Acknowledgment

This is Florida Agricultural Experiment Station Journal Series No. R-04838.

*chapter six*

# Statistical analysis of plant tissue culture data

## Michael E. Compton

Statistics is an important tool for biological research because it provides an objective, nonbiased way to evaluate experimental treatments. In statistics, we use the mathematical probability that an event can occur to determine whether or not treatment effects are real. Statistics measures variation both within and between experimental treatments, allowing scientists to select the best treatment. Therefore, statistics should be included as an important part of experimental design and implementation in order to interpret results. This chapter is a shortened version of a review article published previously on this subject (Compton, 1994) and discusses methodologies associated with scientific research, experimental designs and methods of comparing treatment means, providing examples of how to present data in tables and graphs.

## Methodologies associated with scientific research

In science we use a logical and objective approach to problem solving. We examine what is already known by reading literature and formulate a hypothesis that can be tested experimentally. The hypothesis is tested objectively and either accepted or rejected according to the results obtained. The "null" hypothesis is typically used in statistics and states that all experimental treatments elicit a similar response. Rejection of the null hypothesis denotes that the imposed treatments vary in the response that they cause.

Before designing an experiment, the scientist should clearly and concisely define the problem to be solved in order to formulate questions that will lead to a solution. A set of experimental objectives is formulated to help solve the problem. Next, a series of treatments that allows the scientist to study the problem are established. The treatments should be based on prior knowledge of the subject matter and evoke a wide range of response. Experimental materials used in plant tissue culture studies, such as culture vessels, culture medium, growth environment, etc., are selected in accordance with experimental objectives when establishing treatments. Knowledge of previous experiments, whether successful or not, is essential when selecting treatments and experimental materials. After establishing the treatments and experimental materials, the experimental design, along with the observational units, number of replicates per treatment and randomization scheme, is chosen.

Explants typically are selected as the observational unit in plant tissue culture experiments. Selection of explant type and source may influence the outcome. In most tissue culture experiments the experimental unit, or replicate, is the culture vessel. This is because a culture vessel houses explants, which are the unit of observation. When there is one explant per culture vessel, both the vessel and explant are considered a replicate. When there are multiple explants per culture vessel, the vessel is considered the unit of replication. The number of replicates per treatment affects statistical precision. Generally, precision increases as the number of replicates is increased. However, a plateau is reached where adding replicates does not improve precision.

When designing an experiment, each replicate is assigned to a treatment. When assigning replicates to treatments, it is important to make assignments in a manner such that all have an equal chance of receiving a given treatment. This is called randomization. Randomization schemes vary according to the experimental design employed. After selecting the treatments and experimental design, one should determine the data to be collected. The data should properly evaluate and explain treatment effects according to experimental objectives.

An analysis of variance (ANOVA) summary table with all the degrees of freedom (DF) and associated sources of variation (treatment, experimental error, etc.), along with graphs and/or tables that show the theoretical results, should be constructed to evaluate the effectiveness of the proposed experiment in answering the desired questions. Consulting a knowledgeable colleague or statistician at this time should determine if the experiment was well conceived.

Data must be statistically analyzed and interpreted as planned before the experiment, according to the hypothesis, experimental conditions and previous facts. Results that differ from those obtained previously may be due to mistakes in preparation of experimental materials, data entry and analysis, fluctuation of environmental conditions during the experiment, etc. There is always some degree of uncertainty regarding the conclusions from an experiment. For this reason it is important to repeat the experiment to confirm your results.

## Experimental designs

The following section will discuss the application of completely randomized designs (CRD), randomized complete block designs (RCBD) and split-plot designs in plant tissue culture research.

### Completely randomized design

A majority of plant cell and tissue culture studies employ a CRD, because cell cultures are generally grown in environmental chambers that accurately control light, temperature and humidity. In the CRD, treatments are assigned to experimental units at random. The numbers of treatments and replicates per treatment that can be tested are not limited by the CRD. It is the easiest design to employ and analyze, even when there are missing data or when treatments are unevenly spaced (Lentner and Bishop, 1986; Little and Hills, 1978). The CRD also is the most precise design because it maximizes the DF used for estimating experimental error. This reduces the F-value required to detect statistical differences among treatments. This is important in small experiments or when outside variation is great.

An experiment designed to identify tomato (*Lycopersicon esculentum* Mill.) cultivars that displayed high rates of shoot regeneration from pedicel explants illustrates use of the CRD in plant tissue culture studies. Explants were cultured in test tubes that contained 15 ml of shoot regeneration medium (Compton and Veilleux, 1991). There were 24 tubes (replicates) per treatment, each with one explant (observational unit), resulting in a total of 264 experimental units. Treatments were assigned to the experimental units completely at random. Experimental data included the number of explants with shoots and the number of shoots per explant.

*Table 6.1* Analysis of Variance Summary Table for the Number of Shoots per Explant from Tomato Shoot Regeneration Study

| Source | DF | Mean square | F-value |
|---|---|---|---|
| Cultivar | 10 | 7.937 | 5.87*** |
| Error | 253 | 1.352 | |
| Corrected total | 263 | | |

*Note:* Data analyzed using procedures for a CRD.

*** Significant F-value at the 0.001 level.

From Compton, M.E. *Plant Cell Tissue Organ Cult.* 37, 217–242, 1994. With permission.

ANOVA was used to analyze the data in order to determine if the eleven genotypes differed in their ability to produce shoots from pedicel explants. This procedure generates the DF, sums-of-squares (SS) and mean square (MS) values for treatment and experimental error used to perform the statistical test. Treatment SS measures the degree of variation associated with the treatments, whereas experimental error SS measures variation associated with experimental units. In the tomato study, treatment (cultivar) F-value was calculated by dividing its MS, 7.937, by the mean square error (MSE), 1.35 (Table 6.1). Significance of the observed F-statistic was determined using 10 and 253 DF. A significant F-statistic indicates that there are differences in shoot regeneration rates among the cultivars, but does not determine which genotypes differ. To determine differences among treatment means, a mean separation procedure must be performed. Mean separation procedures should be chosen prior to ANOVA to avoid personal bias in the test. This is because each mean separation test uses a unique formula when calculating differences among means. The types of mean separation tests will be discussed in greater depth later in this chapter.

The CRD is most efficient when there is little variation among experimental units. Efficiency and precision are reduced in situations where there is a high degree of variation not associated with treatment effects. This is because nontreatment variation is assigned to experimental error, resulting in a large MSE and fewer rejections of the null hypothesis. In situations where there is a high degree of outside error and the source can be identified, designs that employ blocking should be used (Lentner and Bishop, 1986; Little and Hills, 1978).

## Single factor CRD with subsampling

Sometimes it is more efficient to culture several explants in a single culture vessel. In this situation the response of each explant is measured, resulting in multiple measurements for each vessel, i.e., subsampling (Lentner and Bishop, 1986). This type of design is more efficient than the normal CRD with one observational unit per experimental unit when there is a wide range in response among explants. Statistical precision is improved by making several measurements per culture vessel, thus reducing variation among replicates of the same treatment. Subsampling may be used to study variability among explants, which may be useful in future experiments.

An experiment that examined the ability of watermelon (*Citrullus lanatus* [Thunb.] Matsum. & Nakai) cotyledon explants to produce adventitious shoots illustrates the use of subsampling. The objective of this experiment was to identify genotypes with high level adventitious shoot organogenesis. Cotyledons from seedlings of 11 cultivars were cultured in petri dishes that contained 25 ml of shoot regeneration medium (Compton and Gray, 1993). Cotyledons from each cultivar were arranged into 6 petri dishes, each with 5 explants, resulting in a total of 66 petri dishes and 330 explants. Explants from each cultivar

*Table 6.2*  Analysis of Variance Summary Table for the Number of
Adventitious Shoots per Explant from Watermelon Shoot
Organogenesis Study

| Source | DF | Mean square | F-value |
|---|---|---|---|
| Genotype | 10 | 38.312 | 6.83[***] |
| Experimental error[a] | 55 | 5.610 | |
| Subsampling error | 319 | 4.644 | |
| Corrected total | 329 | | |

*Note:* Data were analyzed using procedures for a single factor CRD with
subsampling.

[a] Obtained from the nested term replicate within genotype (rep[Genotype]).

[***] Significant F-value at the 0.001 level.

Data obtained from Compton, M.E. and D.J. Gray. *J. Am. Soc. Hort. Sci.* 118,
151–157, 1993.

were assigned to petri dishes at random, making sure that only explants from one cultivar were cultured together. Individual petri dishes were distributed at random in the growth room. The number of explants with adventitious shoots and the number of shoots per explant were recorded.

The ANOVA for CRD experiments with subsampling differs from a CRD with one observational unit per experimental unit in that SS are generated for treatment, experimental error and subsampling error (Table 6.2). A special error term, REP(GENOTYPE), is used for calculating the treatment F-value and measures variation among experimental units. Subsampling SS represents variation caused by culturing multiple explants within the same experimental unit and is not used in computing treatment differences. This improves statistical precision by removing variation among explants from the experimental error. In this example, significance among treatments was detected by dividing the treatment MS by the new MSE, resulting in a significant F-value of 6.83 and indicating that the cultivars displayed varying levels of shoot regeneration. However, a statistical test must be conducted on the treatment means. When conducting mean separation tests in subsampling experiments, the special MSE must be used to calculate values for the statistical test. Failure to specify the correct error term may result in the grouping of means that should be separated and/or separation of similar means.

Subsampling experiments promote efficient use of space and materials and measures variability among observational units. They are most efficient when there is much variation in response among explants and little outside variation. The design is not efficient when there is little variation among explants or when there is a high degree of variability from some other identifiable source. This is because it reduces the DF for experimental error, which means that a higher F-value is required to detect significant differences among treatments. If there is little variation among explants, a CRD without subsampling should be used. If variation from an identifiable outside source exists, a RCBD with subsampling should be used.

## Randomized complete block design

The CRD is only efficient when the experimental units are homogeneous (Lentner and Bishop, 1986). This is because all unrecognized variation, regardless of source, is lumped into the experimental error term. Heterogeneity among experimental units results in a high MSE and reduces statistical precision. In experiments with high heterogeneity, it is best to group experimental units into homogeneous units or blocks. Grouping experimental units into uniform blocks provides a better estimate of treatment effects and improves statistical precision.

***Table 6.3***   Analysis of Variance Summary Table for the Number
of Shoots per Explant from Tomato Shoot Regeneration Study

| Source | DF | Mean square | F-value |
|---|---|---|---|
| Block | 23 | 1.639 | |
| Cultivar | 10 | 7.937 | 5.99[***] |
| Error | 230 | 1.326 | |
| Corrected total | 263 | | |

*Note:* Data were analyzed using procedures for an RCBD.

[***] Significant F-value at the 0.001 level.

From Compton, M.E. 1994. *Plant Cell. Tissue Organ Cult.* 37:217–242. With
permission.

In RCBDs, treatments are grouped into blocks that contain at least one replicate from
each treatment. Experimental units are randomized within blocks, each employing a
separate randomization scheme. This minimizes variability within a block while maximizing variability among blocks. For this reason, experimental units within each block should
be as uniform as possible. The number of treatments should be limited because variability
within each block increases as the number of treatments increase.

The tomato shoot regeneration study described above can be adapted to illustrate an
RCBD by arranging treatments into 24 blocks, each containing one test tube with a single
pedicel explant from each cultivar. The ANOVA for an RCBD differs from that of a CRD
in that block SS is calculated in addition to treatment and experimental error SS. Block SS
identifies variation controlled by blocking. Treatment SS represents variability related to
the imposed treatment. Experimental error SS measures variability among experimental
units. Blocking improves statistical precision by removing variability controlled by blocking from the experimental error. In this example, the F-value for cultivar (5.99) was
obtained by dividing the cultivar MS (7.937) by the MSE (1.326) and significance determined using 10 and 230 DF (Table 6.3). A means separation procedure must be performed
to determine which treatment means differ.

An RCBD is useful when some variation is caused by something other than the
imposed treatment(s). If little difference among blocks occurs, an RCBD will not improve
statistical precision and a CRD should be used. Kempthorne (1973) provides a formula that
determines if blocking has improved precision. In contrast, if there is much variation
among explants, an RCBD with subsampling should be used (see review by Compton,
1994).

## Split-plot design

This design is typically used when two treatment factors are superimposed on each other.
One treatment is assigned to "main plots" that contain all levels of the second factor
(Lentner and Bishop, 1986). Main plot treatments may be randomized in any of the
experimental designs described above. Treatments assigned to subplots are randomized
within each main plot. Each subplot is randomized differently. The advantage of employing a split-plot design is that it identifies experimental error associated with each treatment
factor. Treatments with large experimental error rates are assigned to main plots and those
with reduced experimental error to subplots. Dividing experimental error into main plots
and subplots improves statistical precision. The split-plot design is most efficient when
treatment factors have different experimental error rates. Statistical precision is lost when
there is no difference in experimental error between main plots and subplots. For a more
detailed discussion on the use of split-plot designs in plant tissue culture see the review by
Compton (1994).

*Table 6.4*  Analysis of Variance Summary Table for
Tobacco Callus Growth Study

| Source | DF | Callus dry weight (mg) | |
| --- | --- | --- | --- |
| | | Mean square | F-value |
| BA | 2 | 9.367 | 5.65** |
| IBA | 2 | 12.209 | 7.36*** |
| BA × IBA | 4 | 4.361 | 2.63* |
| Error | 98 | 1.761 | |
| Corrected total | 106 | | |

*Note:* Data were analyzed using procedures for a CRD with treatments arranged in a 3 × 3 complete factorial.

\*,\*\*,\*\*\* Significant F-value the 0.05 (\*), 0.01 (\*\*) or 0.001 (\*\*\*) level, respectively.

Data obtained from Compton, M.E. M.S. Thesis. Southern Illinois University, Carbondale, IL, 1986.

## Factorial experiments

In complete factorial experiments, treatments are arranged such that all possible combinations of two or more factors are examined simultaneously. Factorial experiments are useful because they allow the scientist to examine treatment interdependencies and are more powerful than multiple single factor experiments. Any of the aforementioned treatment randomization schemes can be used for factorial treatment designs. A factorial treatment design was employed in an experiment examining the effects of benzyladenine (BA) and indole-3-butyric acid (IBA) on tobacco (*Nicotiana tabacum* L.) callus growth (Compton, 1986). Treatments were three levels (0, 1 and 10 µM) of BA and IBA arranged in a 3 × 3 complete factorial. Pieces of tobacco callus (1 cm³) were transferred to test tubes that contained 15 ml of test medium. There were six replicate test tubes per treatment that were randomized in a CRD. Callus height, diameter and dry weight were recorded 1 month after culture initiation.

In the analysis of factorial experiments, SS are generated for each treatment and their interaction, and for experimental error. F-values for BA, IBA and their interaction were obtained by dividing the MS for each by the MSE (Table 6.4). Significance for the main effects (BA and IBA) was determined using 2 and 98 DF, whereas significance for the BA by IBA interaction was tested at 4 and 98 DF. ANOVA indicated that callus growth in the form of dry weight was influenced by BA and IBA simultaneously. This was determined by the significant BA by IBA interaction, indicating that BA affected callus dry weight differently at each level of IBA and IBA affected callus dry weight differently at each BA concentration. Interaction means for this type of data are presented in a line graph with treatment differences estimated using SE (Figure 6.1). When treatment interactions are significant, the influence of each treatment separately should not be discussed (Lentner and Bishop, 1986; Zar, 1984).

Factorial experiments can become prohibitive with relatively few factors. For example, the above 3 × 3 factorial experiment had 9 treatments. If a third factor with three levels was added the number of treatment combinations would increase to 27. Add to that 10 replicates and there would be 270 observational units. Factorial experiments require a more complicated analysis model and more care in performing mean separation analyses, especially when there are missing data. Interpretation of higher order interactions may be difficult in any factorial experiment. However, the information to be gained often compensates for the extra work.

**Figure 6.1**  Graph of tobacco callus dry weight as affected by various combinations of BA and IBA. Graph demonstrates the presentation of interaction means when treatments were arranged in a 3 × 3 complete factorial. Vertical lines represent standard errors. □ = 0 BA; Δ = 1 μM BA; ◊ = 10 μM BA. Data obtained from Compton, M.E. M.S. Thesis, Southern Illinois University, Carbondale, IL, 1986. Figure from Compton, M.E., 1994. *Plant Cell Tissue Organ Cult.* 37:217–242. With permission.

## Methods for comparing treatment means

Once a significant F-test, with a small P-value, is obtained the scientist must elucidate specific differences among treatments. This is not a problem when there are only two treatments because treatment means are simply presented in a table with the associated F-test significance level. When there are more than two treatments, a post-ANOVA analysis is necessary. The easiest way to compare treatment means is to rank them in ascending or descending order and pick the best treatment. The problem with this method is that natural variation that occurs within a treatment is not considered. There are many mean separation procedures that account for within treatment variation. These are standard error of the mean (SE), multiple comparison and multiple range tests and regression analysis.

### Standard error of the mean

SE is obtained by dividing the sample standard deviation by the square root of the number observations for that treatment (Zar, 1984). Hence, the size of SE depends on the magnitude of the data. When using SE for mean separation purposes, treatment means are ranked along with their respective SE and the difference between paired values calculated. Treatments are declared different if the collective values for the paired treatments do not overlap. Using SE provides an indication of variability within treatments and allows readers to make their own comparisons. One disadvantage of using SE for mean separation purposes is that it is considered conservative and may only detect differences between means ranked far apart.

### Multiple comparison and multiple range tests

Multiple comparison or multiple range tests should only be used when treatments are unrelated, e.g., different growth regulators, genotypes, etc. (Lentner and Bishop, 1986). The terms *multiple comparison test* and *multiple range test* are used interchangeably. However, the two differ in their method of making comparisons. Multiple comparison tests use the same critical value to compare adjacent and nonadjacent means. Examples of multiple comparison tests are Bonferoni, Fisher's least significant difference (LSD), Scheffe's, Tukey's honestly significant difference test (Tukey's HSD), and Waller-Duncan K-ratio T test (Waller-Duncan). Multiple range tests employ different critical values to compare adjacent and nonadjacent means, which protects against committing a type I error (instance where treatments are declared different that truly are the same). Examples of multiple range tests include Duncan's new multiple range test (DNMRT), Ryan-Einot-Gabriel-Welsh multiple F-test (REGWF), Ryan-Einot-Gabriel-Welsh multiple range test (REGWQ), and Student-Newman-Kuels (SNK).

It is best to use multiple range or multiple comparison tests only after obtaining a significant ANOVA (Lentner and Bishop, 1986). Most mean separation procedures may be used in situations when a significant ANOVA has not been obtained if specific comparisons were planned prior to data analysis. However, there is a higher probability of creating a type I error, especially when comparing nonadjacent means. DNMRT, REGWQ, SNK, Tukey's HSD, and Waller-Duncan are the most commonly recommended mean separation procedures for unrelated treatments (Lentner and Bishop, 1986; SAS Institute, 1988).

## Simple linear regression

Regression, or trend, analysis is used in experiments with quantitative treatments where the primary objective is to develop a model that quantifies the relationship between response variables and treatment levels (Mize and Chun, 1988; Kleinbaum and Kupper, 1978). The model is used to identify which trend (e.g., linear, quadratic or cubic) best describes the data and may be used for predicting optimum treatment levels. It is assumed that there is a continuous response of the observational unit to the imposed treatment when using quantitative levels. Trend analysis is most appropriate when there are three or more evenly spaced, equally replicated treatments. A forward step procedure is most often used to identify the best model. The simplest model (linear) is tested first and more complicated models (quadratic and cubic) tested after rejection of lower order models. A backward step procedure may be used but is less common. Lack-of-fit (LOF), T and r-square ($r^2$) values are used to determine the best model. A significant T-value and nonsignificant LOF signifies that the model accurately describes the relationship between treatment and response variables. A significant LOF value indicates that the model does not fit the data and other models should be tested. The $r^2$ value estimates variation described by the model. A high $r^2$ suggests that much of the variability is described by the model. If a low $r^2$ is obtained, it may be advantageous to check other models. The most appropriate model has a significant T-value, nonsignificant LOF, and high $r^2$. It is important that extrapolation be limited to within treatment boundaries as the model equation may not accurately describe treatment effects beyond the tested parameters.

Regression analysis was used to determine the effect of sucrose on the growth of grape (*Vitis vinifera* L. 'Thompson Seedless') embryogenic cultures (Compton and Gray, 1995). Embryogenic cells and somatic embryos (heart and globular stage) were incubated on medium containing 60, 90, 120, 150 or 180 g/L sucrose for 3 months. Embryogenic cultures were subcultured monthly to fresh medium of the same composition. At the end of the experiment embryogenic cultures were dried in an oven at 70 C for 72 h, and dry weight was determined.

In this example, ANOVA determined that the sucrose concentration in the medium influenced the growth of grape embryogenic cultures. Regression and LOF analyses indicated that the data best fit the cubic model as indicated by a significant T-value and nonsignificant LOF value ($-1774.06 + 57.92X - 0.4837X^2 + 0.001X^3$) (Table 6.5). Optimum

*Table 6.5* Analysis of Variance Summary Table for Trend and Lack of Fit Analyses on Dry Weight of Grape Embryogenic Cultures Grown on Medium with Varying Sucrose Concentrations

| MODEL 3 (cubic) | | Analysis of variance | | | |
|---|---|---|---|---|---|
| Variable | DF | Parameter estimates | Standard error | T for HO: Parameter = 0 | Significance |
| Intercept | 1 | **−1774.057** | 231.967 | −7.648 | 0.0001 |
| Sucrose | 1 | **57.928** | 6.675 | 8.677 | 0.0001 |
| Sucrose$^2$ | 1 | **−0.483** | 0.059 | −8.166 | 0.0001 |
| Sucrose$^3$ | 1 | **0.001** | 0.0001 | 7.508 | **0.0001** |
| | Root MSE | 37.225 | $r^2$ | **0.8815** | |

Lack of fit test

| Source | DF | Sum of squares | Mean square | F-value | Significance |
|---|---|---|---|---|---|
| Sucrose | 0 | — | — | — | — |
| Sucrose$^2$ | 0 | — | — | — | — |
| Sucrose$^3$ | 0 | — | — | — | — |
| LOF | 1 | 3860.17 | 3860.17 | 3.07 | **0.0956** |

Data from Compton, M.E. and D.J. Gray. *Vitis*. In press, 1996.

growth was obtained when embryogenic cultures were incubated on medium with 90–120 g/L sucrose (Figure 6.2). The high (0.8815) $r^2$ obtained for this model indicates that much of the variation in the data was explained by the model. The terms high and low $r^2$ are relative to the type of data analyzed. High $r^2$ values for biological data may range between 0.50 and 0.90, whereas a low $r^2$ for nonbiological data may be 0.90 (Kleinbaum and Kupper, 1978).

**Figure 6.2** Effect of sucrose concentration on dry weight of grape somatic embryogenic cultures. Dotted line represents the predicted regression equation. Treatment means are represented by +.

*Table 6.6*  Percent of Maize Somatic Embryos that Produced Chlorophyll (Greening),
Roots, Coleoptiles and Leaves after Growth on Three Different Embryo
Maturation Media Followed by CRHD for 2 Weeks

| RH | Medium | No. of embryos | Percent embryos with | | | |
|---|---|---|---|---|---|---|
| | | | Greening | Roots | Coleoptiles | Leaves |
| Control[a] | 1[b] | 43 | 58.1 ± 7.5 | 25.6 ± 6.7 | 25.6 ± 6.7 | 20.9 ± 6.2 |
| | 2 | 147 | 81.0 ± 3.2 | 32.7 ± 3.9 | 31.3 ± 3.8 | 46.3 ± 4.1 |
| | 3 | 100 | 72.0 ± 4.5 | 56.0 ± 5.0 | 31.0 ± 4.6 | 47.0 ± 5.0 |
| 90% | 1 | 43 | 0.0 ± 0.0 | 0.0 ± 0.0 | 0.0 ± 0.0 | 0.0 ± 0.0 |
| | 2 | 71 | 0.0 ± 0.0 | 0.0 ± 0.0 | 0.0 ± 0.0 | 0.0 ± 0.0 |
| | 3 | 119 | 0.0 ± 0.0 | 0.0 ± 0.0 | 0.0 ± 0.0 | 0.0 ± 0.0 |
| 70% | 1 | 38 | 0.0 ± 0.0 | 0.0 ± 0.0 | 0.0 ± 0.0 | 0.0 ± 0.0 |
| | 2 | 72 | 0.0 ± 0.0 | 0.0 ± 0.0 | 0.0 ± 0.0 | 0.0 ± 0.0 |
| | 3 | 82 | 12.2 ± 3.6 | 14.6 ± 3.9 | 9.8 ± 3.3 | 13.4 ± 3.8 |

*Note:* ± values represent the standard error of the mean.

[a] Control embryos were transferred directly to germination medium without CRHD.

[b] Medium 1 = N6 plus 4.5 µM 2,4-D, 100 mg/L casamino acids, 25 mM proline, and 2 g/L Phytagel; Medium 2 = 2 weeks on MS plus 0.45 µM 2,4-D, 0.1 µM ABA, and 8 g/L agarose; Medium 3 = 2 weeks on medium 2 and 2 weeks on N6 plus 60 g/L sucrose and 2 g/L Phytagel.

Data from Compton et al. *In Vitro Cell. Devel. Biol.* 28P, 197–201, 1992. Copyright © 1992 by the Tissue Culture Association. Reprinted with permission.

## Categorical data

Categorical data are often based on counts of individuals that can be placed into one of two or more groups. This occurs when data consist of yes/no values (e.g., yes the explant responded or no the explant failed to respond) or when explant response is placed into a category (e.g., explants produced roots, shoots, somatic embryos or did not respond). Categorical data are not continuous and not normally distributed and, therefore, cannot be analyzed with the procedures mentioned above. Chi-square or Maximum Likelihood are used to analyze categorical data (Lentner and Bishop, 1986; Zar, 1984).

An example of categorical data is an experiment where the investigators were interested in identifying the best dehydration treatment for maize (*Zea mays* L. B73 × A188-19 hybrid) somatic embryos (Compton et al., 1992). In this experiment, somatic embryos were reared on medium without ABA (medium 1), transferred to medium with 0.1 µM ABA (medium 2) for 2 weeks, or transferred to medium 2 for 2 weeks before transfer to medium with 60 g/L sucrose and no ABA (medium 3) prior to controlled relative humidity dehydration (CRHD) at 70 or 90 % relative humidity for 2 weeks. Embryo survival was determined 2 weeks after transfer to germination medium by the ability of somatic embryos to produce chlorophyll (green), roots, coleoptiles or leaves. Control embryos obtained from each of the above media were transferred directly to germination medium without CRHD. Surviving embryos were assigned a 1, whereas those that failed to survive were assigned 0 (no response).

According to ANOVA, culture medium and dehydration treatment simultaneously influenced the ability of maize somatic embryos to survive dehydration as determined by the significant treatment by medium interaction contrast. The ability of maize somatic embryos to survive dehydration depended on the pretreatment medium and RH level used during CRHD. SE was used to compare treatment means (Table 6.6). Categorical data can be transformed using arc sine prior to ANOVA and converted back to the original scale for demonstration in tables or graphs (Compton, 1994).

# Conclusions

This chapter has outlined the steps of experimentation and many of the procedures used in the analysis of plant cell culture data. It is important to remember that any well-conceived research project should be planned carefully on paper before the experiment is conducted. This helps to ensure that the experiment will answer the desired questions. Planning is probably the most important step in scientific research. Experimental design and analysis provides an objective, nonbiased way to accurately evaluate treatment differences. Used properly, statistics is a valuable tool to the scientific researcher and it allows the investigator to declare experimental results with confidence.

# Acknowledgments

This is Florida Agricultural Experiment Station Journal Series No. N-00961.

# Literature cited

Compton, M.E. 1986. Stage I exudation, plant growth regulators, and phenolic compounds alter the growth and development of tobacco callus. M.S. Thesis, Dept. of Plant and Soil Science, Southern Illinois University, Carbondale, IL.

Compton, M.E. 1994. Statistical methods suitable for the analysis of plant tissue culture data. *Plant Cell Tissue Organ Cult.* 37:217–242.

Compton, M.E. and D.J. Gray. 1993. Shoot organogenesis and plant regeneration from cotyledons of diploid, triploid and tetraploid watermelon. *J. Am. Soc. Hort. Sci.* 118:151–157.

Compton, M.E. and D.J. Gray. 1996. Effects of sucrose and methylglyoxal bis-(guanylhydrazone) on controlling grape somatic embryogenesis. *Vitis* (in press).

Compton, M.E. and R.E. Veilleux. 1991. Shoot, root and flower morphogenesis on tomato inflorescence explants. *Plant Cell Tissue Organ Cult.* 24:223–231.

Compton, M.E., C.M. Benton, D.J. Gray, and D.D. Songstad. 1992. Plant recovery from maize somatic embryos subjected to controlled relative humidity dehydration. *In Vitro Cell. Devel. Biol.* 28P:197–201.

Kempthorne, O. 1973. *The Design and Analysis of Experiments.* Robert E. Krieger Publishing Co., Malabar, FL.

Kleinbaum, D.G. and L.L. Kupper. 1978. *Applied Regression Analysis and Other Multivariable Methods.* PWS Publishers, Boston.

Lentner, M. and T. Bishop. 1986. *Experimental Design and Analysis.* Valley Book Company, Blacksburg, VA.

Little, T.M. and F.J. Hills. 1978. *Agricultural Experimentation: Design and Analysis.* John Wiley and Sons, Inc., New York.

Mize, C.W. and Y.W. Chun. 1988. Analysing treatment means in plant tissue culture research. *Plant Cell Tissue Organ Cult.* 13:201-217.

SAS Institute, Inc. 1988. *SAS/STAT User's Guide.* Release 6.03. SAS Institute, Inc., Cary, NC.

Zar, J.H. 1984. Biostatistical Analysis. Second ed. Prentice-Hall, Inc., Englewood Cliffs, NJ.

# Part III

## Propagation techniques

*chapter seven*

# Propagation from preexisting meristems

*Michael E. Kane*

Micropropagation is defined as the true-to-type propagation of selected genotypes using in vitro culture techniques. Depending on the species and cultural conditions, in vitro propagation can be achieved by the following four basic methods: (1) enhanced axillary shoot proliferation (shoot culture); (2) node culture; (3) de novo formation of adventitious shoots through shoot organogenesis (chapter eleven); or (4) nonzygotic embryogenesis (chapter sixteen). Currently, the most frequently used micropropagation method for commercial production utilizes enhanced axillary shoot proliferation from cultured meristems (chapter thirty-eight). This method provides genetic stability and is easily attainable for many plant species. Consequently, the shoot culture method has played an important role in development of a worldwide industry that produces more than 250 million plants yearly. Besides propagation, shoot meristems are cultured in vitro for two other purposes: (1) production of pathogen-eradicated plants (chapter thirty-nine); and (2) preservation of pathogen-eradicated germplasm (chapter thirty-three). Concepts related to propagation by shoot and node culture will be discussed in this chapter.

## Shoot apical meristems

It is important to briefly review the general structure of shoot meristems. Shoot growth in mature plants is restricted to specialized regions that exhibit little differentiation and in which the cells retain the embryonic capacity for unlimited division. These regions, called apical meristems, are located in the apices of the main and lateral buds of the plant. Cells derived from these apical meristems subsequently undergo differentiation to form the mature tissues of the plant body. Because of their highly organized structure, apical meristems tend to be genetically stable.

There exist significant differences in the shape and size of shoot apices among different taxonomic plant groups (Fahn, 1974). A typical dicotyledon shoot apical meristem consists of a layered dome of actively dividing cells located at the extreme tip of a shoot and measures about 0.1–0.2 mm in diam. and 0.2–0.3 mm in length. The apical meristem has no vascular connection to the xylem and phloem tissues of the stem. Below the apical meristem, localized areas of cell division and elongation represent sites of newly developing leaf primordia (Figure 7.1). Lateral buds each containing an apical meristem develop within the axils of the subtending leaves. In the intact plant, outgrowth of the lateral buds is usually inhibited by apical dominance of the terminal shoot tip. Organized shoot growth from the apical meristem of plants is potentially unlimited and is said to be indeterminate. However, shoot apical meristems may become committed to the formation of determinate organs such as flowers.

0-8493-9409-0/96/$0.00+$.50

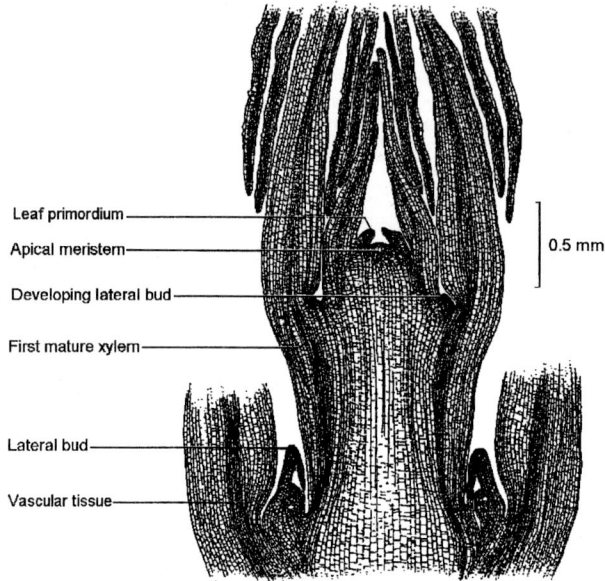

**Figure 7.1** Diagrammatic representation of a dicotyledonous shoot tip. The shoot tip comprises the apical meristem, subtending leaf primordia and lateral buds.

## In vitro culture of shoot meristems

The recognized potential for unlimited shoot growth prompted early, but largely unsuccessful, attempts to aseptically culture isolated shoot meristems in the 1920s. By the middle 1940s, sustained growth and maintenance of organization of cultured shoot meristems through repeated subculture was achieved for several species. Ball (1946), however, provided the first detailed procedure for the isolation and production of plants from cultured shoot meristem tips and the successful transfer of rooted plantlets into soil. Ball is often called the Father of Micropropagation because his shoot tip culture procedure is the one most commonly used by commercial micropropagation laboratories today. Although these studies demonstrated the feasibility of regenerating shoots from cultured shoot tips, the procedures typically yielded unbranched shoots.

Several important findings facilitated application of in vitro culture techniques for large scale clonal propagation from meristems. The discovery that virus-eradicated plants could be generated from cultured meristems led to the widespread application of the procedure for routine fungal and bacterial pathogen eradication as well (Morel and Martin, 1952; Styer and Chin, 1984). Demonstration of rapid production of orchids from cultured shoot tips supported the possibility of rapid clonal propagation in other crops (Morel, 1960; 1965). It should be noted that in vitro propagation in many orchids does not occur via axillary shoot proliferation, but rather the cultured meristems become disorganized and form spheroid protocormlike bodies that are actually nonzygotic embryos.

The final discovery was the elucidation of the role of cytokinins in the inhibition of apical dominance (Wickson and Thimann, 1958). This finding was eventually applied to enhance axillary shoot production in vitro. Application of this method was expedited by development of improved culture media that supported the propagation of a wide diversity of plant species (Murashige and Skoog, 1962; Lloyd and McCown, 1981).

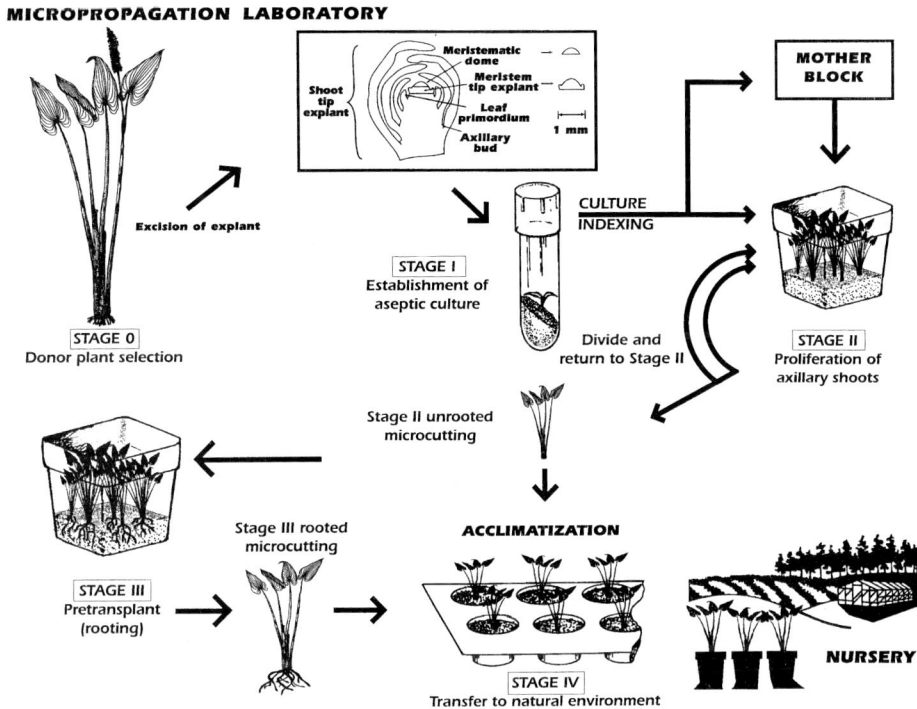

**Figure 7.2** Micropropagation stages for production by shoot culture.

## Meristem and meristem tip culture

Although not directly used for propagation, meristem and meristem tip culture will be briefly described since these procedures are used to generate pathogen-eradicated shoots that subsequently serve as propagules for in vitro propagation. Culture of the apical meristematic dome alone (Figure 7.2) from either terminal or lateral buds, for the purpose of pathogen elimination, is termed *meristem culture*. In reality, true meristem culture is rarely used because isolated apical meristems of many species exhibit both low survival rates and increased chance of genetic variability following callus formation and indirect shoot organogenesis.

Pathogen elimination can often be accomplished by culture of relatively larger (0.2–0.5-mm-long) meristem tip explants excised from plants that have undergone thermo- or chemotherapy. The meristem tip comprises the apical meristem plus one or two subtending leaf primordia (Figure 7.2). This procedure is therefore termed *meristem tip culture*. Caution should be taken when interpreting much of the early published literature of successful *meristem* culture since, in many instances, meristem tip or even larger shoot tip explants were actually used. The term *meristemming* commonly used in the orchid literature is equally ambiguous.

## Shoot and node culture

Although not the most efficient procedure, propagation from axillary shoots has proved to be a reliable method for the micropropagation of a large number of species (Kurtz et al., 1991). Depending on the species, two methods, shoot and node culture, are used. Both methods rely on the stimulation of axillary shoot growth from lateral buds following disruption of apical dominance of the shoot apex. Shoot culture (shoot tip culture) refers

**Figure 7.3**   a. Stage II shoot multiplication is achieved by repeated formation of axillary shoot clusters from explants containing lateral buds. Depending on the species, individual microcuttings or shoot clusters may be rooted and acclimatized ex vitro. Scale bar = 1 cm. b. For maximum survival Stage III rooting may be required prior to acclimatization to ex vitro conditions. c. Rooted and acclimatized Stage IV plantlets.

to the in vitro propagation by repeated enhanced formation of axillary shoots from shoot tips or lateral buds cultured on medium supplemented with plant growth regulators (PGRs), usually a cytokinin (George, 1993). The axillary shoots produced are either subdivided into shoot tips and nodal segments that serve as secondary explants for further proliferation (see Figure 7.3a) or are treated as microcuttings for rooting.

When either verified pathogen-free stock plants are used or when pathogen elimination is not a concern, relatively larger (1–20-mm-long) shoot tip or lateral bud primary explants (Figure 7.2) can be used for culture establishment and subsequent shoot culture. Advantages of using larger shoot tips include greater survival, more rapid growth responses and the presence of more axillary buds. However, these larger explants are more difficult to completely surface disinfest and can potentially harbor undetected latent systemic microbial infection. Compared with other micropropagation methods, shoot cultures (1) provide reliable rates and consistency of multiplication following culture stabilization, (2) are less susceptible to genetic variation, and (3) may provide for clonal propagation of periclinal chimeras.

Node culture, a simplified form of shoot culture, is another method for production from preexisitng meristems. Numerous plants such as potato (*Solanum tuberosum* L.) do not respond well to the cytokinin stimulation of axillary shoot proliferation observed in the micropropagation of many crops. Axillary shoot growth is promoted by the culture of either intact shoots (from meristem tip cultures) positioned horizontally on the medium (in vitro layering) or single or multiple node segments. Typically, single elongated unbranched shoots, comprising multiple nodes, are rapidly produced (see Figure 9.2). These shoots (microcuttings) are either rooted and acclimatized to ex vitro conditions or repeatedly subdivided into nodal cuttings to initiate additional cultures. In some species, modified shoot storage organs, such as miniaturized tubers or corms, may develop from axillary shoots under inductive culture conditions. Although node culture is the simplest method, it is associated with the least genetic variation.

## Micropropagation stages

Murashige (1974) originally described three basic stages (I to III) for successful micropropagation. Recognition of the contamination problems often associated with inoculation of primary explants prompted Debergh and Maene (1981) to include a Stage 0. This stage described specific cultural practices which maintained the hygiene of stock plants and decreased the contamination frequency during initial establishment of primary explants. As a result of our increased information base, currently it is now agreed that there are five stages (Stage 0 to IV) critical to successful micropropagation. These stages not only describe the procedural steps in the micropropagation process, but also represent points at which the cultural environment is altered (Miller and Murashige, 1976). This system has been adopted by most commercial and research laboratories as it simplifies production scheduling, accounting and cost analysis (Kurtz et al., 1991). Requirements for completion of each stage will depend on the plant material and specific method used. Diagrammatic representation of the micropropagation stages for propagation by shoot culture is provided in Figure 7.2.

### Stage 0: Donor plant selection and preparation

Explant quality and subsequent responsiveness in vitro is significantly influenced by the phytosanitary and physiological conditions of the donor plant (Debergh and Maene, 1981; Read, 1988). Prior to culture establishment, careful attention is given to the selection and maintenance of the stock plants used as the source of explants. Stock plants are maintained in clean controlled conditions that allow active growth, but reduce the probability of disease. Maintenance of specific pathogen-tested stock plants under conditions of relatively lower humidity and use of drip irrigation and antibiotic sprays have proved effective in reducing the contamination potential of candidate explants. Such practices also allow excision of relatively larger and more responsive explants often without increased risks of contamination.

Numerous practices are also employed to increase explant responsiveness by modifying the physiological status of the stock plant and include the following: (1) trimming to stimulate lateral shoot growth; (2) pretreating with sprays containing cytokinin or gibberellic acid; and (3) using forcing solutions containing 2% sucrose and 200 mg/L 8-hydroxyquinoline citrate for induction of bud break and delivery of growth regulators to target explant tissues (Read, 1988). Currently, information on the effects of other factors such as stock plant nutrition, light and temperature treatments on the subsequent in vitro performance of meristem explants is lacking.

### Stage I: Establishment of aseptic cultures

Initiation and aseptic establishment of pathogen eradicated and responsive terminal or lateral shoot meristem explants is the goal of this stage. The primary explants obtained from the stock plants may consist of surface sterilized shoot apical meristems or meristem tips for pathogen elimination or shoot tips from terminal or lateral buds (Figure 7.2).

The presence of microbial contaminants can adversely affect shoot survival, growth and multiplication rate. Bacteria and fungal contaminants often persist within cultured tissues that visually appear contaminant free. Consequently, it is essential that Stage I cultures be indexed (screened) for the presence of internal microbial contaminants prior to serving as sources of shoot tip and nodal explants for Stage II multiplication.

The following factors may affect successful Stage I establishment of meristem explants: (1) explantation time, (2) position of the explant on the stem, (3) explant size, and (4) polyphenol oxidation. Time of explantation can significantly affect explant response in vitro. In deciduous woody perennials, shoot tip explants collected at various times during the spring growth flush may vary in their ability for shoot proliferation. Shoot tips collected during, or at the end of, the period of most rapid shoot elongation exhibit weak proliferation potential. Explants collected before or after this period are capable of strong shoot proliferation in vitro (Brand, 1993). Conversely, the best results are obtained with herbaceous perennials that form storage organs, such as tubers or corms, when explants are excised at the end of dormancy and after sprouting.

Explants also exhibit different capacities for establishment in vitro depending on their location on the donor plant. For example, survival and growth of terminal bud explants are typically greater than lateral bud explants. Often similar lateral meristem explants from the top and bottom of a single shoot may respond differently in vitro. In woody plants exhibiting phasic development, juvenile explants typically are more responsive than those obtained from the often nonresponsive mature tissues of the same plant. Sources of juvenile explants include the following: (1) root suckers, (2) basal parts of mature trees, (3) stump sprouts, and (4) lateral shoots produced on heavily pruned plants.

The excision of primary explants often promotes the release of polyphenols and stimulates polyphenol oxidase activity within the damaged tissues. The polyphenol oxidation products often blacken the explant tissue and medium. Accumulation of these polyphenol oxidation products can eventually kill the explants. Procedures used to decrease tissue browning include (1) use of liquid medium with frequent transfer; (2) addition of antioxidants, such as ascorbic acid or polyvinylpyrrolidone (PVP); and (3) culture in reduced light or darkness.

There clearly is not one universal culture medium for establishment of all species. However, modifications to the Murashige and Skoog (Murashige and Skoog, 1962) basal medium formulation are most frequently used (see chapter two). Cytokinins and/or auxins are most frequently added to Stage I media to enhance explant survival and shoot development (Hu and Wang, 1983). The types and levels of PGRs used in Stage I media are dependent on the species, genotype and explant size.

Knowledge of the specific sites of hormone biosynthesis in intact plants provides insight into the relationship between explant size and dependence on exogenous PGRs in the medium. Endogenous cytokinins and auxins are synthesized primarily in root tips and leaf primordia, respectively. Consequently, smaller explants, especially cultured apical meristem domes, exhibit greater dependence on medium supplementation with exogenous cytokinin and auxin for maximum shoot survival and development (Shabde and Murashige, 1977). Larger shoot tip explants usually do not require the addition of auxin in Stage I medium for establishment. Rapid adventitious rooting of shoot tip explants often provides a primary endogenous cytokinin source. Most Stage I media are agar solidified and supplemented with at least a cytokinin (Wang and Charles, 1991). The most frequently used cytokinins are $N^6$-benzyladenine (BA), kinetin (Kin) and $N^6$-(2-isopentenyl)-adenine (2-iP). Because of its low cost and high effectiveness, the synthetic cytokinin BA is most widely used.

Many types of auxins are used. The naturally occurring auxin indole-3-acetic acid (IAA) is the least active, whereas the stronger and more stable compounds α-naphthalene acetic acid (NAA), a synthetic auxin, and indole-3-butyric acid (IBA), a naturally-occurring auxin, are more often used. Stage I medium PGR levels and combinations are selected that promote explant establishment and shoot growth, but limit formation of callus and adventitious shoot formation.

A commonly held misconception is that primary explants exhibit immediate and predictable growth responses following inoculation. For many species, particularly herbaceous and woody perennials, consistency in growth rate and shoot multiplication is achieved only after multiple subculture on Stage I medium. Physiological stabilization may require from 3–24 months and four to six subcultures. Failure to allow culture stabilization, before transfer to a Stage II medium containing higher cytokinin levels, may result in diminished shoot multiplication rates or production of undesirable basal callus and adventitious shoots. With some species, the time required for stabilization can be reduced by initial culture in liquid medium.

In many commercial laboratories, stabilized cultures, verified as being specific pathogen tested and free of cultivable contaminants, are often maintained on media that limit shoot production to maintain genetic stability. These cultures, called "mother blocks" (see chapter thirty-eight), serve as sources of shoot tips or nodal segments for initiation of new Stage II cultures (Figure 7.2).

## Stage II: Proliferation of axillary shoots

Stage II propagation is characterized by repeated enhanced formation of axillary shoots from shoot tips or lateral buds cultured on a medium supplemented with a relatively higher cytokinin level to disrupt apical dominance of the shoot tip. A subculture interval of 4 weeks with a three- to eightfold increase in shoot number is common for many crops propagated by shoot culture. Given these multiplication rates, conservatively, more than $4.3 \times 10^7$ shoots could be produced yearly from a single starting explant.

Stage II cultures are routinely subdivided into smaller clusters, individual shoot tips or nodal segments that serve as propagules for further proliferation (Figure 7.3a). Additionally, axillary shoot clusters may be harvested as individual unrooted Stage II microcuttings or multiple shoot clusters for ex vitro rooting and acclimatization (Figure 7.2). Clearly, Stage II represents one of the most costly stages in the production process.

Both source and orientation of explants can affect Stage II axillary shoot proliferation. Subcultures inoculated with explants that had been shoot apices in the previous subculture often exhibit higher multiplication rates than lateral bud explants. Inverting shoot explants in the medium can double or triple the number of axillary shoots produced on vertically oriented explants per culture period in some species.

The number of subcultures possible before initiation of new Stage II cultures from the mother block is required depends on the species or cultivar and its inherent ability to maintain acceptable multiplication rates while exhibiting minimal genetic variation and off-types (Kurtz et al., 1991). Some species can be maintained with monthly subculture from 8–48 months in Stage II. In contrast, in some ferns (*Nephrolepsis*), as few as three subcultures may only be possible before the frequency of off-types increases to unacceptable levels.

Increased production of off-types is often attributed to production of adventitious shoots, particularly those that arise through an intermediary callus stage (see chapter eleven). Stage II cultures, originally regenerating from axillary shoots, often begin producing adventitious shoots at the base of axillary shoot clusters after a number of subcultures on the same medium. These so-called mixed cultures can develop without any morphological differences being apparent. Selecting only terminal shoots of axillary origin for subculture, instead of shoot bases, decreases the frequency of off-types including the segregation of periclinal chimeras.

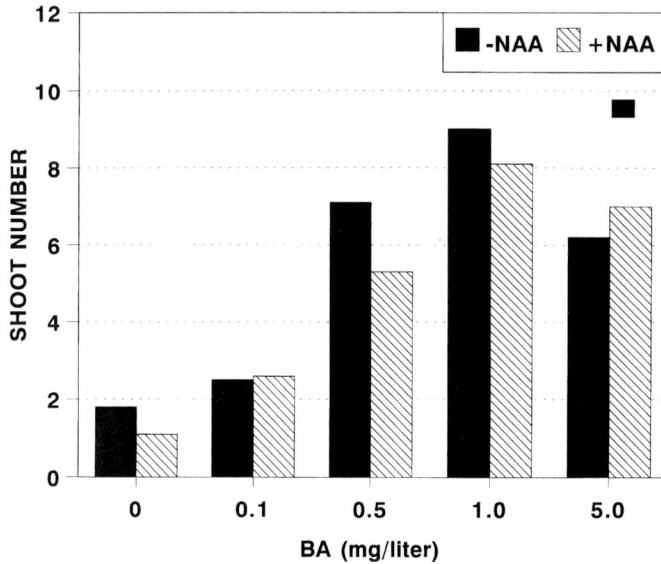

**Figure 7.4**   Effect of BA concentration on Stage II axillary shoot proliferation produced from two-node explants of *A. arbutifolia* after 28-day culture in presence and absence of 0.1 mg/l NAA.

Selection of Stage II cytokinin type and concentration is made on the basis of shoot multiplication rate, shoot length, and frequency of genetic variation. Although shoot proliferation is enhanced at higher cytokinin concentrations, the shoots produced are usually smaller. Depending on the species, exogenous auxins may or may not enhance cytokinin-induced axillary shoot proliferation (Figure 7.4). Addition of auxin in the medium often mitigates the inhibitory effect of cytokinin on shoot elongation, thus increasing the number of usable shoots of sufficient length for rooting (Figure 7.5). This benefit must be weighed against the increased chance of callus formation.

The possibility of adverse carryover effects on the survivability and rooting of plantlets in Stage IV should be evaluated when selecting a Stage II cytokinin. For example, with some tropical foliage plant species, the use of BA in Stage II can significantly reduce the number of plantlets that survive transplanting to as low as 10% (Griffis et al., 1981). Use of Kin or 2-iP instead of BA yields survival rates in excess of 90%. This adverse effect of BA on Stage IV survival has been attributed to abnormal stomate function. However, an inhibitory carryover effect of BA on Stage IV rooting is also possible. Conceivably, switching the Stage II cytokinin from BA may also eliminate the requirement for Stage III rooting in some species.

## Stage III: Pretransplant (rooting)

This step is characterized by preparation of Stage II shoots or shoot clusters for successful transfer to soil. The process may involve (1) elongation of shoots prior to rooting, (2) rooting of individual shoots or shoot clumps, (3) fulfilling dormancy requirements of storage organs by cold treatment or (4) prehardening cultures to increase survival. Where possible, commercial laboratories have developed procedures to transfer Stage II microcuttings to soil, thus bypassing Stage III rooting (Figure 7.2).

There are several reasons for eliminating Stage III rooting. Estimated costs for Stage III range from 35–75% of the total production costs. This reflects the significant input of labor and supplies required to complete Stage III rooting. Considerable cost saving can be

**Figure 7.5** Inclusion of auxin may reduce the inhibitory effect of BA on axillary shoot elongation. Data shown for shoots generated from two-node explants of *A. arbutifolia* after 28-day culture in presence and absence of 0.1 mg/L NAA.

realized if Stage III is eliminated. Furthermore, it is often observed that in vitro–formed root systems are largely nonfunctional and die following transplanting. This results in a delay in transplant growth prior to production of new adventitious roots.

For various reasons, however, it may not be feasible to transplant Stage II microcuttings directly to soil. Given the aforementioned limitations of Stage III rooting, Debergh and Maene (1981) proposed using Stage III solely to elongate Stage II shoot clusters prior to separation and rooting ex vitro. Elongated shoots may be further pretreated in an aqueous auxin solution prior to transplanting. Usually, Stage III rooting of herbaceous plants can be achieved on medium in the absence of auxins. However, with many woody species the addition of an auxin (IBA or NAA) in Stage III medium is required to enhance adventitious rooting (Figure 7.3b). Optimum auxin concentration is determined based upon percent rooting, root number and length (Table 7.1). It is critical that the roots not be allowed to elongate to prevent root damage during transplanting. Care must be taken when selecting an auxin. For example, use of NAA for Stage III rooting has been shown to decrease survival rates or suppress posttransplant growth (Connor and Thomas, 1981).

## Stage IV: Transfer to natural environment

The ultimate success of shoot or node culture depends on the ability to transfer and reestablish vigorously growing plants from in vitro to greenhouse conditions (Figure 7.3c). This involves acclimatizing or hardening-off plantlets to conditions of significantly lower relative humidities and higher light intensities. Even when acclimatization procedures are carefully followed, poor survival rates are frequently encountered. Micropropagated plants are difficult to transplant for two primary reasons: (1) a heterotrophic mode of nutrition and (2) poor control of water loss.

Plants cultured in vitro in the presence of sucrose and under conditions of limited light and gas exchange exhibit no or extremely reduced capacities for photosynthesis. Reduced photosynthetic activity is associated with low RubPcase activity. During acclimatization,

***Table 7.1***   Addition of Auxin Promotes Stage III Rooting of Shoot Cuttings

| Treatment IBA (mg/L) | % Rooting | Root number | Root length (mm) |
|---|---|---|---|
| 0 | 37 | 2.4 | 23.2 |
| 0.05 | 43 | 3.5 | 18.1 |
| 0.1 | 55 | 4.1 | 14.2 |
| 0.5 | 71 | 5.7 | 6.5 |
| 1.0 | 84 | 7.1 | 4.3 |

*Note:* Effects of IBA on Stage III Rooting of 10-mm Shoots of *Aronia arbutifolia* after 4 Weeks Culture Are Shown.

there is a need for plants to rapidly transition from the heterotrophic to photoautotrophic state for survival (Preece and Sutter, 1991). Unfortunately, this transition is not immediate. For example, in cauliflower, no net increase in $CO_2$ uptake is achieved until 14 days after transplantation. This occurs only following development of new leaves since the leaves produced in vitro in the presence of sucrose never develop photosynthetic competency. Interestingly, before senescencing, these older leaves function as "lifeboats" by supplying stored carbohydrate to the developing and photosynthetically competent new leaves. This is not the rule with all micropropagated plants since the leaves of some species become photosynthetic and persist after acclimatization.

A composite of anatomical and physiological features, characteristic of plants produced in vitro under 100% relative humidity, contributes to the limited capacity of micropropagated plants to regulate water loss immediately following transplanting. These features include reductions in leaf epicuticular wax, poorly differentiated mesophyll, abnormal stomate function and poor vascular connection between shoots and roots.

To overcome these limitations, plantlets are transplanted into a well-drained "sterile" growing medium and maintained initially at high relative humidity and reduced light (40–160 $\mu mol \cdot m^{-2} \cdot sec^{-1}$) at 20–27 C. Relative humidity may be maintained with humidity tents, single tray propagation domes, intermittent misting or fog systems. However, use of intermittent mist often results in slow plantlet growth following waterlogging of the medium and excessive leaching of nutrients. Transplants are acclimatized by gradually lowering the relative humidity over a 1–4-week period. Plants are gradually moved to higher light intensities to promote vigorous growth.

## Conclusion

Propagation from preexisting meristems through shoot and node culture is the most reliable and widely used procedure. However, the need for multiple subcultures on different media makes shoot and node culture extremely labor intensive. Total labor costs, typically ranging from 50 to 70% of production costs, limit expansion of the micropropagation industry. Current application of the technology is restricted to high value horticultural crops such as ornamental plants. Expansion of the industry to include production of vegetable, plantation and forest crops depends on development of more-efficient micropropagation systems. Cost-reduction strategies including elimination of production steps and development of reliable automated micropropagation systems will facilitate this expansion.

## Literature cited

Ball, E.A. 1946. Development in sterile culture of shoot tips and subjacent regions of *Tropaeolum majus* L. and of *Lupinus albus* L. *Am. J. Bot.* 33:301–318.

Brand, M. 1993. Initiating cultures of *Halesia* and *Malus*: influence of flushing stage and benzyladenine. *Plant Cell Tissue Organ Cult.* 33:129–132.

Conner, A.J. and M.D. Thomas. 1981. Re-establishing plantlets from tissue culture: a review. *Comb. Proc. Intl. Plant Prop. Soc.* 31:342–357.

Debergh, P.C. and L.J. Maene. 1981. A scheme for commercial propagation of ornamental plants by tissue culture. *Sci. Hort.* 14:335–345.

Debergh, P.C. and P.E. Read. 1991. Micropropagation. In: *Micropropagation Technology and Application.* P.C. Debergh and R.H. Zimmerman, Eds., pp. 1–13. Kluwer Academic Publishers, Boston.

Fahn, A. 1974. *Plant Anatomy.* Pergamon Press, New York.

George, E.F. 1993. *Plant Propagation by Tissue Culture.* Part 1. *The Technology.* Exegetics Ltd., London.

Griffis, J.L., Jr., G. Hennen and R.P. Oglesby. 1981. Establishing tissue-cultured plant in soil. *Comb. Proc. Intl. Plant Prop. Soc.* 33:618–622.

Hu, C.Y. and P.J. Wang. 1983. Meristem, shoot tip and bud cultures. In: *Handbook of Plant Cell Culture.* Vol. 1. *Techniques for Propagation and Breeding.* D.F. Evans, W.R. Sharp, P.V. Ammirato and Y. Yamada, Eds., pp. 177–227. Macmillan Publishing, New York.

Kurtz, S., R.D. Hartmann and I.Y.E. Chu. 1991. Current methods of commercial micropropagation. In: *Scale-Up and Automation in Plant Propagation. Cell Culture and Somatic Cell Genetics of Plants*, Vol. 8. I.K. Vasil, Ed., pp. 7–34. Academic Press, San Diego.

Lloyd, G. and B. McCown. 1981. Commercially feasible micropropagation of mountain laural, *Kalmia latifolia*, by use of shoot-tip culture. *Intl. Plant Prop. Soc. Proc.* 30:421–427.

Miller, L.R. and T. Murashige. 1976. Tissue culture propagation of tropical foliage plants. *In Vitro.* 12:797–813.

Morel, G. and C. Martin. 1952. Guerison de dahlias atteints d'une maladie a virus. *C. R. Acad. Sci. Ser. D.* 235:1324–1325.

Morel, G. 1960. Producing virus *Cymbidium. Am. Orchid Soc. Bull.* 29:495–497.

Morel, G. 1965. Clonal propagation of orchids by meristem culture. *Cymbidium Soc. News.* 20:3–11.

Murashige, T. and F. Skoog. 1962. A revised medium for rapid growth and bioassays with tobacco tissue cultures. *Physiol. Plant.* 15:473–497.

Murashige, T. 1974. Plant propagation through tissue culture. *Annu. Rev. Plant Physiol.* 25:135–166.

Preece, J.E. and E.G. Sutter. 1991. Acclimatization of micropropagated plants to greenhouse and field. In: *Micropropagation Technology and Application.* P.C. Debergh and R.H. Zimmerman, Eds., pp. 71–93. Kluwer Academic Publishers, Boston.

Read, P.E. 1988. Stock plants influence micropropagation success. *Acta Hort.* 226:41–52.

Shabde, M. and T. Murashige. 1977. Hormonal requirements of excised *Dianthus caryophyllus* L. shoot apical meristem in vitro. *Am. J. Bot.* 64:443–448.

Styer, D.J. and C.K. Chin. 1984. Meristem and shoot-tip culture for propagation, pathogen elimination, and germplasm preservation. *Hort. Rev.* 5:221–277.

Wang, P.J. and A. Charles. 1991. Micropropagation through meristem culture. In: *Biotechnology in Agriculture and Forestry*, Vol. 17, *High-Tech and Micropropagation I.* Y.P.S. Bajaj, Ed., pp. 32–52. Springer-Verlag, Berlin.

Wickson, M. and K.V. Thimann. 1958. The antagonism of auxin and kinetin in apical dominance. *Physiol. Plant.* 11:62–74.

*chapter eight*

# Micropropagation of Syngonium *by shoot culture*

*Michael E. Kane*

One of the most successful and reliable commercial applications of micropropagation has been the production of tropical foliage plants (house plants) using shoot culture (Henny et al., 1981). In fact, most of the foliage plants purchased in garden shops are produced using micropropagation. Shoot culture is characterized by the establishment of Stage I cultures using isolated surface disinfested meristem tips, shoot tips or lateral buds as primary explants (chapter seven). Shoots that develop from these explants usually are first screened (indexed — see also chapter thirty-nine) for microbial contamination. Aseptic cultures, once physiologically adapted to culture, are then subdivided into nodal or shoot tip secondary explants and transferred to a medium supplemented with a cytokinin to pro-mote axillary shoot proliferation. The axillary shoots produced are either subcultured for repeated proliferation (Stage II) or rooted as microcuttings in vitro (Stage III) or ex vitro (Stage IV) to produce plantlets (George, 1993) — see chapter thirty-eight.

*Syngonium podophyllum* Schott., commonly known as nephthytis or arrowhead vine, is one of the most commercially important tropical foliage plants propagated in vitro. Plants produce arrowhead-shaped variegated leaves in the juvenile state and climbing vines with three-parted leaves when mature. Each year more than 6 million plants of the various *Syngonium* cultivars are produced in the United States. Like other members of the Araceae, *Syngonium* is susceptible to infection by numerous systemic pathogens that negatively affect plant vigor and quality. However, use of meristem tip culture followed by shoot culture has enabled rapid production of highly branched specific pathogen-eradicated plants. It is conservatively estimated that 5000 *Syngonium* plants can be produced yearly from a single primary shoot tip explant (Miller and Murashige, 1976).

The primary objective of these laboratory exercises is to illustrate the sequential steps required for the micropropagation of *Syngonium* by shoot culture. The laboratory exercises will familiarize students with the specific procedures used to (1) establish Stage I cultures using isolated shoot tip explants, (2) index Stage I cultures for cultivable bacterial and fungal contaminants, (3) rapidly propagate plants by axillary shoot proliferation (Stage II) and (4) successfully root and acclimatize microcuttings ex vitro (Stage IV). These exercises also offer the opportunity to exemplify several important concepts: (1) relationship be-tween initial shoot tip explant size, response and culture contamination; (2) requirement for physiological adaption of Stage I cultures prior to Stage II axillary shoot proliferation; and (3) direct rooting and acclimatization of microcuttings. The two laboratory exercises outlined below include maintenance of stock plants, media preparation, procedures and anticipated results. Refer to Figure 8.1 for a diagrammatic representation of the procedures used to complete both exercises.

0-8493-9409-0/96/$0.00+$.50

**Figure 8.1** Diagrammatic representation of the *Syngonium* shoot culture laboratory exercise protocol. A. Nodal and terminal vine tip explants are obtained from potted plants and surface sterilized. B. Lateral and terminal bud scales are aseptically removed by making shallow horizontal and vertical excisions (dotted lines). C. Successive inner leaf layers are removed to isolate the meristem and shoot tip explants. D. Primary explants are inoculated onto SEM ($D_1$) and SMM ($D_2$). Growth responses are compared after 6 weeks of culture. E. Thin basal stem cross sections, excised from 6-week-old shoots generated on SEM, are cultured in both liquid and agar-solidified AC Broth for 7 days to screen for the presence of cultivable contaminants. F. Shoots are transferred to SEM. G. Cultures screening positive for contaminants are discarded. Sterile cultures are transferred to SMM and cultured for 4 weeks. H. Microcuttings are rooted and acclimatized ex vitro (Stage IV).

## General considerations

### Maintenance of Syngonium *stock plants*

Shoot tip explants excised from lateral buds and terminal shoot tips from *Syngonium* vines (Figure 8.2a) are placed in culture. It is recommended that vining stems be used because the lateral bud scales are usually larger and easier to remove without damaging the underlying meristem tips. However, if vines are not available, terminal shoot tips and lateral buds on nonvining plants can be used. Potted plants can be inexpensively purchased from most retail garden shops; however, larger plants will produce vines more rapidly. Most *S. podophyllum* cultivars, such as 'White Butterfly', produce climbing vines, and *S. wendlandii* Schott, another nephthytis species, is an especially fast vine producer.

**Figure 8.2** Shoot tip culture of *Syngonium*. a. *Syngonium* vine used as the source of excised meristem tip and shoot tip explants from surface sterilized nodal segments (lower center) and terminal tips (lower right). Scale bar = 1 cm. b. Exposed shoot meristem tip within a lateral bud prior to excision and inoculation. Distance between arrows is 3 mm. c. Left: Stage I shoot culture established from a primary shoot tip explant cultured on SEM for 6 weeks. Center: Inhibition of shoot growth and promotion of basal shoot organogenic callus following 6-week culture of primary shoot tip (arrow) explanted directly on SMM. Right: Typical shoot multiplication response after 4 weeks on SMM from a shoot explant initially established on SEM for 6 weeks. Scale bar = 1 cm. d. Rooted acclimatized plantlet 3 weeks following transfer to ex vitro conditions.

Plants can be grown in most soilless potting mixes contained in 20-cm-diam. hanging baskets. Plants should be maintained in a shaded (70–80% light reduction) greenhouse or similar environment where they will receive about 150 $\mu mol \cdot m^{-2} \cdot sec^{-1}$ light at 22–30 C. Plants should be fertilized twice weekly (150 mg/L N) to promote vining. Depending on growing conditions and plant size, vines will be produced within 6–8 weeks. Two plants should produce sufficient numbers of explants for each student or team of students.

## Materials

The following are needed for each student or student team:

* *Syngonium* vines to provide 15 nodal explants (including terminal vine tip)
* Metal microspatula
* 20% commercial bleach
* Ten culture tubes each containing 12 ml *Syngonium* Stage I Establishment Medium (SEM)
* Ten culture tubes each containing 12 ml *Syngonium* Stage II Multiplication Medium (SMM)
* Incubator or a fluorescent light on a bench top providing approximately 30 $\mu mol \cdot m^{-2} \cdot sec^{-1}$ for a 16-h photoperiod at 25 C

## Composition and preparation of the media

Two culture media, SEM and SMM, are required. These media are identical to those described by Miller and Murashige (1976) with the exception that both media are solidified with Phytagel. The basal medium of both consists of Murashige and Skoog (1962) inorganic salts (see chapter two for composition) supplemented with 170 mg (1.25 mM) $K_2PO_4 \cdot H_2O$, 30 g (87.6 mM) sucrose, 100 mg (0.55 mM) myo-inositol and 0.4 mg (1.2 µM) thiamine HCl, solidified with 2 g/L Phytagel. SEM also contains 3 mg (14.8 µM) 2-iP and 1 mg/L (5.7 µM) IAA whereas, SMM is supplemented with 20 mg/L (98.4 µM) 2-iP alone. Both SEM and SMM can be purchased prepackaged without Phytagel from Sigma Chemical, St. Louis, MO (M8400 and M8525, respectively). Media are adjusted to pH 5.7 before addition of Phytagel and dispensed as 12 ml volumes into $150 \times 25$ mm culture tubes and autoclaved. The autoclaved media should be cooled and solidified as 45° slants. For exercise 1, five culture tubes of each medium are prepared for each student or team of students. Additional SEM and SMM will be needed immediately before exercise 2 is performed.

The AC Broth indexing medium used in exercise 2 is manufactured by Difco Laboratories (Detroit, MI) and consists (grams per liter) of 20 Proteose Peptone No. 3, 3 Bacto Beef Extract, 3 Bacto Malt Extract, 3 Bacto Yeast Extract, 5 Bacto Dextrose and 0.2 ascorbic acid. The medium is distributed through Fisher Scientific Supply (Norcross, GA). The solid indexing medium is prepared by the addition of 10 g/L agar. Both media are prepared without pH adjustment and dispensed as 10-ml volumes into 20-ml glass scintillation vials. Culture tubes can be substituted. However, using relatively deeper $150 \times 25$ mm culture tubes will make it more difficult to perform the stab and streak indexing procedure. At least six vials of each medium for each student should be prepared and dispensed immediately before exercise 2 is performed. After autoclaving, the agar-supplemented medium is cooled and solidified as 45° slants.

## Exercises

### Experiment 1. Stage I establishment of shoot tip meristems

The purpose of this exercise is to familiarize students with the procedures used to successfully establish Stage I cultures from excised *Syngonium* meristem tip and shoot tip explants. The goal of Stage I is to establish axenic, responsive shoot cultures. Cultures are indexed for specific pathogens or microbial contamination and allowed to become physiologically adapted to in vitro culture conditions before being clonally multiplied by repeated proliferation of axillary shoots in Stage II (see Figure 7.2). Before proceeding, the structure of a generalized shoot apex should be briefly reviewed (see Figure 7.1).

The size of the primary meristem explant will depend on the purpose of culture establishment. For example, when virus elimination is a goal, excision and culture of the apical meristematic dome alone (meristem culture) may be attempted. In reality, cultured apical meristems of many species exhibit low survival rates in vitro and their use may increase the probability of genetic variability due to the tendency to form organogenic callus (Styer and Chin, 1984) — see chapter eleven. Consequently, relatively larger (0.2–1.0 mm) and more-responsive meristem tip explants (meristem tip culture) are usually used. When pathogen eradication is not a concern, relatively larger (1–2-mm-long) terminal or lateral bud shoot tip explants are also used to establish cultures.

#### Procedure

*Syngonium* vines consist of a terminal shoot apex and numerous lateral buds each enclosed by a sheathing petiole (Figure 8.2a). The leaves are removed with a scalpel to expose the terminal and lateral buds, and any adventitious roots are excised. The defoliated vines are cut to yield 15 nodal segments, each about 1 cm long and consisting of a single bud. It can be noted that the smoothly cutinized epidermis of *Syngonium* should facilitate effective surface disinfestation of the explants. Nodes exhibiting bud break should not be used. Rinse nodal segments in tap water for 15 min and then surface disinfest in 20% commercial bleach with constant agitation for 10 min followed by three 5-min rinses in sterile deionized water.

In the transfer hood, under 15–30× magnification provided by a dissecting microscope, a 1–5-mm-long meristem tip or shoot tip explant is excised from each surface-disinfested bud with a sterile scalpel. To accomplish this, remove the outer bud scales by making a very shallow, but continuous, incision around the base of the bud and then a continuous median longitudinal incision down the front and rear of the bud scale (see Figure 8.1B). Remove bud scale sections with forceps. The inner leaves are successively removed to expose the meristem tip (Figure 8.2b). Excisions should be performed in a sterile petri dish containing a small volume of sterile water to prevent tissue desiccation. The potential for cross contamination is decreased by using a clean petri dish for each nodal segment. Students should attempt to isolate meristem tip and shoot tip explants of various size ranging from 1–5 mm in length. Final explant length measurements are made using a thin plastic ruler positioned underneath the petri dish.

Use a stainless steel, tapered microspatula to transfer a single meristem tip or shoot tip explant into five culture tubes, each containing either SEM or SMM. The cut base of each explant should be positioned in the center of the slant in firm contact with the medium, but not completely embedded. Each tube should be labeled with the size of the explant and maintained in an incubator or under a fluorescent light on a bench top at 25 C and provided with approximately 30 μmol·m$^{-2}$·sec$^{-1}$ for 16 h.

Make weekly observations and note differences in the growth responses of the various size explants cultured on SEM and SMM. After 6 weeks, determine the following: (1) percentage of cultures with visible contamination, (2) percentage of responsive explants on each medium, (3) the mean number of axillary shoots produced per responsive explant on each medium, and (4) presence of shoot organogenic callus at the base of each explant. Data should be collected without removing the plants from the culture vessels. Individual student or student team data should be complied to provide more representative treatment responses.

### Anticipated results
*Syngonium* primary explant size should significantly affect culture survival and responsiveness. Meristem tip explants about 2 mm in length or smaller usually exhibit phenolic browning and high mortality when cultured on either SEM or SMM. When cultured on SEM, larger explants enlarge rapidly and produce leaves and adventitious roots during the 6-week culture period (Figure 8.2c). However, cultures inoculated with larger explants may exhibit a greater frequency of visible contamination. It is important to note that the absence of visible contamination is not an assurance that a culture is aseptic or axenic.

It is often asked why primary explants are not directly inoculated onto Stage II multiplication medium to promote axillary shoot proliferation and save time. The first and most obvious reason for this is the danger of mass propagation of contaminated cultures. The second reason involves a requirement for physiological adjustment of Stage I explants to in vitro culture conditions. Often this physiological adaptation must be completed before enhanced clonal multiplication at higher cytokinin levels is possible. For some plants, repeated subculture on Stage I medium is required before physiological stabilization is achieved.

The requirement for Stage I stabilization in *Syngonium* is exemplified by the poor response of the meristem tips explanted directly on SMM. Unlike explants cultured on SEM, shoot development from primary explants on SMM should be inhibited, whereas formation of shoot organogenic callus should be promoted at the base of each explant

(Figure 8.2c). This response is commercially undesirable since indirect shoot organogenesis may lead to genetic variability in the plants produced and will definitely result in separation of plant chimeras. It is interesting to note that both axillary shoot and adventitious shoot development may occur simultaneously in the same culture. These are often referred to as mixed cultures. Prevention of mixed cultures is usually controlled by the type, level and combinations of plant growth regulators (PGRs) used.

*Questions*

* If your donor plant was virus infected, which explant tip source (lateral bud or vine tip) do you predict might yield the greatest number of virus eradicated Stage I cultures?
* For commercial micropropagation (see chapter thirty-eight), why is it unwise to attempt to save time by eliminating Stage I and immediately culturing surface disinfested primary explants (shoot tips) on SMM?
* In their study on *Syngonium* micropropagation, Miller and Murashige (1976) reported that both primary explant establishment and subsequent Stage II shoot multiplication were greatest when explants were cultured in liquid rather than on solid media. Give possible explanations for these observations.

## *Experiment 2. Procedures for Stage I culture indexing, Stage II shoot multiplication and Stage IV acclimatization*

Rapid in vitro (contaminant-free) production of pathogen-eradicated plants is a fundamental goal of the micropropagation process. The surfaces of donor plants are normally colonized and/or inhabited with diverse bacteria and fungi, most of which are nonpathogenic, and surface disinfestation procedures are used to specifically eliminate this microflora. The presence of microbial contaminants in culture can adversely affect growth, multiplication rate and survival of shoot cultures. Bacterial and fungal contaminants often persist within cultured shoots that visually appear contaminant free (Knauss, 1976). Consequently, it is essential that Stage I shoot cultures be indexed (screened) for the presence of internal infections before being multiplied in Stage II (see indexing chapter thirty-nine).

The first objective of this exercise is to familiarize students with a general indexing procedure used to screen for cultivable bacterial and fungal contaminants in the Stage I *Syngonium* shoot cultures generated in exercise 1. This indexing exercise can be completed during the same laboratory period that the final data collection for exercise 1 is made. The second objective is to examine the response of secondary shoot explants excised from 6-week-old established and indexed Stage I shoot cultures following culture on SMM for 4 weeks. Lastly, procedures for successful rooting and Stage IV acclimatization of Stage II microcuttings will be examined.

*Materials*

The following supplies are needed for each student or team of students:

* Six scintillation vials or 150 × 25 mm culture tubes containing liquid AC Broth
* Six scintillation vials or 150 × 25 mm culture tubes containing agar-solidified AC Broth (slants)
* Five culture tubes each containing 12 ml freshly prepared SEM
* Five culture tubes each containing 12 ml freshly prepared SMM
* Plug trays with soilless growing medium and propagation dome or mist bench

### Procedure

Each student or team should first consecutively number each of their five Stage I *Syngonium* cultures established on SEM in exercise 1. One scintillation vial each of the liquid and solidified indexing medium should be labeled "Control". The other corresponding pairs of indexing media should be labeled "Syngonium #1–5". Starting with culture #1, remove the shoot (see Figure 8.2c) from the culture vessel and place in a sterile petri dish. Trim roots from the shoot and make several thin cross sections of the stem base using a sterile scalpel. Inoculate the liquid AC Broth in the scintillation vials with one or two tissue sections. Using forceps, inoculate the solidified AC Broth slant by gently stabbing the tissue partially into the medium at the base of the slant. Lift the tissue from the medium, drag it upward along the medium surface and then position it at the top of the slant. Trim off leaves and transfer the shoot onto freshly prepared SEM and label the tube "Syngonium #1". Repeat the indexing procedure with the other Stage I *Syngonium* cultures. Stage I subcultures are maintained under the same culture conditions as described in exercise 1.

In the event that all shoot cultures index negative for contamination, several visibly contaminated cultures should also be indexed. This can be accomplished, without removing the contaminated tissue from the tube, by dipping the end of the forceps into the contaminated medium and then inoculating the indexing media as described above. Cultures containing sporulating fungal contaminants should not be used. Vials or tubes containing inoculated indexing media are cultured at 25 C. After 7 days, examine the indexing media for signs of microbial growth. Stage I shoot cultures that index positive for contamination should be discarded. Transfer the axenic Stage I cultures onto SMM and maintain cultures under the same culture conditions as described previously.

After 4 weeks, Stage II shoot clusters are divided into single microcuttings and rooted directly in soilless growing medium contained in plug trays or any other small container placed under intermittent mist (5 sec every 10 min) for 3 weeks under lightly shaded conditions. Alternatively, if a mist system is not available, plastic propagation domes can be placed over the plug trays to maintain humidity. After 10 days, the domes are gradually removed over a period of 7 days to acclimatize the plantlets to lower humidity.

### Anticipated results

Rapid clouding of the inoculated liquid AC Broth is a positive indication of the presence of cultivable contaminants in the tissue sample. The presence of contaminants on solid AC broth is confirmed by development of colonies on the surface of the medium and/or development of a halo in the medium where the tissue sample was stabbed. The uninoculated controls should not display microbial growth. Although AC Broth will promote the growth of many microorganisms, it is important to note that growth of all bacteria and fungi will not be supported on this medium. Verified contaminated Stage I cultures are discarded. After 4 weeks, each Stage II *Syngonium* culture should consist of three to five axillary shoots with no or minimal development of basal organogenic callus (Figure 8.2c). This is in contrast to the inhibition of shoot growth and promotion of organogenic callus observed on the primary explants inoculated directly onto SMM in exercise 1. *Syngonium* microcuttings should readily root and become acclimatized to ex vitro conditions without the need for Stage III rooting. Rooted and vigorously growing acclimatized plantlets will be produced in 3 weeks (Figure 8.2d).

### Questions

* Why is it important to both stab into and streak the Stage I tissue sample across the surface of the agar-solidified indexing medium?
* How is it possible to have a contaminated Stage I shoot culture that indexes negative for contamination?

* Provide a possible explanation for the differences in growth response of primary explants cultured first on SEM and then SMM vs. those inoculated directly onto SMM.
* What are the two primary environmental factors that must be regulated during acclimatization?

## *Literature cited*

George, E.F. 1993. *Plant Propagation by Tissue Culture.* Part 1. *The Technology.* Exegetics Ltd., Edington.

Henny, R.J., J.F. Knauss and A. Donnan. 1981. Foliage plant tissue culture. In: *Foliage Plant Production,* J.N. Joiner Ed., pp. 137–178. Prentice-Hall, Englewood Cliffs, NJ.

Knauss, J.F. 1976. A tissue culture method for producing *Dieffenbachia picta* cv. Perfection free of fungi and bacteria. *Proc. Fla. State Hort. Soc.* 89:293–296.

Miller, L.R. and T. Murashige. 1976. Tissue culture propagation of tropical foliage plants. *In Vitro* 12:797–813.

Murashige, T. and F. Skoog. 1962. A revised medium for rapid growth and bioassays with tobacco cultures. *Physiol. Plant.* 15:473–497.

Styer, D.J. and C.K. Chin. 1984. Meristem and shoot-tip culture for propagation, pathogen elimination, and germplasm preservation. *Hort. Rev.* 5:221–277.

*chapter nine*

# Micropropagation of potato by node culture and microtuber production

*Michael E. Kane*

The potato (*Solanum tuberosum* L.), a crop of worldwide importance, is an integral part of the diet of a large proportion of the population of the world. Potato production ranks fourth among major food crops and, unlike other major food crops, potatoes are vegetatively propagated and, unfortunately, susceptible to infection by many viral and bacterial pathogens. At least 23 viruses that decrease tuber quality and yield are known to infect potato. Reliance on vegetative propagation from field-produced seed tubers results in multiplication and spread of infected tubers. Routine production of pathogen eradicated seed tubers is necessary to maintain adequate yield (Miller and Lipschultz, 1984). This is accomplished using meristem tip culture to produce specific pathogen-free tested (SPT) potato tuber seed stocks (Cassells and Long, 1982). Micropropagation techniques are subsequently used to clonally propagate these pathogen-eradicated stock plants (Dodds et al., 1992). For a discussion concerning indexing plants and pathogen elimination, see chapter thirty-nine.

Two very simple, but reliable, micropropagation techniques can be demonstrated using potato. The first technique, called single node or multiple node culture, involves production of shoots from cultured single or multiple nodes (in vitro layering) positioned horizontally on the medium. In potato, the initial source of these nodes are SPT shoots established in vitro from meristem tips (Figure 9.1). Interestingly, potato cannot be multiplied by using the shoot tip culture method (see chapter seven) because axillary shoot proliferation is not promoted by the addition of cytokinin to the medium. Typically, single elongated unbranched shoots comprising multiple nodes are rapidly produced. These shoots (microcuttings) are either rooted and acclimatized to field conditions or are subdivided into single node cuttings to initiate additional cultures (Figure 9.1). Although not the most efficient method, node culture is the most reliable micropropagation procedure to produce plants that are true to type.

The second, and perhaps more fascinating, micropropagation technique involves production of miniaturized tubers (microtubers) on potato shoots cultured in vitro (Figure 9.1). Microtuber production is a very useful method to propagate and store valuable potato stocks and may be adaptable for automated commercial propagation and large-scale mechanized field planting (McCown and Joyce, 1991). Depending on the cultivar, potato microtuber development may be promoted under short-day photoperiods and following addition of cytokinin and high levels of sucrose to the medium (Seabrook et al., 1993). Under ideal planting and early growth conditions, field-planted microtubers perform as well as regular seed tubers (McCown and Wattimena, 1987).

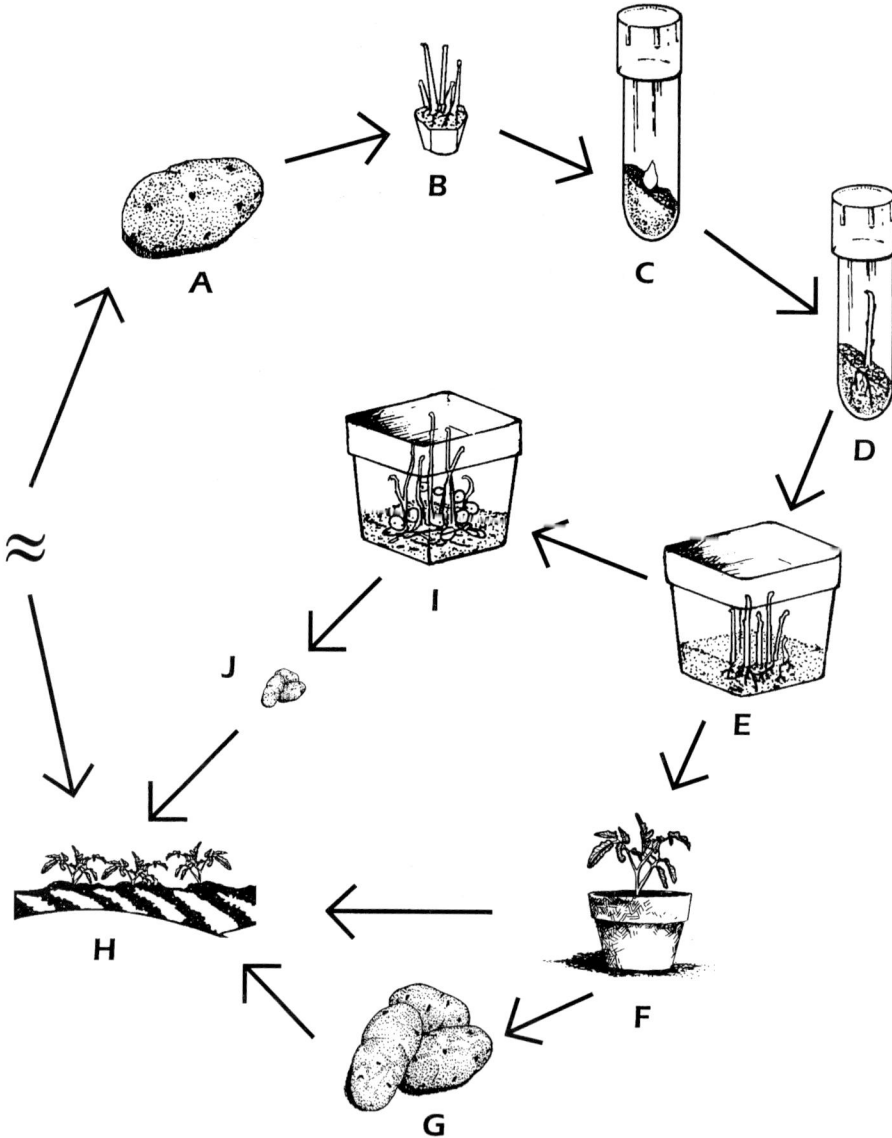

**Figure 9.1**  Diagrammatic representation of potato micropropagation by node culture and microtuber production. A. Tuber obtained from a multigeneration pathogen-infected field potato population. B. Meristem tip explants excised from sprouted tuber sections. C. Establishment of meristem tip in culture. D. Culture is SPT and indexed for other contaminants. E. Shoots determined to be specific pathogen free are propagated by node culture to increase numbers. F. Microcuttings are rooted and acclimatized ex vitro. G. Tubers produced ex vitro from STP plants are used as planting stock. H. Field planting of STP seed tuber I. Shoots may be transferred to a microtuber induction medium and promotive conditions. J. Microtubers produced are stored or used as STP seed tubers for field planting.

The primary objective of this laboratory exercise is to illustrate micropropagation of potato by node culture and microtuber production. This exercise will familiarize students with the following concepts: (1) efficient shoot production is possible from propagules with preexisting shoot meristems in the absence of cytokinin; (2) potato shoot meristems that normally produce elongated shoots in vitro can be induced to form miniature tubers following alteration of environmental and cultural conditions; and (3) microtuberization is a valuable method for potato propagation, storage and distribution. The laboratory exercise is detailed below, including establishment and maintenance of stock shoot cultures, media preparation, procedures and anticipated results.

## General considerations

### Establishment and maintenance of stock potato shoot cultures

Students will use potato nodal cuttings and intact shoots from shoot cultures established from surface disinfested shoot tips excised from sprouted tuber sections. This eliminates the need to perform an often unsuccessful surface disinfestation step and decreases culture response time. Most potato cultivars (genotypes) can be induced to form tubers in vitro; however, the capacity for microtuberization is influenced by genotype. Potato cultivars such as 'Kennebec', 'Superior', 'Red Pontiac' or 'Russet Burbank' have proved to be responsive (Leclerc et al., 1994) and are available in most produce stores. Use only fresh potato tubers that are free of decay.

Wash potato tubers under flowing tap water and then soak for 10 min in 20% (v/v) commercial bleach (1.05% NaOCl). Cut tubers into 5–10 pieces, each about 2 cm$^2$ and then soak for 1 h in 100 mg/L (0.29 mM) gibberellic acid (GA$_3$) to break bud dormancy and promote sprouting. Place treated tuber sections on moist vermiculite in an incubator maintained at approximately 25 C under a 16-h photoperiod provided by cool white fluorescent lamps at about 30 μmol·m$^{-2}$·sec$^{-1}$. Excise sprouted shoots when they are approximately 3–5 cm in length. Large shoot tip explants 5–10 mm long should be used to initiate stock cultures. Rinse excised shoot tips in tap water for 10 min and then surface disinfest in 20% commercial bleach with constant agitation for 10 min followed by three five-min rinses in sterile distilled water. Trim away bleached tissues and inoculate shoot tip explants into separate 150 × 25 mm glass culture tubes containing 15 ml Murashige and Skoog (1962) basal salts (see chapter two for composition) supplemented with 30 g (88 mM) sucrose, 100 mg (0.55 μM) myo-inositol and 0.4 mg (1.2 μM) thiamine HCl, solidified with 8 g/L agar. This medium can be purchased prepackaged without the sucrose and agar from Sigma Chemical, St. Louis, MS (M-6899). Cultures are maintained under the previously described culture conditions. Shoots should be about 4–6 cm long after 30–40 days culture. Subculture nodal cuttings from sterile cultures onto the same medium at about 4–6-week intervals to generate additional shoot cultures.

### Materials

The following items are needed for each team of students:

* Twenty-four 5-week-old stock potato shoot cultures (in culture tubes)
* Long (29-cm) forceps
* Four GA-7 vessels each containing 60 ml Node Culture Medium (NCM)
* Eight GA-7 vessels each containing 60 ml Microtuber Induction Medium (MIM)
* Plug trays with soilless growing medium and propagation dome or mist bench
* Incubator or a fluorescent light on a bench top providing approximately 30 μmol· m$^{-2}$·sec$^{-1}$ for a 16-h photoperiod and 20 C
* Incubator set to maintain 20 C and provide approximately 30 μmol·m$^{-2}$·sec$^{-1}$ for a 8-h photoperiod (for microtuberization exercise)

**Figure 9.2**   Typical development of potato shoots from two-node cuttings cultured for 7 days in a Magenta GA-7 vessel containing NCM. Scale bar = 10 mm. Corresponds to part E of Figure 9.1.

## Composition and preparation of media

Two culture media, NCM and MIM (see above), are required for the laboratory exercise. Magenta GA-7 vessels (Magenta Corp., Chicago, IL), each containing 60 ml medium should be used if available (Figure 9.2). However, baby food jars containing 30 ml medium can be substituted. Each student team will require four GA-7 vessels containing NCM and eight GA-7 vessels containing MIM. The NCM is identical in composition to the stock shoot culture medium described previously. The MIM contains the same components as the NCM except that the sucrose concentration is increased to 80 g/L (234 mM) and the medium is supplemented with 5 mg/L (22.2 µM) benzyladenine (BA). Both media are adjusted to pH 5.7 before addition of agar and autoclaving.

## Exercises

### Experiment 1. Potato micropropagation by node culture

The objective of this exercise is to demonstrate that node culture, the simplest micropropagation method, can be used to generate large numbers of plants in a short period of time. Each shoot produced in vitro consists of multiple nodes that potentially can be used as a propagule to produce another shoot. However, some potato cultivars may be less responsive or produce smaller shoots in vitro when single-node cuttings are used. During routine subculture of the stock shoot cultures it may be useful to compare the response of single-node and two-node cuttings. The typical response of two-node cuttings after only 7 days culture is shown in Figure 9.2.

#### Procedure

Remove intact shoots from stock cultures and place in sterile petri dish. Add a small volume of sterile water to prevent tissue desiccation and subdivide into single-node cuttings using a sterile scalpel. Trim leaf blades from each nodal cutting, leaving only the petiole bases. Orient four single-node cuttings horizontally onto NCM medium in each of the four GA-7 vessels. Place cultures in an incubator or on a laboratory bench top under fluorescent lighting providing approximately 30 µmol·m$^{-2}$·sec$^{-1}$ for a 16-h photoperiod at

20 C. The percent responsive nodal cuttings, average number of shoots and total nodes produced per nodal cutting are determined after 3 weeks.

After data collection, the potato shoots produced may be treated as microcuttings and rooted directly in soilless growing medium contained in plug trays or any other small container and placed under intermittent mist (5 sec every 10 min) for 2 weeks under lightly shaded conditions. Alternatively, if a mist system is not available, plastic humidity domes can be placed over the plug trays to maintain humidity. After 10–12 days, the domes are gradually removed over a period of 7 days to acclimatize the plantlets.

### Anticipated results

Shoot growth from the nodes should occur rapidly. Typically, an unbranched shoot approximately 5 cm long and consisting of three to five nodes, plus the shoot tip, will be produced from a nodal cutting after 3 weeks. Determine the average number of new nodes (plus the shoot tip) produced per nodal cutting after 3 weeks. Using this value, estimate the number of shoots generated from the original nodal cutting after 1 year. This is accomplished by calculating $Y^X$, where $Y$ = node number (including shoot tip) produced per nodal cutting after the 3-week culture cycle and $X$ = 17.3, the number of culture cycles per year. This yearly production rate can be quite large, and its calculation provides an opportunity to illustrate the power of this simple micropropagation technique. Potato microcuttings should quickly root and acclimatize to ex vitro (Stage IV) conditions.

### Questions

* Given a 3-week culture cycle and based upon your calculated average total nodes produced per nodal cutting, how many potato shoots can be produced from a single node cutting in a year?
* Why is node culture considered the most reliable method to in vitro propagate plants that are true to type?

## Experiment 2. Microtuberization of potato

### Procedure

Transfer four 4-cm-long microcuttings with shoot tips intact into each of the eight GA-7 vessels containing MIM. Position the microcuttings vertically by embedding the cut end of each about 5 mm into the medium. Place four cultures in an incubator set to provide 30 $\mu mol \cdot m^{-2} \cdot sec^{-1}$ fluorescent lighting for an 8-h photoperiod at 20 C and the remaining cultures in an incubator set to maintain a 16-h photoperiod at the same light level and temperature. Usually, microtuberization is promoted at temperatures ranging from 15–20 C. However, higher culture temperatures can be used if lower temperature control is not available. Depending on culture response, the number of microtubers produced per shoot and mean microtuber fresh weight can be determined after 6–10 weeks.

### Anticipated results

Potato tubers normally form underground but are produced in vitro from axillary meristems along the shoot (Figure 9.3). Usually only one microtuber fully develops on each shoot. During the tuberization process, there is apparently strong competition between axillary meristems along the shoot, such that the first meristem to tuberize inhibits tuberization of the other meristems (McCown and Joyce, 1991). This has been a factor limiting efficient microtuber production and makes it difficult to design exercises examining factors affecting microtuber production per propagule. However, it is important to point out that each microtuber, being a modified stem, contains numerous shoot meristems, the so-called eyes of the potato. Microtuber meristems exhibit varying degrees of

**Figure 9.3**  Typical microtuber formation from cultured shoots 10 weeks after culture initiation. Note the presence of only one primary microtuber on each shoot. Scale bar = 10 mm. Corresponds to part H of Figure 9.1.

dormancy that can be broken by storing at a lower temperature for several weeks. Depending on the cultivar used, microtubers produced under an 8-h photoperiod usually weigh more than those produced under a 16-h photoperiod. One variation to this exercise would be to examine the influence of sucrose concentration (0, 60, 120, 180 and 240 mM) on size and fresh weight of microtubers produced under only an 8-h photoperiod.

*Questions*

* What are several advantages of producing potato in vitro via formation of microtubers?
* Provide a possible reason why so few microtubers develop on each shoot.

## Literature cited

Cassells, A.C. and R.D. Long. 1982. The elimination of potato viruses X, Y, S and M in meristem and explant cultures of potato in the presence of Virazole. *Potato Res.* 25:165–173.

Dodds, J.H, D. Silva-Rodriguez and P. Tovar. 1992. Micropropagation of potato (*Solanum tuberosum* L.). In: *Biotechnology in Agriculture and Forestry*, Vol. 19. *High Tech and Micropropagation III*, Y.P.S. Bajaj, Ed., pp. 91-106. Springer-Verlag, New York.

Leclerc, Y., D.J. Donnelly and J.E.A. Seabrook. 1994. Microtuberization of layered shoots and nodal cuttings of potato: the influence of growth regulators and incubation periods. *Plant Cell Tissue Organ Cult.* 37:113–120.

McCown, B.H. and P.J. Joyce. 1991. Automated propagation of microtubers of potato. In: *Scale-Up and Automation in Plant Propagation*, I.K. Vasil, Ed., pp. 95–109. Academic Press, San Diego.

McCown, B.H. and G.A. Wattimena. 1987. Field performance of micropropagated potato plants. In: *Biotechnology in Agriculture and Forestry.* Vol. 3, Y.P.S. Bajaj, Ed., pp. 80–88. Springer-Verlag, New York.

Miller, S.A. and L. Lipschultz. 1984. Potato. In: *Handbook of Plant Cell Culture.* Vol. 3. *Crop Species*, P.V. Ammirato, D.A. Evans, W.R. Sharp and Y. Yamada, Eds., pp. 291–326. Macmillan, New York.

Murashige, T. and F. Skoog. 1962. A revised medium for rapid growth and bioassays with tobacco cultures. *Physiol. Plant.* 15:473–497.

Seabrook, J.E.A., S. Coleman and D. Levy. 1993. Effect of photoperiod on in vitro tuberization of potato (*Solanum tuberosum* L.). *Plant Cell Tissue Organ Cult.* 34:43–51.

*chapter ten*

# Micropropagation of lilacs

*Deborah D. McCown and Andrew J. Daun*

Collectively, the species of the genus *Syringa* are some of the more easily micropropagated woody plants. The French hybrid lilac, *S. vulgaris* L., is among the easiest to establish in culture and has a relatively high multiplication rate. Other species, generally identified as shrub lilacs (e.g., *S. microphylla* and *S. julianae* Schneid.), are slightly more recalcitrant. The tree lilacs, such as *S. reticulata* and *S. pekinensis* Rupr., can be successfully micropropagated, but generally are more difficult to establish and acclimatize to the culture system (McCown and McCown, 1987). However, once established, they perform much like the shrub lilacs, but with slightly lower multiplication rates.

For the horticultural industry, rapid introduction of new superior cultivars is important. Desirable traits may include disease resistance, improved flower color or more compact growth habit. For commercial purposes, successful isolation (or culture) of vegetative tissue in late winter or early spring can result in significant numbers of propagules the following year. We have routinely produced several thousand rooted microcuttings of a number of clones in a single year; lilacs have excellent axillary shoot proliferation potential in culture.

These laboratory exercises have the following two objectives: (1) isolation and culture of vegetative shoots of a woody plant in the microculture system and (2) exploration of the effect of cytokinin type and concentration on axillary shoot multiplication and development. Cytokinins make up a specific class of plant growth regulator (PGR).

The specific cultivar of *S. vulgaris* used in these exercises is probably a minor concern, as the number in commercial propagation is approaching 100. There is, however, one important caveat — a few lilacs are apparently chimeras, notably 'Sensation' and 'Primrose' (McCulloch, 1989). There are reports that these cultivars may be unstable in culture and yield plants that are not identical to the parental type; thus, avoid known chimeral lilacs. Some suggested cultivars that are readily available through nurseries and are known to be easily micropropagated include 'Charles Joly', 'Ludwig Spaeth', 'Madame Lemoine' and 'President Grevy'.

## Exercises

### Experiment 1. Isolation and culture of terminal and axillary buds

#### Materials

The tissue available for isolation for microculture (tissue culture) is an important factor and, along with a number of other variables, can influence success rates. The ideal tissue is obtained from a small vegetative plant maintained in a "relatively" clean environment such as a greenhouse. If using greenhouse plants, maintain them in soilless potting mixture and fertilize with a liquid preparation of 20-20-20 + micronutrient solution on a regular basis. Supplemental lighting in the greenhouse may be necessary to maintain long days (greater than 16-h daylength). Cutting plants back on a regular basis will encourage production of new shoots. Vegetative shoots from plants growing outside can also be successfully used to initiate cultures, but the number of fungi and bacterial spores from rain/soil splash etc. makes surface disinfestation more difficult. In either case, the tissues should be vegetative and actively growing.

The following items are needed for each student or student team:

* Actively growing vegetative shoots (ten pieces per student or team)
* 15% commercial bleach (0.79% NaOCl) in distilled water (tap water may reduce bleach activity) plus 1 or 2 drops of liquid dishwashing detergent
* Two 200-ml baby food jars of sterile water
* Ten 125-ml baby food jars or Magenta GA-7 vessels containing culture medium

The amount of bleach solution and number of vessels of rinse water and culture medium are dependent on the number of shoots or nodal sections available for culture. For both the bleach and rinse water, the volume should be sufficient to fully immerse the tissue.

The number of vessels for liquids or culture medium will depend both on the amount of tissue to be disinfested and on the number of different cultivars being cultured. If more than one cultivar is used, then each will require a separate set of vessels, each carefully labeled with the cultivar name. It is nearly impossible to distinguish cultivars once the shoots have been prepared for culture or even after shoots have begun to proliferate. Each student or team of students should have a minimum of ten pieces of tissue (terminal buds and/or nodal sections) to work with. Baby food jars with B-caps (Magenta Corp., Chicago, IL) are conveniently sized vessels; large jars are for bleach and rinse water, and smaller jars for culture medium.

The basal culture medium is composed of Woody Plant Medium (WPM) basal salts (Lloyd and McCown, 1980) — see chapter two for composition. This medium and all other chemicals listed can be purchased from Sigma Chemical Company, St. Louis, MO. The basic medium is amended with 0.4 g (5 mM) $NH_4NO_3$, 1.3 g (6 mM) D-gluconic acid (hemicalcium salt), 20 g (59 mM) sucrose, 4 g agar and 1.4 g/L Phytagel™ (5 g/L AgarGel™, a mixture of the above two gelling agents, can be substituted). The gluconic acid is added to help prevent hyperhydric growth that can be a problem with lilacs. Approximately 30 ml of medium is used per small baby food jar.

### Tissue preparation
When disinfesting tissue from greenhouse-grown plants, vegetative shoots are generally cut into small segments approximately 2–3 cm long. This would include the terminal (apical) bud and several segments including a single node each. Using a razor blade (see Figure 10.1), remove the leaf blade, but keep a small section of the petiole that is subtending the bud. Removing the entire petiole and leaving the bud exposed results in much greater injury and consequent bud death in the bleach solution. The trade-off is that the petiole–stem–bud junction may harbor spores that are protected from the disinfesting action of the bleach solution; however, we find this to be the lesser problem of the two.

### Disinfestation
Once the shoots are cut into sections and the leaves trimmed, place them into the bleach solution. This operation and all subsequent steps of the procedure are completed in a laminar flow hood using aseptic techniques. Disinfest the tissue in 15% bleach–soap

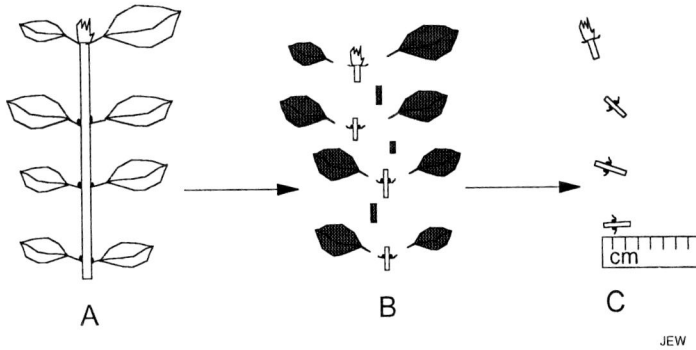

JEW

**Figure 10.1** Diagram illustrating how to cut shoots (A) into single-node segments 2–3 cm long, removing the leaves (B) and keeping a small part of the petiole subtending the bud (C). Shaded sections are discarded.

solution for 15 min with some agitation to break the surface tension and to ensure that all portions of the explant are exposed. The concentration of bleach vs. the time in solution is a dose effect, and both variables could be altered if successful disinfestation is not achieved. Remove the tissue pieces from the bleach with sterile forceps and place into sterile water contained in the first rinse jar. Agitate gently for a few minutes, then transfer shoots to the second rinse vessel. Following the second rinse, tissue sections are ready to be transferred to solidified, sterile culture medium contained in the small baby food jars.

*Initial culture establishment and expected results*

Ideally, each student (or team) will have a minimum of ten pieces of tissue to work with. All tissue segments containing buds should be placed in vessels containing medium without PGRs. A single piece of tissue (a node or terminal bud) per jar should be gently pushed, but not submersed, into the surface of the medium. Cultures should be placed under 24-h artificial light (25 $\mu mol \cdot m^{-2} \cdot sec^{-1}$) at a temperature of 26–27 C. If a culture room is not available, a single 40-W fluorescent light 25 cm above a shelf or bench will provide adequate light. Lilacs like warm temperatures, and growth will be much slower if the temperature is lower. Cultures should be checked daily for contamination. Fungal growth is generally apparent within a few days, but bacterial contamination may take a week to 10 days to be visible. Discard contaminated cultures — it is hoped they will be few!

New growth from terminal bud explants will be apparent in approximately 2 weeks; axillary buds from nodal segments are somewhat slower, but new shoots should form in about 4 weeks. Once the explant tissue has produced new shoots, the experiment to investigate the effects of different cytokinins and concentration on axillary shoot proliferation can begin.

## Experiment 2. Shoot production from axillary buds

Each student or student team will need ten jars of the basal medium described above — two jars of each of the five PGR concentrations listed below:

1. No PGRs
2. 0.22 mg/L (1 $\mu$M) benzyladenine (BA)
3. 1.0 mg/L (5 $\mu$M) BA,
4. 0.22 mg/L (1 $\mu$M) zeatin and
5. 1 mg/L (5 $\mu$M) zeatin

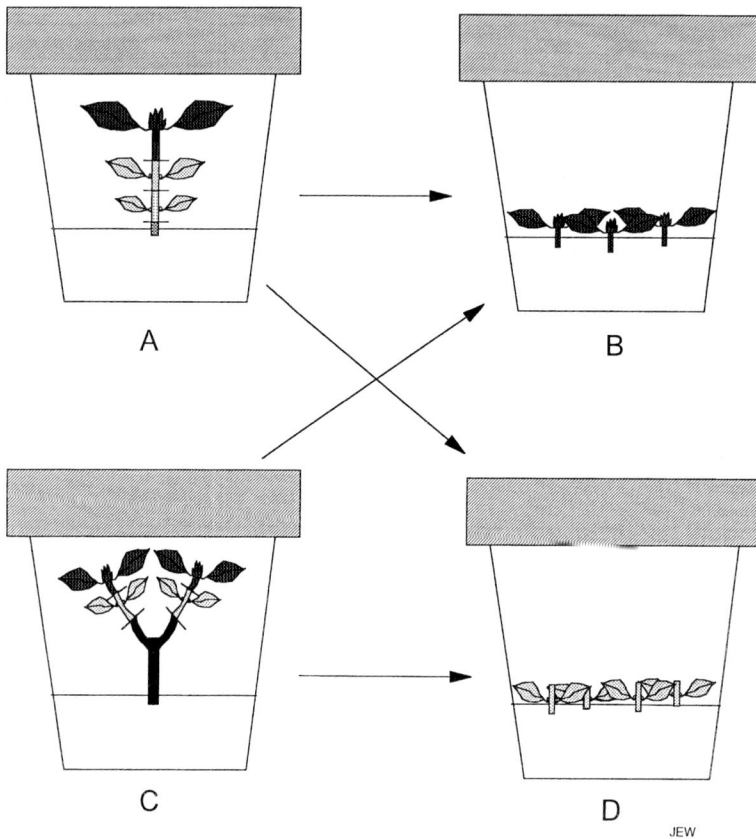

**Figure 10.2**  Diagram illustrating subculturing the new isolates and separating terminal buds and nodal sections. Jars labeled A and C are the new isolates, and jars labeled B and D are fresh culture medium. Terminal buds are placed in one set of jars (B); nodal sections are placed in another set (D).

Be sure to have a system for labeling different PGRs and their concentrations. We use a different colored marker for each PGR; a black zero equals no PGR, a green one or five equals either 1 or 5 μM BA and a purple one or five equals either 1 or 5 μM zeatin (another type of cytokinin).

The objective is to obtain good shoot proliferation from axillary buds without promoting adventitious shoots (see chapter eleven). The PGR concentrations provided are not definitive, but are only meant as beginning guidelines for these experiments. For commercial production, a more sophisticated experimental matrix consisting of other cytokinins and more concentrations is usually designed. The medium without PGRs is included as a control, and shoot proliferation would be expected to be relatively poor.

As new shoots are produced from the initial isolated buds, they should be transferred to the above media. Shoots should be cut as before, so that the terminal bud is in one baby food jar and the nodal sections are in other jars (Figure 10.2). With ten original explants, the shoots in two jars can be subcultured onto each of the five different PGR-containing media. Therefore, you should be able to generate ten new jars — two for each treatment or medium. If some of the initial cultures have been lost to contamination, distribute the remainder among the ten new jars as equally as possible. At this point, only one piece of tissue per jar is needed, but work carefully to avoid contamination!

At the warm incubation temperatures suggested above, lilacs can be subcultured about every 3 weeks. Shoots should be maintained on consistent PGR type and concentration

**Table 10.1** Examples of Axillary Shoot Production of *Syringa vulgaris* 'Little Boy Blue' (LBB), 'Marie Frances' (MF) and 'Arch McKean' (AM) Cultivars at the End of a 3.5-Week Growing Cycle

| Jar | Initial nodal segments | | | Usable cuttings generated | | |
|---|---|---|---|---|---|---|
| | LBB | MF | AM | LBB | MF | AM |
| 1 | 22 | 17 | 16 | 34 | 25 | 27 |
| 2 | 21 | 17 | 13 | 17 | 23 | 10 |
| 3 | 14 | 13 | 16 | 18 | 35 | 12 |
| 4 | 17 | 13 | 9 | 19 | 16 | 11 |
| 5 | 19 | 15 | 10 | 21 | 16 | 20 |
| 6 | 17 | 15 | 21 | 31 | 28 | 43 |
| 7 | 22 | 12 | 16 | 27 | 9 | 22 |
| 8 | 22 | 16 | 12 | 20 | 15 | 19 |
| 9 | 14 | 14 | 15 | 30 | 10 | 22 |
| 10 | 20 | 18 | 21 | 8 | 19 | 29 |
| Ave. | 19 | 15 | 15 | 20 | 20 | 21 |
| Multiplication factor | | | | 1.1 | 1.3 | 1.4 |

(e.g., if cultured on medium containing 1 μM BA, then they should be subcultured to the same medium) and divided so that terminal buds and single-node sections are in separate jars. As shoots grow and proliferate, jars will be filled with more tissue and you will begin to accumulate more than two jars per treatment. Six to ten pieces of tissue per jar will be adequate for data collection. For commercial purposes, individual jars will contain around 20 shoots each. At each subculturing, notes should be taken on the number of axillary shoots produced. Apical dominance will keep the terminal shoots unbranched, but you should see differences in the number of axillary shoots produced from the nodal segments among the different treatments. Axillary shoot production from single-node sections is the primary method we use for multiplication of lilacs in commercial tissue culture.

Construct a table for each cultivar. You will probably want to include the name of the cultivar, PGR type and concentration, date (and perhaps the number of times the shoot has been subcultured) and the number of axillary shoots produced per node segment or terminal bud.

*Anticipated results*

At Knight Hollow Nursery, we specialize in introducing new and often proprietary cultivars to the market. We have found that 1 μM zeatin provided the best results in terms of axillary bud proliferation in the dozen lilac cultivars we have micropropagated. Many environmental conditions can influence axillary shoot proliferation, and you may find that, under your laboratory conditions or for your particular cultivar, another medium gives improved multiplication. That is part of the fun of experiments.

Data on three cultivars that have been maintained on 1 μM zeatin are presented in Tables 10.1 and 10.2. They should be used only for comparison with the results of your experiment. Table 10.1 shows the axillary shoot proliferation at 3.5 weeks or at the point we would begin to subculture again. If cultures are allowed to grow an additional 2 weeks, the number of usable (or rootable) microcuttings increases significantly (Table 10.2). Thus, the last subculture round prior to rooting is generally slightly longer than the usual subculture period.

The experiment discussed above can be completed in a normal semester of 16 weeks, but it will be critical to start immediately. If 4 weeks are allowed following initial isolation of tissues for new shoot development and then shoots are subcultured at 3 week intervals, there will be a maximum of five subculture cycles. This experiment works using French

*Table 10.2*   Examples of Axillary Shoot Production of *Syringa vulgaris*
'Little Boy Blue' (LBB), 'Marie Frances' (MF) and 'Arch McKean'
(AM) Cultivars at the End of a 5.5-Week Growing Cycle

| Jar | Initial nodal segments | | | Usable cuttings generated | | |
|-----|------|------|------|------|------|------|
|     | LBB  | MF   | AM   | LBB  | MF   | AM   |
| 1   | 22   | 17   | 12   | 59   | 35   | 39   |
| 2   | 17   | 15   | 19   | 45   | 47   | 40   |
| 3   | 20   | 19   | 15   | 63   | 47   | 52   |
| 4   | 22   | 13   | 16   | 42   | 48   | 35   |
| 5   | 21   | 19   | 14   | 46   | 62   | 36   |
| 6   | 14   | 20   | 18   | 53   | 44   | 58   |
| 7   | 21   | 14   | 14   | 56   | 41   | 39   |
| 8   | 22   | 15   | 12   | 42   | 30   | 32   |
| 9   | 20   | 20   | 12   | 61   | 54   | 30   |
| 10  | 18   | 17   | 18   | 43   | 28   | 32   |
| Ave. | 20  | 17   | 15   | 51   | 44   | 36   |
| Multiplication factor | | | | 2.6 | 2.6 | 2.4 |

hybrid lilacs because they are so easily established in culture, which is unusual for woody plants. We have found that *Celastris* (both American bittersweet and Chinese bittersweet) behaves in a similar manner and can be substituted for lilacs if they are unavailable.

All lilac shoots produced in culture can be successfully rooted ex vitro with no additional treatment. We do not know of any commercial tissue culture laboratory that currently roots lilacs in vitro; culture space is too valuable and better and more functional root systems are produced ex vitro (McCleland and Smith, 1988). Lilac microcuttings generally root at nearly 100% so long as the following conditions are met. First, the cuttings should be actively growing and not hyperhydric. Second, high humidity must be maintained until cuttings are sufficiently rooted, then humidity can be decreased gradually. Third, while the humidity is decreased, the light level should be increased. The rooting and acclimization process generally takes approximately 8 weeks.

*Questions*

* * If the 5 µM cytokinin concentration yields the highest axillary shoot proliferation, what follow-up experiment would you design? What problems might be encountered?
* * How would you distinguish axillary shoots from adventitious shoots?
* * In a commercial operation, terminal buds and nodal sections are often kept in separate vessels. Can you explain the reason for this practice?
* * If nodal sections were left as two or three node pieces, would you predict the same or different axillary shoot proliferation? What effect would cytokinin concentration have on this response? How?
* * Would orientation of the nodal pieces in the medium (vertical, lateral, upside down) change the response? How?

# Literature cited

Lloyd, G. and B. McCown. 1980. Commercially feasible micropropagation of mountain laurel, *Kalmia latifolia*, by use of shoot-tip culture. *Proc. Intl. Plant Prop. Soc.* 30:421–427.

McCleland, M.T. and M.A.L. Smith. 1988. Response of woody plant microcuttings to in vitro and ex vitro rooting methods. *Proc. Intl. Plant Prop. Soc.* 38:593–599.

McCown, D.D. and B.H. McCown. 1987. North American hardwoods. In: *Cell and Tissue Culture in Forestry*, Vol. III. J.M. Bonga and D. J. Durzan, Eds., pp. 247–260. Martinus Nijhoff, Boston.

McCulloch, S. 1989. Tissue culture of French hybrid lilacs. *Proc. Intl. Plant Prop. Soc.* 39:105–114.

*chapter eleven*

# Propagation from nonmeristematic tissues — organogenesis

*Otto J. Schwarz and Robert M. Beaty*

The ability of plant tissues to form various organs de novo (i.e., organogenesis) has long been an object of interest and practical utility. The ancient Chinese successfully cloned selected genotypes of forest trees through the organogenic process of adventitious rooting of woody cuttings. In a recent historical review of adventitious rooting research, the practice of "cuttage" or rooting of cuttings was said to predate the time of Aristotle (384–322 B.C.) and Theophrastus (371–287 B.C.) (Haissig and Davis, 1994). The process of organogenesis provides the basis for asexual plant propagation largely from nonmeristematic somatic tissues. One needs only to visit a local garden center in the early spring or fall to witness the many varieties of woody and herbaceous plants offered for sale to the gardening public. Many of these plants are propagated exclusively via asexual means. The cloning of these selected genotypes probably included the de novo organ formation of roots on cuttage, the in vitro multiplication of preexisting shoot meristems followed by de novo root formation on the resultant microshoots, or possibly de novo shoot regeneration on in vitro cultured explant tissues followed by adventive rooting of the microshoots. Whether accomplished in vivo or in vitro using either meristematic or nonmeristematic tissues, the de novo genesis of plant organs is broadly defined here as organogenesis. As stated above, our focus is specifically on "organogenesis in vitro" on nonmeristematic plant tissues. To further understand the problems to be encountered in obtaining an all-encompassing definition of the terms associated with plant organ regeneration in vitro or in vivo, a reading of the reviews by Dore (1965) and Haissig and Davis (1994) is recommended.

The material presented in this chapter represents only a small portion of the information available in the scientific literature concerning the developmental process of organogenesis. It is hoped that the concepts selected for discussion provide a model framework upon which to build an understanding of the morphogenic events that accompany organogenesis and, in addition, to begin to gain insight into the complex process that directs the morphogenetic process.

## Organogenesis as a developmental process

In its broadest sense, development is the process that results in a functional, mature organism. According to Fosket (1994), it includes "all of those events during the life of a plant or animal that produce the body of the organism and give it the capacity to obtain food, to reproduce and to exploit the opportunities and deal with the hazards of its environment." Organogenesis is a developmental process that is in some ways unique to plants. Animal cells follow developmental paths that normally involve irreversible differentiation toward a specific cell or tissue type. In other words, animal cells and tissues remain structurally and functionally committed to an initial developmental end point. Plant cells, however, may retain the ability to dedifferentiate from their current structural and functional state and to begin a new developmental path toward a number of other morphogenetic endpoints.

Plant tissues in vitro may produce many types of primordia, including those that will eventually differentiate into embryos, flowers, leaves, shoots and roots. For the purpose of this chapter, embryogenesis will not be discussed — see chapter sixteen. Although the botanical definition of an "organ" is quite restrictive, in this discussion we regard the development of any of the aforementioned primordia to be encompassed by the defintion of "organogenesis". These primordia originate de novo from a cellular dedifferentiation process followed by initiation of a series of events that results in their formation. The cell or cells thought to be the direct progenitors are somehow stimulated to undergo a number of rapid cell divisions leading to the formation of a meristemoid. Meristemoids are characterized as being an aggregation of meristemlike cells (i.e., smaller, isodiametric, thin-walled, microvacuolated cells with densely staining nuclei and cytoplasm) (Thorpe, 1978; 1982). Early in their development, meristemoids are thought to be morphogenetically plastic and capable of developing into a number of different primordia cells (i.e., root, shoot, etc.). This unique developmental flexibility has been widely used by plant propagators. Two organogenic events constitute a common approach to in vitro plant propagation. The first event is to regenerate multiple shoot meristems followed by their growth and development into microshoots of a suitable size for the second event, which is the induction of de novo root meristem production. It is believed that under in vitro culture conditions, these organogenic events can be the result of two differing ontogenic pathways.

According to Hicks (1980), two developmental sequences leading to organogenesis emerge from the literature. They differ in the presence or absence of a callus stage in the organogenic sequence of events. A developmental sequence involving an intervening callus stage is termed *indirect* organogenesis.

$$\begin{array}{ccccc} \text{PRIMARY} & & & & \text{ORGAN} \\ & \longrightarrow & \text{CALLUS} \longrightarrow \text{MERISTEMOID} \longrightarrow & \\ \text{EXPLANT} & & & & \text{PRIMORDIUM} \end{array}$$

De novo organ formation via indirect organogenesis may increase the possibility of introducing variation in the chromosomal constitution (e.g., ploidy change) of the cells in the callus stage and, hence, the possibility for both physiological and morphogenic variation in the resulting organs. This type of variation has been termed *somaclonal variation*. Although some plants are more prone to this problem than others, a general rule of thumb for plant propagation is to minimize any stage in the organogenic sequence containing "unorganized" callus growth. For the research scientist concerned with determining the physiochemical mechanism(s) driving the organogenic process, there is a second important reason to avoid or at least minimize this type of developmental event. The presence of a proliferating callus stage has the potential to greatly complicate the analysis of ongoing molecular events that accompany and perhaps drive de novo organ production. Those particular cells or groups of cells that are destined to become organogenic are in relatively small number among the much larger mass of dividing callus cells, reducing the ability to analyze their particular chemistry and physiology.

Direct organogenesis is accomplished without an intervening proliferative callus stage. For de novo organogenesis (i.e., shoots, roots, flowers) the sequence of events is described as follows:

$$\begin{array}{ccccc} \text{PRIMARY} & & & & \text{ORGAN} \\ \text{EXPLANT} & \longrightarrow & \text{MERISTEMOID} & \longrightarrow & \text{PRIMORDIUM} \end{array}$$

The major distinction between these two de novo pathways is, of course, the presence or absence of a discernible callus stage early in the morphogenic sequence of events. The callus tissue, containing cells that have dedifferentiated into a less-determined, morphologically more flexible form, serves as the starting point for de novo organogenesis. In the absence of a callus intermediate stage, cells present within the explant have the capacity to act as the direct precursors of the new primordium. Hicks (1980) presents a substantial listing of in vitro culture systems that provide examples of direct and indirect organogenesis from both meristematic and nonmeristematic explant tissues.

Researchers have utilized the organogenic process as a model system for asking questions about the causal events that direct shoot or root production and, by generalization, other morphogenic processes. In one such model, the process of organogenesis has been broken into several phases, illustrated in the following diagram (adapted from Christianson and Warnick, 1985).

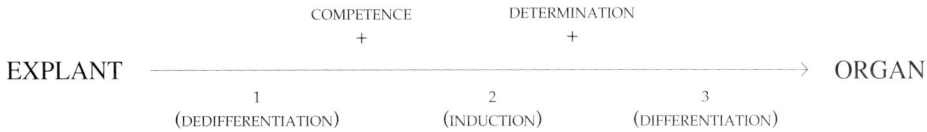

$$\begin{array}{ccccc} & \text{COMPETENCE} & & \text{DETERMINATION} & \\ & + & & + & \\ \text{EXPLANT} & \text{---------------------------------} & & \longrightarrow & \text{ORGAN} \\ & 1 & 2 & 3 & \\ & \text{(DEDIFFERENTIATION)} & \text{(INDUCTION)} & \text{(DIFFERENTIATION)} & \end{array}$$

Key points of interest to developmental biologists are the initial two phases that precede the differentiation phase. These phases encompass events that begin with dedifferentiation, which results in the attainment of "competence", followed by induction, which culminates in the fully "determined" state. The morphological differentiation and development of the nascent shoot or root then proceed, eventually resulting in a functional organ. The first two phases of the model encompass those events that occur prior to morphogenesis. As shown by Hicks (1980), embryos, seedling parts, apical meristems, primordial organs, cell layers from mature organs, complex fragments from mature organs, intact organs and protoplasts have been used for primary explants.

## Dedifferentiation

The process of dedifferentiation involves reversion to a less-committed, more flexible/plastic developmental state that may or may not give rise to callus tissue. In the case of direct organogenesis, competent cells that have not produced callus tissue are thought to be the sole organ progenitors, implying that they undergo the dedifferentiation process and proceed individually or perhaps in small groups to produce a new primordia. Leaf explants of morning glory or bindweed (*Convolvulus arvensis* L.) produce roots or shoots via an indirect organogenic pathway that involves a dedifferentiation process that produces limited callus growth from which the new organs are generated (Christianson and Warnick, 1983). The result of completion of this first phase is that the primary explant acquires a state of competence, which is defined by its ability to respond to organogenic stimuli.

Attainment of tissue competence is not always a single-step process involving, for example, the application of plant growth regulators (PGRs). When cultured for 4 days on relatively high kinetin and low 2,4-dichlorophenoxyacetic acid (2,4-D) concentrations and then removed to PGR-free medium, alfalfa (*Medicago sativa* L.) callus produces roots. Shoots are produced when callus is treated to the same culture regime except that the medium contains high 2,4-D and low kinetin concentrations. However, the PGR pretreatment followed by transfer to basal, PGR-free medium is not sufficient to bring all of the tissue through the second (induction) phase that is required for eventual root or shoot production (Walker et al., 1979). An additional size requirement imposed a lower limit of 105 μm as the minimal diametric size for callus cellular aggregates to be competent for morphogenetic induction. Cellular aggregates allowed to grow to the minimum size became competent for organogenic induction and root or shoot formation.

## Induction

The induction phase occurs between the time the tissue becomes competent and the time it becomes fully determined for primordia production. This phase has been shown to encompass a number of "phenocritical" times postulated to result from the function of an integrated gene pathway that guides the developmental process and precedes morphological differentiation. Landauer (1958) suggested that certain chemical and physical stimuli could interrupt a genetically determined developmental pathway, modifying the morphogenic result, which produces a mutantlike phenotype or "phenocopy". Christianson and Warnick (1984) have identified several chemical agents that intervene during the induction phase of *C. arvensis*. Agents that induce the production of a phenocopy are called phenocopying agents. These phenocopying agents were shown to be only effective during certain experimentally determined times during the induction phase. These phenocritical times were postulated to define key steps in the genetic pathway orchestrating the developmental process. The phenocopying agents, or stage-specific inhibitors, were thought to block in some direct way the action of a gene or gene product of a currently active "orchestrating gene". In the *C. arvensis* system, this meant blocking shoot organogenesis and producing the modified phenotype (i.e., phenocopy), in this case, callus.

The end of the induction process is defined as the point when a cell or group of cells becomes fully committed to the production of shoots or roots. Operationally, this end point is attained when explant tissue (1) can be removed from the root or shoot- inducing medium; (2) can be placed on basal medium without PGRs, containing mineral salts, vitamins, and a source of carbon; and (3) can proceed to produce the desired organs. The tissue has now completed the induction process and can be considered to be fully determined.

There is another term of interest that is used to describe or perhaps measure the extent of developmental commitment toward a particular end point (i.e., bud formation) that a tissue has reached during the induction phase. This term is *canalization*, and it is defined by Waddington (1940) as "the property of developmental pathways of achieving a standard phenotype in spite of genetic or environmental disturbances." *Convolvulus arvensis* explant tissue that has become fully canalized for shoot production can be placed on media designed to induce roots and still form shoots (Christianson and Warnick, 1983). If the explant is removed from the shoot-inducing medium before it becomes fully canalized, shoot production is severely inhibited, with root production becoming the prominent end point. This developmental flexibility illustrates the degree of morphogenic plasticity that can be exhibited by plant tissue under in vitro culture conditions.

The extent that explant tissues become competent and eventually determined is, in part, a function of the physical and chemical environment to which they have been exposed. Under in vitro culture conditions, the properties of the medium and associated culture environment play a central role in delivering these organogenic signals. Skoog and Miller (1957) first demonstrated that a key system for the chemical regulation of in vitro organogenesis was the ratio of auxin to cytokinin present in the culture medium. They summarized their findings by concluding that "these tend to show that quantitative interactions between growth factors, especially between IAA and kinetin (auxins and kinins) and between these and other factors, provide a common mechanism for the regulation of all types of growth investigated from cell enlargement to organ formation". A great many other variables, such as day length, light quality and quantity, explant tissue age and size, genotype and mineral nutrition, to mention a few, have been shown to affect regeneration ability. It is certainly possible that, as Christianson and Warnick (1988) suggest, "the possibility that at least some cases of nonresponsiveness to culture in vitro are simply due to the failure of one 'happy medium' to elicit competence for induction and to induce organogenesis". In its broadest sense, this idea suggests that a lack of tissue organogenic responsiveness under any particular set of experimental conditions is the result of failure of the explant tissues to achieve the state of competence for induction. This is not a particularly comforting situation for an investigator's state of mind, if one calculates all of the possible combinations and permutations presented by testing even a few of these parameters. Until appropriate biochemical/genetic markers are discovered that clearly indicate the current developmental disposition of the primary explant tissue and until methods are discovered that can override the current state of genetic programming of recalcitrant tissue, our approach will have to remain that of "the educated guess".

## Differentiation

Of the three phases proposed by the model, the third phase or differentiation phase is the best documented. It is in this phase of the process that morphological differentiation and development of the nascent organ begins. In describing the initial differentiation events of organ initiation, McDaniel (1984) states that "the general picture that emerges is that organ initiation involves an abrupt shift in polarity followed by a smoothing of this shift into a radially symmetrical organization and the concurrent growth along the new axis to form the bulge characteristic of organogenesis". These initial events and those leading to bud formation are shown in pictorial series in Figures 11.1 to 11.3. There has been a great deal of discussion in the literature concerning these initial differentiation events with respect to the tissue types involved, and perhaps more pointedly, to the precise number of cells that take part in meristem initiation. Again, the subject remains somewhat obscure in that it seems that there is not an absolute answer. Christianson (1987) states, based on his own experiments and those of Marcotrigiano and Gouin (1984a; b) using plant chimeras, that "shoots formed in vitro can arise from more than one cell in an explant, but they usually do not." Based upon extensive cytological investigation of the cellular events surrounding organogenic shoot initiation on immature zygotic embryos of sunflower (*Helianthus annuus* L.), Bronner et al. (1994) concluded that shoots developed from the "simultaneous divisions of several cells" and, therefore, were multicellular in origin. The epidermis and outer layers of the cortex of the immature embryos were found to contribute these germinal cells. The tissue origin and cell number involved in these very early differentiative events seem to depend upon a number of, as yet, poorly understood variables. Recognizable centers of cell division can occur deep within the explant tissues or, as just described for sunflower, be more superficial in origin.

One can follow, through the careful use of microscopy, the internal histological and surface architectural changes that occur during the differentiation phase of the organogenic process (Figures 11.1 to 11.3). The morphological sequence of events that is depicted tracks the de novo production of shoot buds by the process of direct organogenesis on cotyledons of the Central American pine (*Pinus oocarpa* Schiede). This series is presented as a more or less typical example of the organogenic process. Keep in mind that this is a singular example and that all matter of variations of the developmental theme have been described for in vitro cultured plant species. However, the sequence of events that leads to organogenic

**Figure 11.1**  Histological micrographs of the differentiation phase of adventitious bud formation on seedling cotyledonary explants of a Central American pine (*Pinus oocarpa*). a. Transverse section of the cotyledonary explant before exposure to a shoot bud induction medium. The epidermis has a regular, well-ordered appearance on all three sides. The more central mesophyll tissue is compact and has little intercellular space (bar = 46 µm). b. After 3 days on hormone-containing induction medium, the cells of the epistomatic surfaces have become irregular in size and shape, resulting in an uneven cotyledonary surface (bar = 46 µm). c. The presence of meristemoids is easily observed after 12–18 days and is denoted by the darkly stained nuclei in the group of small cells clustered at the top center of the micrograph (bar = 28 µm). d. The meristematic region rapidly increases in size. In this case, initial growth extends the meristemoid downward into the mesophyll tissue (bar = 28 µm). e. Meristematic growth may also occur laterally along the plane of the explant surface (bar = 28 µm). f. The rapidly developing bud meristem begins to protrude above the surrounding epidermis (bar = 46 µm). g. The dividing cells of the meristematic regions sometimes do not involve the epidermis. These regions may or may not develop into functional bud meristems. h. A meristematic dome with several leaf primordia protrudes high above the surrounding epidermis. Note the numerous anticlinal divisions in the epidermis just to the left of the center of the dome and the periclinal divisions just to the right of the center of the dome (bar = 25 µm). i. Fully developed adventitious buds, complete with enclosing outer primordial leaves. Note the procambial strands in the area of the leaf primordia (arrows). The cytohistological zones associated with pine apical meristems are represented as follows: I, apical zone initials; II, central mother cell zone; III, peripheral tissue zone; and IV, rib meristem (bar = 28 µm). (From Bajaj, Y.P.S., Ed., *Biotechnology in Agriculture and Forestry and Trees*, III, Springer-Verlag, p. 312. With permission.)

**Figure 11.2**    SEMs of the differentiation phase of adventitious bud formation on seedling cotyledonary explants of a Central American pine (*Pinus oocarpa*), showing developmental events prior to meristematic dome protrusion. a. SEM of the explant prior to exposure to shoot bud induction medium. Stomata occur in rows of two and three surrounded by small subsidiary cells. Rows of elongated rectangular epidermal cells are interspersed between the rows of stomata (bar = 83 μm). b. The explant after 3-day exposure to the induction medium. The abaxial side with almost no stomata and one of the epistomatic sides with its numerous stomata are just beginning to mirror the changes that have already begun in the subsurface tissue by a slight increase in roughness of the epidermal surface (bar = 114 μm). c. The cells of the epidermis have begun to divide causing an obvious disruption of the regular rows of cells between rows of stomata (bar = 96 μm). d. The explant surface has now become nodular and uneven. Disruption of the regular rows of cells has increased as the meristematic areas increase in size, producing nodules (bar = 192 μm).

shoot bud formation in the cotyledons of this conifer is probably generally applicable to most plant species producing adventitious buds (Thorpe, 1980; Yeung et al., 1981).

The final differentiation phase of organogenesis provides the first opportunity to observe the structural genesis of the new organ made possible by the developmental program that was put in place during the preceding "induction" phase. The sequence of morphogenic events observed in de novo bud formation on cotyledons of *P. oocarpa* included changes in surface morphology, the appearance of meristemoids, vertical and/or horizontal expansion of the meristematic region, protrusion of the meristematic region above the surrounding epidermis, the appearance of an organized meristem with leaf primordia, and full development of an adventitious bud. The explant was placed on induction medium containing the PGR benzyladenine (BA) for 10–14 days. It was then transferred to basal medium without PGR for the remainder of the experiment. In this particular experiment, the organogenic process was observed for approximately 40 days.

Histological sections of day 0, newly cultured cotyledons illustrate the anatomy of untreated control tissue. In transverse section (Figure 11.1a), the cotyledons are triangular in outline and the cells of the epidermis are more or less regular in conformation on all three sides of the cotyledon. The round mesophyll cells surrounding the central vascular

**Figure 11.3**   SEMs of the differentiation phase of adventitious bud formation on seedling cotyledonary explants of a Central American pine (*Pinus oocarpa*), showing meristematic dome protrusion to a well-developed adventitious bud. a. Meristematic domes protrude well above the surrounding epidermis of the explant 2–3 weeks after beginning exposure to the shoot bud induction medium. Note the continuity of the epidermal surface over the entire dome (bar = 83 μm). b. A slightly more differentiated dome with newly formed leaf primordia. "L" indicates leaf primordia. "A" indicates the shoot apex (bar = 71 μm). c. Shoot bud with numerous leaf primordia and elongating leaves about 5 weeks after culture initiation. "L" indicates leaf primordia. "A" indicates the shoot apex (bar = 156 μm). d. A fully developed shoot bud (B), enclosed by outer primordial leaves. Elongation along the central axis, (B) of the bud has begun. Arrow indicates another bud earlier in development (bar = 714 μm).

tissue are compact and exhibit very little intercellular space. Scanning electron microscopy (SEM) micrographs of day 0 control cotyledonary tissue (Figure 11.2a) echo the regularity of the surface and near surface tissues shown in the histological sections. Even ranks of long, smooth-surfaced rectangular cells lie between irregular double and triple rows of stomata. The subsidiary cells surrounding the stomata are noticeably smaller in surface area than the long ranks of cells between the double and triple rows of epidermal cells. After 3 days on induction medium, the epidermal cells of the epistomatic cotyledonary surfaces have become irregular (Figure 11.1b). Small groups of meristemoids, mainly clustered at or just under the epidermis, are evident by 15–18 days (Figure 11.1a, c, d, e, and g). These meristematic zones enlarge both perpendicularly and parallel to the cotyledon surface, taking on a nodular appearance. During this time, as a result of mitotic activity, the epidermal cells lose much of their regularity and become uneven in length, forming small cell clusters (Figure 11.2c–d). These localized areas of meristematic activity begin to form meristematic domes (Figure 11.1f and h) that eventually protrude from the surface of the explant (Figure 11.3a). In this organogenic scheme, the ontogeny of the development of the dome structures involves incorporation of cells of the epidermis. In some conifers, the meristematic domes have been observed to push through the surface epidermal layer, causing the epidermis to rupture. The dome soon achieves a higher degree of organization and begins to produce leaf primordia (Figure 11.3b). Both periclinal and anticlinal divisions occur to the right and left of the central "apical" portion of the newly developed meristem, allowing for the successive production of leaf primordia (Figure 11.1h). The apex continues to grow, giving rise to numerous leaf primordia and culminating in the formation of an adventitious bud (Figure 11.1i and Figure 11.3c).

The cytohistological zones typically associated with conifer seedling apical domes (Sacher, 1954) and apical meristem domes in developing conifer embryos (Fosket and Miksche, 1966) are clearly mirrored in this fully developed shoot bud (see legend, Figure 11.1i) produced in vitro by de novo organogenesis. The simultaneous development of numerous meristemoids on a single explant (Figure 11.3d) is not uncommon in in vitro organogenic culture systems. The degree of differentiation of these nascent bud meristems is highly variable. Buds in early (arrow) to fully mature stages (B) can be present on a single explant (Figure 11.3d).

The series of morphogenic events chronicled for *P. oocarpa* seems to fit the model presented for organogenesis via a "direct" developmental pathway. In practice, well-developed buds are removed from the cotyledons and are placed on a medium formulated to allow their further growth and development. Elongated shoots, larger than 5 mm, are then transferred to a rooting medium, or in some propagation systems, the base of the microshoots are treated with a PGR rooting preparation and placed directly into a commercially available rooting medium under carefully controlled ex vitro greenhouse conditions. In either case, the purpose of these manipulations is to induce roots to form via organogenesis on the distal end of the shoot in order to complete plantlet regeneration.

## Summary

The vast majority of published information concerning in vitro plant organogenesis has been based upon the intuitive application of technical procedures established through the venerable technique called "trial and error". Although by necessity of a descriptive nature, many careful investigations have had as their underlying purpose the elucidation of the "basic causes of cellular differentiation and organized development" (Earle and Torrey, 1965). For the developmental biologist, the investigative goal is therefore no less than the establishment of a "mechanism" for this "inducible" organogenic process. A great many of the almost infinite number of combinations and permutations of experimental variables have been tried, and results have been largely positive. That is, a great many plant species have responded with the de novo production of shoot and/or root meristems. There remains a good number of instances in which experimental trial and error have been to no avail. The key needed to unlock the developmental potential of these recalcitrant explant tissues must await the uncovering of the link between genetics and the production of three dimensional form, i.e., the phenomenon of morphogenesis.

## Literature cited

Bronner, R., G. Jeannin and G. Hahne. 1994. Early cellular events during organogenesis and somatic embryogenesis induced on immature zygotic embryos of sunflower (*Helianthus annuus*). *Can. J. Bot.* 72:239–248.

Christianson, M.L. 1986. Causal events in morphogenesis. In: *Plant Tissue and Cell Culture*. pp. 45–55. Alan R. Liss, Inc., New York.

Christianson, M.L. and D.A. Warnick. 1983. Competence and determination in the process of in vitro shoot organogenesis. *Dev. Biol.* 95:288–293.

Christianson, M.L. and D.A. Warnick. 1984. Phenocritical times in the process of in vitro shoot organogenesis. *Dev. Biol.* 101:382–390.

Christianson, M.L. and D.A. Warnick. 1985. Temporal requirement for phytohormone balance in the control of organogenesis in vitro. *Dev. Biol.* 112:494.

Christianson, M.L. and D.A. Warnick. 1988. Organogenesis in vitro as a developmental process. *Hort. Sci.* 23:515.

Dore, J. 1965. Physiology of regeneration in carmophytes. In: *Encyclopedia of Plant Physiology*, Vol. XV/2, W. Ruhland, Ed., pp. 1–91. Springer-Verlag, Berlin.

Earle, E.D. and J.G. Torrey. 1965. Morphogenesis in cell colonies grown from *Convolvulus* cell suspensions plated on synthetic media. *Am. J. Bot.* 52:891–899.

Fosket, D.E. 1994. *Plant Growth and Development — A Molecular Approach.* 580 pp. Academic Press, New York.

Fosket, D.E. and J.P. Miksche. 1966. A histochemical study of the seedling shoot meristem of *Pinus lambertiana. Am. J. Bot.* 53:694–702.

Haissig, B.E. and T.D. Davis. 1994. A historical evaluation of adventitious rooting research to 1993. In: *Biology of Adventitious Root Formation*, T.D. Davis and B.E. Haissig, Eds., pp. 275–332. Plenum Press, New York.

Hicks, G.S. 1980. Patterns of organ development in tissue culture and the problem of organ determination. In: *The Botanical Review*. Vol. 46, A. Cronquist, Ed., pp. 1–23. New York Botanical Garden, New York.

Landauer, W. 1958. On phenocopies, their developmental physiology and genetic meaning. *Am. Nat.* 92:201–213.

Marcotrigiano, M. and F.R. Gouin. 1984a. Experimentally synthesized plant chimeras. 1. In vitro recovery of *Nicotiana tabacum* L. chimeras from mixed callus cultures. *Ann. Bot.* 54:503–511.

Marcotrigiano, M. and F.R. Gouin. 1984b. Experimentally synthesized plant chimeras. 2. A comparison of in vitro and in vivo techniques for the production of interspecific *Nicotiana* chimeras. *Ann. Bot.* 54:513–521.

McDaniel, C.N. 1984. Competence, determination, and induction in plant development. In: *Pattern Formation, a Primer in Developmental Biology*. G.M. Malacinski and S.V. Bryant, Eds., pp. 393–411. Macmillan Publishing Co., New York.

Sacher, J.A. 1954. Structure and seasonal activity of the shoot apicies of *Pinus lambertiana* and *Pinus ponderosa. Am. J. Bot.* 41:749–759.

Skoog, F. and C.O. Miller. 1957. Chemical regulation of growth and organ formation in plant tissues cultured in vitro. *Symp. Soc. Exp. Biol.* 11:118–140.

Thorpe, T.A. 1978. Physiological and biochemical aspects of organogenesis in vitro. In: *International Congress of Plant Tissue and Cell Culture*. 4th ed., pp. 49–58. International Association for Plant Tissue Culture. University of Calgary, Calgary.

Thorpe, T.A. 1980. Organogenesis in vitro: Structural, physiological, and biochemical aspects. In: *Intl. Rev. Cytol. Suppl.* 11a, I. K. Vasil, Ed., pp. 71–111. Academic Press, New York.

Thorpe, T.A. 1982. Physiological and biochemical aspects of organogenesis in vitro. In: *Proc. 5th Intl. Cong. Plant Tissue and Cell Culture*, pp. 121–124.

Waddington, C.H. 1940. The genetic control of wing development of *Drosophila. J. Genet.* 41:75–139.

Walker, K.A., M.L. Wendeln and E.G. Jaworski. 1979. Organogenesis in callus tissue of *Medicago sativa*, the temporal separation of induction processes from differentiation processes. *Plant Sci. Lett.* 16:23.

Yeung, E.C., J. Aitken, S. Biondi and T.A. Thorpe. 1981. Shoot histogenesis in cotyledon explants of *radiata* pine. *Bot. Gaz.* 142:494–501.

*chapter twelve*

# Direct shoot organogenesis from leaf explants of chrysanthemum

*Robert N. Trigiano and Roger A. May*

Chrysanthemums or "mums" (*Dendranthema grandiflora* Tzvelev. synonymous with *Chrysanthemum morifolium* Ramat.) are among the most popular floricultural species and are grown as garden and pot crops or as cut flowers. Mums originated in the Far East and are now a major floricultural crop in many countries throughout the world. The majority of the hundreds of different cultivars are propagated commercially by stem cuttings, but many have been successfully micropropagated also by adventitious shoot formation (organogenesis) from a variety of tissue and callus cultures (e.g., Lu et al., 1990). Some of these regeneration protocols can be used in eliminating viruses from cultivars (meristem tip cultures), in rapidly increasing the number of plants from unique sports, in recovering mutations induced by either chemical or irradiation processes and genetically transforming selected genotypes for flower color and disease resistance. Although infrequently encountered in exercises for plant tissue culture classes, chrysanthemums can be used to demonstrate many methodologies and concepts in regeneration protocols that are based on direct shoot organogenesis.

The following laboratory exercises illustrate the technique of direct shoot organogenesis using leaf sections of chrysanthemum. Specifically, the laboratory experiences will familiarize students with the following concepts: (1) differences in the ability of genotypes to regenerate shoots and roots; (2) effects of plant growth regulator (PGR) combinations on shoot and root formation; and (3) pulse treatments for initiation and growth of shoots. These exercises are adapted primarily from a publication by Trigiano and May (1994).

## General considerations

### Growth of plants

All of the cultivars of chrysanthemum mentioned in the experiments may be obtained from Yoder Brothers, Barberton, OH as unrooted or rooted cuttings. We suggest growing the following cultivars for the exercises: 'Adorn', 'Goldmine', 'Hekla', 'Iridon' and 'Rave'. However, if laboratory or greenhouse space is limited, grow only the cultivars 'Goldmine' and 'Iridon'. Plant three to five cuttings in each of at least ten 10-cm-diam. plastic pots containing any soilless medium 3 months prior to the beginning of the exercises. This should provide adequate materials for classes of about 15–20 students. A good estimate of the number of plants needed per class is to provide a minimum of one plant per student or student team per exercise. The conditions for growing the plants used for explants is very important — if possible, cultivate the plants in the laboratory or growth room illuminated with about 100 $\mu$mol·m$^{-2}$·sec$^{-1}$ fluorescent and incandescent light for 16 h per day at 22–25 C. Alternately, plants grown in the greenhouse under shade and with cooling will provide suitable materials. Supplemental lighting in the greenhouse will be necessary to maintain long days and inhibit flowering if the plants are grown in the spring and fall. Pinch longer shoots often to encourage branching and initiation of new leaves. Fertilize regularly (twice weekly with 300 ppm N) to maintain vigorous growth (see May and Trigiano [1991] for details of growing conditions).

## Explant preparation and basal medium

For satisfactory and reproducible results, select only young, partially expanded, light green, 2–5-cm-long leaves for explant tissue. Surface disinfest whole leaves (do not remove the petiole) using 5% commercial bleach (0.26% NaOCl) for 5–10 min (or 10% bleach for material grown in the greenhouse) with constant agitation followed by three rinses with sterile distilled water. Place a single leaf with the top surface up in a sterile 100 × 15 mm plastic petri dish, and drain excess water by tilting and gently tapping the dish on the flow hood table. The top surface of the leaf is easily identified by its darker color and abundant trichomes. Excise four or five $^1/_2$–1-cm$^2$ sections from the midrib area of the leaf (Figure 12.1) using a number 10 scalpel blade. Place the sections with the abaxial surface (under-side) of the leaf in contact with the agar medium.

The basal culture medium is composed of Murashige and Skoog (MS) (1962) basal salts (see chapter two for composition) amended with 30 g (88 mM) sucrose, 1 g (0.55 mM) myo-inositol, 5 mg (2.9 $\mu$M) thiamine HCl, pH 5.7 and 8 g of agar per liter. After sterilization, the medium is poured into 60 × 15 mm plastic petri dishes. Typically, one or two leaf sections are placed in each of the dishes. PGR requirements will be provided for each variation of the experiment. We usually prepare at least four dishes of each treatment for each student or student team.

## Exercises

### Experiment 1. Differences between genotypes in the ability to regenerate shoots and roots

An important basic concept in the tissue culture of plants is that not all genotypes (in this case, genotypes are identified as cultivars) will respond equally under similar cultural conditions. In fact, there are some cultivars of chrysanthemum that will not produce either shoots or roots in vitro using any treatment. This experiment was developed to illustrate this concept and is probably the easiest of the following exercises to conduct with large classes. We suggest that the exercise be completed by teams of two or three students. This experiment also provides the basic protocol for the other two experiments in this chapter. Refer to Figure 12.1 for a diagrammatic representation of this exercise. Experiment 1 will require approximately 12 weeks for completion.

#### Materials
The following items are needed for each student or team of students:

* 100 ml of 5% commercial bleach per cultivar
* Four 60 × 15 mm petri dishes containing MS + 0.25 mg/L (1 $\mu$M) BA + 2 mg/L (11.5 $\mu$M) IAA per cultivar
* Four 60 × 15 mm petri dishes containing MS medium with 11.5 $\mu$M IAA only
* Four 60 × 15 mm petri dishes containing MS medium with 1.0 $\mu$M BA only
* Sixteen 60 × 15 mm petri dishes containing MS medium without PGRs per cultivar

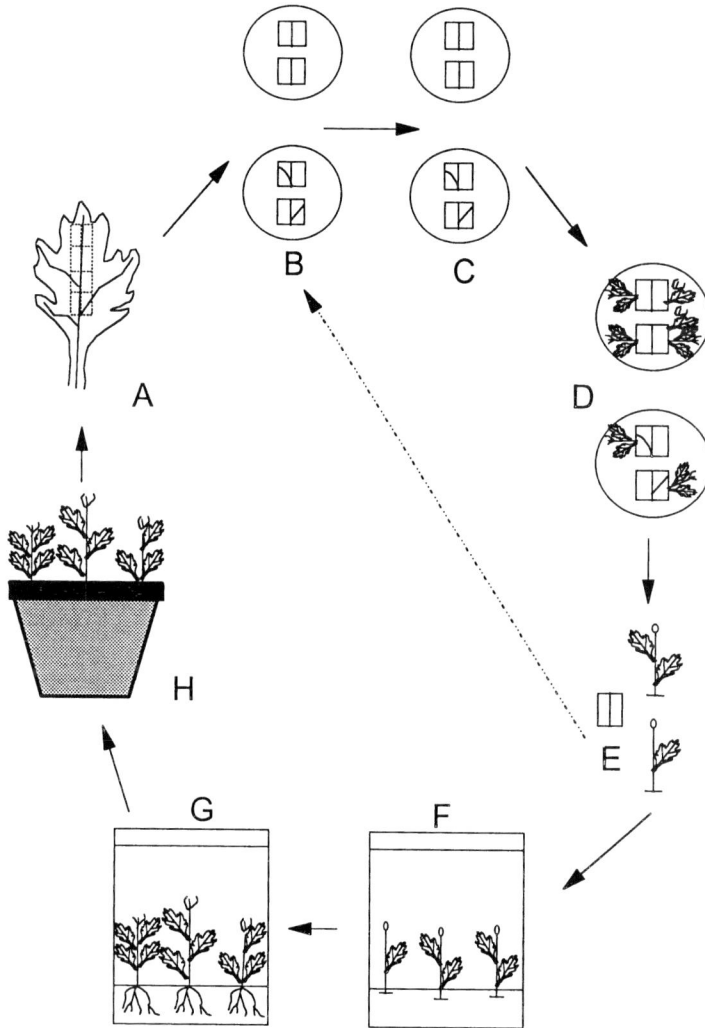

**Figure 12.1** Diagrammatic representation of the micropropagation protocol for chrysanthemum. A. Leaf after surface disinfestation and excision of leaf sections (dotted lines). B. Two leaf sections incubated on shoot initiation medium. C. Leaf sections transferred to shoot elongation medium after 2 weeks. D. Shoots formed on cut edges of the explants 5 weeks after culture initiation. E. Shoots greater than 2 cm are excised and large lower leaves removed. Leaves produced in vitro that may be used as explants are returned to initiation medium to produce more shoots (dotted line). F. Shoots are transferred to Magenta vessels for rooting. G. Shoots have elongated and produced roots after 3 weeks. H. Plantlets are transferred to soilless medium and acclimatized to growing conditions.

* Two to five GA-7 Magenta culture vessels or other similar size vessel (baby food jars) containing 75 ml MS medium without PGRs
* Incubator providing approximately 25 $\mu mol \cdot m^{-2} \cdot sec^{-1}$ and 25 C
* Cell packs with soilless medium, and mist bed or plastic wrap or plastic bag to construct a moist chamber for high humidity

*Procedure*

After the leaves have been surface disinfected, culture two leaf sections in each 60 × 15 mm petri dish containing MS medium supplemented with 11.5 $\mu$M IAA and 1.0 $\mu$M BA, IAA only, BA only, and MS basal medium lacking PGRs. This is the shoot initiation phase. The experiment may be considered as a completely randomized design with treatments consisting of the different PGRs replicated four times

Incubate petri dishes at 25 C with 25 μmol·m$^{-2}$·sec$^{-1}$ of fluorescent light for 2 weeks. Alternatively, if a growth room or cabinet is not available, the laboratory bench top should provide adequate conditions for the experiment to be conducted successfully. Transfer all explants to MS medium lacking PGRs and incubate under the same conditions as described previously for 3 weeks (shoot elongation phase). The number of shoots and/or roots per experimental unit (the petri dish) should be counted for each of the four treatments and all cultivars used in the study. Student teams may share data and each team should analyze the data (e.g., determine standard deviation or conduct an ANOVA etc.; see chapter six).

After the 3-week shoot elongation phase, excise shoots greater than 2 cm from the leaf margins and remove any large leaves that are positioned near the cut surface of the shoot axes. Insert the cut end of the shoots into the MS basal medium without PGRs contained in GA-7 Magenta vessels. Three rows of five shoots can be placed into each vessel. Incubate as before for 3 weeks (root initiation phase).

After 3 weeks, gently remove any agar from the roots with running tap water. Plantlets, with any large lower leaves removed, may be transferred to soilless medium in cell packs or individual 1 × 1 in. cells and placed under intermittent mist for 2–3 weeks (acclimatization phase). Alternatively, if a mist bed is not available, a plastic wrap or bag may be used to form a "tent" around the tray containing the cell packs. After a week, make several small holes in the tent with a sharpened pencil. Add a few more holes to the tent every 2–3 days. The plants should be acclimatized in about 2–3 weeks and now can be grown in the greenhouse or the laboratory.

### Anticipated results

After the first 2 weeks of culture, leaf explants exposed to IAA and BA should have distorted shapes, and sparse, light green crystalline callus may have formed around the cut edges of all cultivars. Darker green meristematic zones or small buds may also be apparent on 'Hekla', 'Iridon' and 'Rave' explants. This ends the initiation phase of the exercise — Stage 0 (selection of materials) and Stage I (establishment of cultures) of traditional micropropagation.

Three weeks after transfer to medium without PGRs, shoots and some roots should be formed on leaf sections of 'Hekla', 'Iridon' and 'Rave' that were initially cultured in medium containing PGRs. In this case, shoots and roots are formed directly from mesophyll cells without an intervening callus phase or direct organogenesis. Shoots should have formed primarily on the cut margins of the explant. The most shoots (and least number of roots) should be on 'Iridon' explants and decreasing numbers on 'Hekla' and 'Rave'. 'Adorn' and 'Goldmine' explants may have produced a few roots and maybe an occasional shoot. None of the explants of any of the cultivars should have produced appreciable amounts of callus, and those initially incubated on either only IAA, BA, or medium lacking PGRs should have formed very few shoots or roots — 'Adorn' and 'Goldmine' may have formed some roots. This is the "shoot elongation" phase of the experiment or Stage II of micropropagation.

Shoots that were excised from leaves should have elongated to almost twice the original length and produced roots (Stage III of micropropagation) in the Magenta vessels containing MS medium without PGRs. Typically, three to five adventitious roots are formed per shoot. These shoots are easily acclimatized to normal growing conditions within 3 weeks using the described procedures (Stage IV of micropropagation).

Questions

* What purpose do the treatments containing individual PGRs (IAA or BA only) and without any PGRs serve in the initiation phase of the experiment? Why are they necessary in the design of the experiment?
* Why is it necessary to transfer the explants from initiation medium (with PGRs) to shoot elongation medium (without PGRs)?
* Why can't the plants be transferred directly from the Magenta vessels to normal growing conditions found in the laboratory or greenhouse?

## Experiment 2. The effects of PGRs on formation of shoots and roots

One of the first steps in developing a micropropagation protocol of any plant is to determine the type (or combination) and concentration of PGRs that will provide the desired response — in our case, shoot formation. Leaf tissue of some cultivars of chrysanthemum are very responsive to both the type and concentration of cytokinin in the initiation medium.

This experiment uses the cultivars 'Adorn' and 'Iridon', or if time and/or materials are limited, the experiment may be completed with 'Iridon' only. Prepare several modifications of MS medium, each containing 11.5 µM IAA combined with the following concentrations and types of cytokinin or cytokininlike compounds: 0.01, 0.1, 1.0 and 5.0 µM kinetin, 2-iP, BA and thidiazuron (TDZ). There are a total of 17 treatments including IAA only.

Since there are 17 treatments, it is impossible to include all treatments in every block without using at least four or five different leaves. The inherent variability for the ability to form shoots between different leaves from the same stock plant would probably necessitate increasing the replications to some very large number in order to detect significant differences between treatments. Therefore, from our experience, this experiment works best when designed as a randomized incomplete block design with eight replications. Each incomplete block is composed of four entries or the number of explants obtained from one leaf, thus limiting the variability within an individual block to that of an individual leaf. As a matter of fact, explants from within an individual leaf are incredibly uniform in their ability to produce shoots. A more complete treatment of this experimental design with chrysanthemum may be found in Kuklin et al. (1993). A design for this exercise can be found in Table 12.1.

An associated exercise to demonstrate that explants from an individual leaf are equally capable of forming shoots is easily prepared and executed. Culture the four midvein explants from ten leaves on MS medium supplemented with 1.0 µM BA and 11.5 µM IAA for 2 weeks, then transfer to MS medium without PGRs for 3 weeks. Simply compare the number of shoots produced by each explant of an individual leaf — they should be more or less equal.

### Materials

For each cultivar in the experiment:

* Eight 60 × 15 mm petri dishes for each treatment
* Eight 60 mm × 15 mm petri dishes containing MS medium without PGRs

### Procedure

The procedure is essentially the same as that given in experiment 1 except that the experimental design in Table 12.1 should be followed. Treatments (PGR types and concen-

*Table 12.1* An Incomplete Block Design with 17 Treatments and 8 Replications

| Block | Treatment number | | | | Block | Treatment number | | | |
|---|---|---|---|---|---|---|---|---|---|
| 1 | 8 | 17 | 6 | 1 | 18 | 8 | 9 | 7 | 14 |
| 2 | 1 | 13 | 11 | 17 | 19 | 8 | 10 | 12 | 2 |
| 3 | 8 | 13 | 7 | 17 | 20 | 2 | 7 | 14 | 6 |
| 4 | 15 | 8 | 3 | 17 | 21 | 3 | 14 | 14 | 4 |
| 5 | 1 | 2 | 17 | 9 | 22 | 7 | 5 | 12 | 10 |
| 6 | 11 | 10 | 1 | 5 | 23 | 6 | 15 | 3 | 9 |
| 7 | 16 | 17 | 12 | 4 | 24 | 12 | 7 | 6 | 2 |
| 8 | 9 | 12 | 15 | 11 | 25 | 3 | 4 | 11 | 12 |
| 9 | 11 | 12 | 3 | 9 | 26 | 14 | 10 | 13 | 15 |
| 10 | 13 | 1 | 11 | 4 | 27 | 9 | 5 | 15 | 14 |
| 11 | 9 | 14 | 2 | 1 | 28 | 10 | 13 | 16 | 9 |
| 12 | 15 | 7 | 4 | 13 | 29 | 8 | 16 | 2 | 14 |
| 13 | 15 | 1 | 11 | 12 | 30 | 5 | 6 | 4 | 16 |
| 14 | 10 | 2 | 14 | 17 | 31 | 13 | 5 | 2 | 16 |
| 15 | 16 | 6 | 13 | 3 | 32 | 7 | 15 | 4 | 6 |
| 16 | 16 | 11 | 7 | 5 | 33 | 5 | 16 | 10 | 8 |
| 17 | 11 | 5 | 17 | 3 | 34 | 6 | 2 | 8 | 10 |

trations) should be represented by a number from 1 through 17. Depending on the size of the class, each student or team should be assigned to complete specific blocks, not treatments, for each cultivar. After surface disinfestation, one leaf section should be placed in each of four 60 mm petri dishes (treatments) contained within a block. Explants should remain on medium with PGRs for 2 weeks and then transferred to MS medium without PGRs for 3 weeks. The number of shoots and roots should be counted at the end of 5 weeks. The experiment only requires 5 weeks to complete since regeneration of plants is not necessary to demonstrate the effects of different PGRs.

*Anticipated results*

After incubation for 2 weeks on shoot initiation medium and an additional 3 weeks on shoot elongation medium, shoots and roots should be formed. Most treatments, except IAA only, will support some shoot formation on explants of 'Iridon' (Figure 12.2a); only roots should form on explants of 'Adorn' regardless of the combination of PGRs employed (Figure 12.2b) Shoots produced on medium containing higher concentrations of TDZ may not elongate after transfer to medium without PGRs. Also, most shoots produced on these treatments will exhibit hyperhydricity. Students should determine which treatment produced (statistically) the most shoots and/or roots.

*Questions*

* Which PGRs would you use to initiate shoots for a regeneration protocol and why?
* Are there other factors not considered by this experiment that might influence the number of shoots produced per experimental unit?
* What advantages does an incomplete block design offer over other designs, such as completely randomized or randomized complete block designs?
* Some treatments allowed for the production of many shoots; however, a similar or greater number of roots were also formed. What are some of the disadvantages of forming roots during the shoot's initiation and elongation phases of micropropagation of chrysanthemum? Are there any advantages?

**Figure 12.2** Shoots of chrysanthemum 'Iridon' (a) and roots from 'Adorn' (b) formed on the cut edges of explants after an initial incubation of 2 weeks with PGRs followed by transfer to medium without PGRs for 3 weeks. Numbers represent μM concentrations.

## Experiment 3. Pulse treatment with PGRs for initiation and growth of shoots

Most adventitious shoot micropropagation protocols are based on a treatment of explants with PGRs for a specific period of time (initiation period) followed by incubation on medium that either lacks or has reduced concentrations of PGRs. We have used this technique in each of the previous two exercises, and such treatments are usually referred to as "pulse" treatments. The practical reasons for utilizing pulse treatments are easily demonstrated using 'Iridon' and varying the time the explants are incubated on initiation medium. The objective of this experiment is to determine the optimum time that explants should remain on initiation medium to form the maximum number of shoots.

### Materials
For each student or student team:

* Sixteen $60 \times 15$ mm petri dishes containing MS medium with 11.5 µM IAA + 1.0 µM BA
* Twenty $60 \times 15$ mm petri dishes containing MS medium without PGRs

### Procedure
Surface disinfect leaves of the chrysanthemum cultivars and place two leaf sections into each $60 \times 15$ mm petri dish containing MS medium with PGRs as described in experiment 1. The treatments, arranged in a completely randomized design for this experiment, are 0, 1, 3, 5, 7, 10, 14, 21 and 35 days of culture on initiation medium followed by MS medium without PGRs for the remainder of the 35-day period. The 0-day treatment, should be initiated on MS medium without PGRs and then transferred to the same medium after 14 days.

### Anticipated results
Few, if any, shoots should form on explants from treatments 0, 1, 3, and 5 days, whereas 10–17 days of incubation on medium with PGRs should foster the development of the greatest number of shoots. Fewer shoots are formed on explants from the 21-day treatment, and although some shoots develop on explants from the 35-day treatment, they usually are stunted and/or exhibit hyperhydricity.

### Questions

* What is hyperhydricity, and why did it occur with shoots produced on the 35-day treatment in this experiment?
* Why were shoots not formed from explants included in the 0, 1, 3 and 5-day treatments?

## Literature cited

Kuklin, A.I., R.N. Trigiano, W.L. Sanders and B.V. Conger. 1993. Incomplete block design in plant tissue culture research. *J. Tiss. Cult. Meth.* 15:204–209.

Lu, C.-Y., G. Nugent and T. Wardley. 1990. Efficient, direct plant regeneration from stem segments of chrysanthemum (*Chrysanthemum morifolium* Ramat. cv. Royal Purple). *Plant Cell Rpt.* 8:733–736.

May, R.A. and R.N. Trigiano. 1991. Somatic embryogenesis and plant regeneration from leaves of *Dendranthema grandiflora*. *J. Am. Soc. Hort. Sci.* 116:366–371.

Murashige, T. and F. Skoog. 1962. A revised medium for rapid growth and bioassays with tobacco tissue cultures. *Physiol. Plant.* 15:473–497.

Trigiano, R.N. and R.A. May. 1994. Laboratory exercises illustrating organogenesis and transformation using chrysanthemum cultivars. *HortTechnology.* 4:325–327.

*chapter thirteen*

# Shoot organogenesis from cotyledon explants of watermelon

*Michael E. Compton and Dennis J. Gray*

Watermelon, *Citrullus lanatus* (Thunb.) Matsum. and Nakai, is a frost tender, vining annual vegetable crop with sweet-tasting fruit that is often consumed as a cool dessert. The species originated in tropical Africa. In the United States, watermelon is cultivated mainly in the south, southwestern and central states, although some early-maturing cultivars are available for northern climates. The fruit of watermelon varies in size, shape, rind color and pattern, flesh and seed color and maturity date, depending on the cultivar. Most fruit have a striped rind with red flesh and small black seeds, and weigh 20–30 lbs at maturity. However, seedless varieties, which have a slightly sweeter taste, also are available. Watermelons are susceptible to a number of bacterial, fungal and viral diseases, requiring annual field rotation, frequent chemical sprays and disease-resistant cultivars.

In light of both its importance and susceptibility to diseases, watermelon is an excellent candidate for improvement through genetic engineering. Insertion of bacterial, fungal and virus resistant genes through recombinant DNA technology would facilitate the development of new disease-resistant genotypes without significantly altering the genetic composition of cultivars with high fruit quality. In addition, the production of polyploids through somaclonal variation in tissue culture could be exploited to produce new parental breeding lines, which are useful in developing new seedless watermelon cultivars.

Recent cell and tissue culture research has resulted in an efficient plant regeneration system for watermelon (Compton and Gray, 1993; Dong and Jia, 1991; Srivastava et al., 1989). The published protocols use seedling cotyledons as a source of explants, from which adventitious shoots can be obtained by the process of organogenesis. The shoots can be rooted to form complete plants in 6–10 weeks from culture initiation. These protocols have been used both to obtain model transgenic plants and to obtain tetraploid plants, which can be used to breed improved triploid watermelon cultivars (Compton and Gray, 1994). The purpose of the laboratory exercises in this chapter is to familiarize students with the basic procedures of adventitious shoot regeneration from watermelon cotyledon explants and to illustrate differences in shoot regeneration competence from different regions of the cotyledon and among various watermelon cultivars.

## General considerations

### Explant preparation and culture media

The basic protocol described in Figure 13.1 is a modification of that described previously by Compton and Gray (1993). Seeds are imbided for 4–15 h before extracting the embryos.

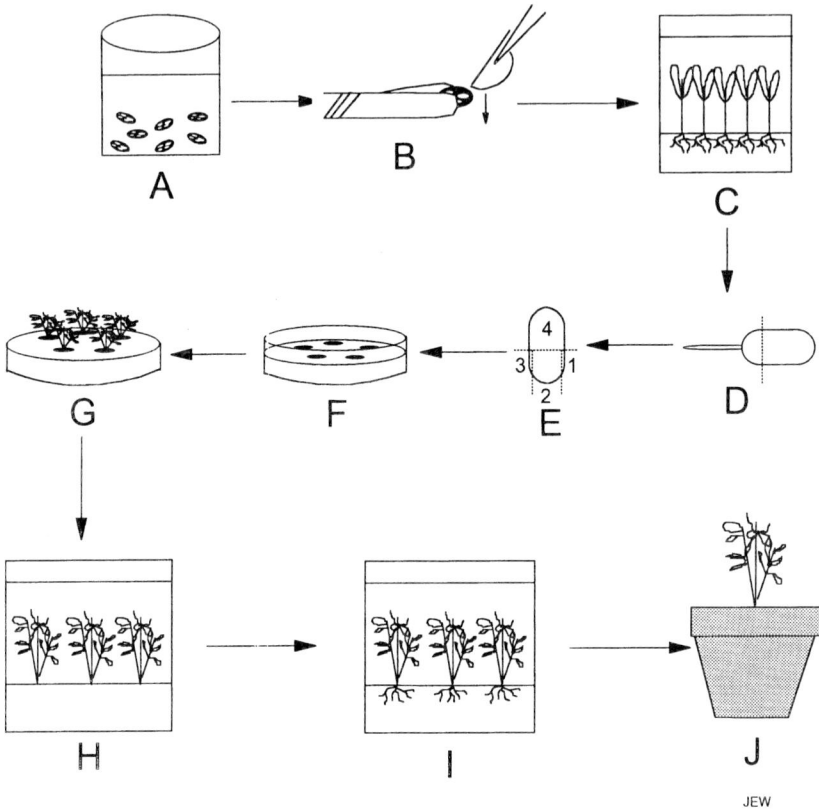

JEW

**Figure 13.1** Flow chart of the experimental protocol. A. Imbibe seeds in distilled water for 4–15 h. B. Remove seed coats using special forceps and scalpel with #10 blade. C. After surface disinfestation, germinate embryos in vitro for 5–6 days. D. Remove cotyledons from seedlings using a scalpel and a #11 blade. E. Prepare explants by making three longitudinal cuts (1, 2 and 3) and a final cross-sectional cut (4). F. Culture explants in $100 \times 15$ mm petri dishes with regeneration medium. G. Shoots appear in 3–4 weeks. H. Transfer shoot cultures to Magenta vessels containing elongation medium. I. Excise 1–2-cm-long shoots and transfer to rooting medium. J. Transfer plants to soil and acclimatize.

The seed coat is removed from imbibed seeds using a scalpel with a #10 blade and, preferably, a special toothed forceps, which is used to firmly hold the seeds. Seeds are held at the pointed end using the toothed forceps and positioned on their side. The seed coat is cut by applying a sharp downward motion at the rounded end. Several cuts may be required to make an opening large enough so that the seed coat can easily be removed. The embryo is removed by gently working it side-to-side using the blunt side of the scalpel, being careful not to damage the embryo. Embryos are stored in distilled water, and after all have been prepared they are surface disinfested in 100 ml of a 25% bleach solution (1.25% NaOCl plus 2 drops surfactant) for 25 min with constant agitation. Embryos are rinsed three to six times with sterile distilled water. Disinfested embryos are transferred to Magenta vessels containing germination medium (Table 13.1) using long-handled forceps. Six embryos are cultured per vessel.

After 5–7 days on germination medium, the embryos have germinated, greened and developed into 2–5-cm-long seedlings. Seedlings with their cotyledons still in close contact with each other are at the best stage. Seedlings are removed from the culture vessel and transferred to sterile plastic petri dishes for dissection. Seedlings can be easily handled after removing the root and half of the hypocotyl. The cotyledons are removed by making

*Table 13.1* Amendments to MS Medium Used for Regeneration
and Rooting of Watermelon Shoots from Cotyledon Explants

| Constituents | Media (amounts in mM) | | | |
|---|---|---|---|---|
| | G[a] | SR[b] | E[c] | R[d] |
| Sucrose | 87.6 | 87.6 | 87.6 | 58.4 |
| BA | — | 0.005 | — | — |
| IBA | — | — | — | 0.001 |

*Note:* Adjust pH of each medium to 5.7 before adding 7 g/L TC agar and autoclaving (see chapter two for composition of MS medium).

[a] Germination medium.

[b] Shoot regeneration medium.

[c] Shoot elongation medium.

[d] Rooting medium.

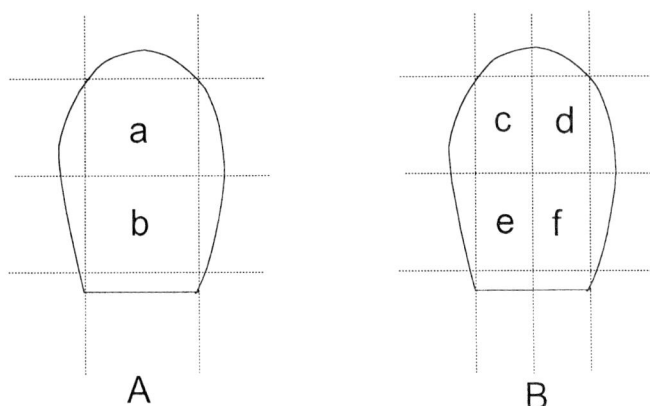

**Figure 13.2** Cotyledon explants to be cultured for experiment 1.

a shallow cut across the width of the cotyledon about 1–2 mm above the point of attachment to the hypocotyl and gently "flicked" to detach them from the hypocotyl. A scalpel equipped with a #11 blade works best. Care must be taken not to include the shoot apex. Cotyledons have been properly removed when the shoot tip can be seen remaining on the embryo axis after detaching the cotyledons. Cotyledons are placed abaxial side (i.e., the "bottom" side) up and cut as diagrammed in Figure 13.1F or Figure 13.2, depending on the experiment. Explants are cultured abaxial side down in $100 \times 15$ mm plastic petri dishes containing 25 ml of regeneration medium (Table 13.1). Five explants are randomly distributed to each petri dish of regeneration medium and cultured abaxial side down. Plates are sealed with parafilm, labeled and transferred to the growth chamber.

After 4 weeks on regeneration medium, explants with shoots and buds are transferred to Magenta vessels containing shoot elongation medium (Table 13.1). This encourages shoot elongation and facilitates rooting. All brown leaves and nonresponding portions of the cotyledon are trimmed away before transferring shoot cultures to vessels containing elongation medium. After 3–4 weeks, shoots greater than 2 cm in length are excised from the primary explant and all expanded leaves removed, leaving the shoot apex and first true leaf. Shoots are transferred to Magenta vessels containing rooting medium (Table 13.1). Shoots are ready for acclimatization when they have formed roots that are at least 1 cm long (10–14 days). For acclimatization, remove plantlets from the culture vessels and gently wash away any agar surrounding the roots. Plants are transplanted to cell packs containing autoclaved soilless medium (equal parts Pro Mix BX and coarse vermiculite). Cell packs with plants are placed in a plastic flat containing approximately 300 ml of sterile distilled water and covered with a clear plastic lid. Flats are sealed with wide masking tape and placed in a well lit area ($\geq 50$ $\mu mol \cdot m^{-2} \cdot sec^{-1}$ and 16-h photoperiod). Plants should remain covered for about 1 week before removing the tape and gradually removing the lid.

## Exercises

### Experiment 1. Difference in regeneration potential among distinct regions of watermelon cotyledons

Use of the proper explant type is an important factor in successful regeneration. Not all explants from the same plant or cotyledon possess an equal regeneration potential. To develop an efficient plant regeneration protocol it is necessary to identify the best region of the plant from which to isolate explants. In watermelon, the best explant source is the cotyledon of in vitro–germinated seedlings. However, not all regions of watermelon cotyledons possess an equal regeneration potential. In this experiment, the students will be culturing explants from different regions of the watermelon cotyledon to identify the best explant for adventitious shoot regeneration.

#### Materials

The following will be needed for each student or team:

* 30 'Minilee' watermelon seeds
* Toothed forceps (#6-134 Miltex Instrument Company, Inc., Lake Success, NY recommended)
* Sterile plastic bottles with caps for sterilizing seeds (Nalgene® 125-ml wide-mouth polypropylene bottle. Lids may be modified for easy rinsing of seeds by cutting a 2-cm-diam. hole and inserting a 1000-$\mu$m mesh screen)
* 100 ml of 25% bleach solution with 1–2 drops surfactant
* Six Magenta GA-7 vessels containing 50 ml of germination medium
* Twenty-four $100 \times 15$ mm plastic petri dishes containing 25 ml of regeneration medium
* Magenta GA-7 vessels containing 50 ml of shoot elongation medium
* Magenta GA-7 vessels containing 50 ml of rooting medium
* Growth chamber providing 25–50 $\mu mol \cdot m^{-2} \cdot sec^{-1}$ irradiance and 25 C
* Cell packs with autoclaved soilless medium (equal parts Pro Mix BX and coarse vermiculite), $28 \times 53$ cm plastic flat without holes and clear plastic dome lid

#### Procedure

For this experiment, use seeds of the watermelon cultivar 'Minilee' (see next experiment for source). Extract and germinate embryos as described previously. Excise cotyledons as outlined and cut into six different types of explants as depicted in Figure 13.2. Each group should test all of the explant types. Culture 20 explants (5 per dish) of each cotyledon region in $100 \times 15$ mm petri dishes with regeneration medium. Seal and label dishes and transfer them to the growth chamber. Arrange dishes from all groups in a completely randomized design with subsampling (see chapter six). After 4 weeks record the number of explants from each region that have shoots or vegetative buds. Transfer explants with shoots and/or buds to Magenta vessels containing elongation medium for an additional

4 weeks. Remove all shoots that are at least 2 cm long from the primary explant and transfer them to Magenta vessels with rooting medium. Acclimatize plants as described above. Analyze data using analysis of variance for a completely randomized design with subsampling and compare means using standard error of the mean or another multiple comparison procedure as described in chapter six.

### Anticipated results
Within 3–5 days of culture initiation all explant types typically enlarge to about four to five times their original size. Buds, leaves and shoots should be visible within 3–4 weeks after culture initiation. Explant types b, e and f (Figure 13.2) should produce the highest response frequency (50–90%). Those from regions a, c or d may respond but at a much lower frequency (<10%).

### Questions

* Which region produced the most explants with shoots and/or buds?
* Why would some regions of the cotyledon respond more readily than others?
* Were there any noticeable abnormalities among the regenerated shoots? If so, were they associated with a particular treatment?

## Experiment 2. Effect of plant genotype on the ability of watermelon cotyledon explants to produce shoots

The genotype of the plant often is one of the most important factors in determining explant competence for organ regeneration. Within a species there are often one or more genotypes that readily respond to the imposed plant regeneration protocol, whereas others fail to respond to any devised treatment. It is extremely important to select highly regenerative genotypes for genetic transformation to be successful. Therefore, it is beneficial to test regeneration rates of as many cultivars as feasibly possible. In watermelon, most cultivars tested to date are competent for shoot regeneration from cotyledon explants. However, the regeneration potential often varies among cultivars. In this experiment the ability of cotyledon explants from four watermelon cultivars to produce shoots will be examined. The cultivars to be tested include 'Crimson Sweet' (Stokes Seeds, Inc., Buffalo, NY), 'Minilee' (Willhite Seed Co., Poolville, TX), 'Sangria' (Seedway, Inc., Elizabethtown, PA) and 'Yellow Doll' (Otis S. Twilley Seed Co., Inc., Trevose, PA.). Other cultivars may be substituted, but those listed should evoke a wide range of responses.

### Materials
The following items are needed for each student or team:

* 15–18 watermelon seeds (a different cultivar should be used by each student or team)
* Toothed forceps
* Sterile plastic bottles with caps for sterilizing seeds
* 100 ml of 25% bleach solution with 1–2 drops surfactant
* Three Magenta GA-7 vessels containing 50 ml of germination medium
* Five 100 × 15 mm plastic petri dishes containing 25 ml of regeneration medium
* Magenta GA-7 vessels containing 50 ml of shoot elongation medium
* Magenta GA-7 vessels containing 50 ml of rooting medium
* Growth chamber providing 25–50 $\mu mol \cdot m^{-2} \cdot sec^{-1}$ irradiance and 25 C

* Cell packs with autoclaved soilless medium (equal parts Pro Mix BX and part coarse
  vermiculite), $28 \times 54$ cm plastic flat without holes and clear plastic dome lid

### Procedure

For this exercise, each student or team should work with seeds from a different cultivar.
The seeds should be imbibed, surface disinfested, and germinated as described in Figure
13.1. Basal cotyledonary explants should be prepared as described in Figure 13.1D and E
and randomly distributed in $100 \times 15$ mm petri dishes containing regeneration medium.
Each student should culture five petri dishes, each with five explants. Seal and label the
petri dishes and transfer to the growth chamber. Arrange the petri dishes in a completely
randomized design (see chapter six).

After 9–12 days, look at the explants using a stereomicroscope. Some should have
developed buds along their cut edges. Record the number of explants with buds.

After 4 weeks on regeneration medium, record the number of explants that produced
shoots and/or buds and calculate the percentage of explants that responded. Trim away
all brown leaves and nonresponding portions of the cotyledon before transferring shoot
cultures to Magenta vessels containing elongation medium.

After 4 weeks in elongation medium, some shoots should elongate to more than 2 cm.
Remove these shoots from the original explant and remove all fully expanded leaves,
leaving the shoot apex and youngest leaf. Transfer shoots to culture vessels containing
rooting medium. Up to nine shoots can be cultured per vessel. Shoot cultures can be
subcultured to fresh elongation medium at this time to produce more shoots.

When shoots have formed roots $\geq 1$ cm in length (10–14 days), record the number that
rooted and calculate the rooting percentage. Transfer plants to cell packs containing
soilless medium and acclimatize as described previously. Record the number of plants
surviving acclimatization and calculate the survival frequency 1 week after completely
removing the lid. Analyze the data using analysis of variance for a completely randomized
design with subsampling and compare means using the standard error of the mean or an
appropriate multiple comparison procedure (see chapter six).

### Anticipated results

Most explants will have expanded four to five times their original size after 3–5 days on
regeneration medium. After 9–12 days, some explants should have developed buds along
their cut edges (Figure 13.3a). Some of these buds develop into 0.5–1-cm-long shoots after
4 weeks on regeneration medium (Figure 13.3b). Explants from the cultivars should
respond differently. 'Minilee', 'Yellow Doll' and 'Crimson Sweet' respond better than
'Sangria'. Transferring explants with shoots to elongation medium without BA allows the
shoots to elongate (Figure 13.3c) so that they can be easily rooted. Approximately 75–95%
of shoots will normally root on medium with IBA. Not all plants will survive acclimatization. Typically 60–90% of watermelon plants from tissue culture will survive acclimatization. Once the plants have been acclimatized they can be transplanted to 10-cm-diam. pots
filled with Pro Mix BX and grown in the greenhouse (Figure 13.3d) or moved to the garden
when conditions are suitable.

### Questions

* Were there any differences in the frequency of explants with shoots, shoots that
  rooted or acclimatized plants among the cultivars tested?
* Why should explants from some cultivars respond better than those from others?
* Did you notice any abnormal looking plants? Describe the abnormalities.

**Figure 13.3** Organogenesis from cotyledons of watermelon. a. Shoot buds formed on the cut edge of watermelon cotyledon incubated on shoot regeneration medium for 9–12 days. b. Leaves, buds and shoots formed on watermelon cotyledon 4 weeks after culture initiation. c. Elongated shoots on watermelon cotyledon explant 4 weeks after transfer to shoot elongation medium. d. Acclimatized plant transferred to the greenhouse.

## Acknowledgment

This is Florida Agricultural Experiment Station Journal Series No. R-04840.

## Literature cited

Compton, M.E. and D.J. Gray. 1993. Shoot organogenesis and plant regeneration from cotyledons of diploid, triploid, and tetraploid watermelon. *J. Am. Soc. Hort. Sci.* 118:151–157.

Compton, M.E. and D.J. Gray. 1994. Regeneration of tetraploid plants from cotyledons of diploid watermelon. *Proc. Fla. State Hort. Soc.* 107:107–109.

Dong, J.Z. and S.R. Jia. 1991. High efficiency plant regeneration from cotyledons of watermelon (*Citrullus vulgaris* Schrad.). *Plant Cell Rpt.* 9:559–562.

Srivastava, D.R., V.M. Andrianov, and E.S. Piruzian. 1989. Tissue culture and plant regeneration of watermelon (*Citrullus vulgaris* Schrad. cv. Melitopolski). *Plant Cell Rpt.* 8:300–302.

## chapter fourteen

# Direct and indirect shoot organogenesis from leaves of Torenia fournieri

*Mark P. Bridgen*

Plant regeneration in tissue culture is a crucial aspect of plant biotechnology as it facilitates the production of genetically engineered plants, the release of disease-free plants from meristem cultures and the rapid multiplication of difficult to propagate species. Plant regeneration through tissue culture can be accomplished by enhanced axillary branching from shoots or lateral buds (micropropagation), somatic embryogenesis, embryo culture or organogenesis (Dodds and Roberts, 1985). Organogenesis is the initial and subsequent de novo growth of shoots, roots, leaves or flowers from a tissue and is especially easy to demonstrate with *Torenia fournieri* L. (Bridgen et al., 1994).

*Torenia fournieri* is commonly called the wishbone flower, bluewings, or torenia. Wishbone flower refers to the shape and position of the stamens that arch together, similar to a chicken wishbone, and fuse at the anther. Torenia has opposite leaves, a rounded growth habit, attains a height of approximately 30 cm and is well adapted to pot conditions. It is a plant that is easy to obtain from seed and easy to grow under greenhouse, field or classroom conditions.

Torenia has been used since 1985 to study the process of somaclonal variation (Brand and Bridgen, 1989; Hadi, 1986; Hadi and Bridgen, 1990) and the white-flowering cultivar 'UConn White' was developed through this process (Brand and Bridgen, 1989). Significant work had been previously completed on the organogenesis of torenia from stem segments (Tanimoto and Harada, 1984) and leaf sections (Bajaj, 1972).

In this exercise, techniques will be described to demonstrate direct and indirect shoot organogenesis from leaf sections of torenia; leaf sections will develop shoots and callus very quickly. These exercises can be completed by the end of a 10-week term or 14-week semester. The primary objective of this exercise is to demonstrate the processes of direct and indirect organogenesis with *T. fournieri*. The secondary objective is to give the student practice with aseptic technique and medium preparation. Details of the experiments follow and are adapted primarily from a publication by Bridgen et al. (1994).

## General considerations

### Growth of plants.

Torenia can be propagated by shoot cuttings of established plants or from seed that can be obtained from most commercial seed companies. Germination will occur in 7–14 days if seeds are not covered and remain under moist conditions at 21–23 C. Plants grow easily in any soilless medium under greenhouse or classroom conditions with adequate moisture, humidity, light (minimum 100 $\mu mol\cdot m^{-2}\cdot sec^{-1}$), and night temperatures of 13–15 C. When grown outside, the plants flourish when given shade, cool temperatures and moist growing media. Plants should be fertilized regularly with a soluble fertilizer to maintain vigorous growth and healthy, green leaf development. Allow the plants to grow for at least 8 weeks after germination before leaf sections are removed. Depending on environmental conditions, flowering should begin 12 weeks after seed germination. Plan on at least one plant per student; a few extra plants for the classroom may be desirable. These plants are especially susceptible to white fly infestation and can be damaged if control measures are not taken. If pesticides are used for control, follow all label recommendations for rates and safety.

## Explant preparation and basal medium

Select and remove young leaves from the second and third nodal regions from the top of established torenia plants. Young and vigorously growing plants will respond more quickly in vitro and will give better results than older leaves. Clean leaves gently in tepid running tap water. From this point on, all work should be completed under aseptic conditions. This exercise can be completed with or without the use of a laminar flow hood. In either situation, aseptic procedures should be explained and demonstrated to the students prior to this laboratory. Clean all work surfaces with 70% ethanol and allow to dry when ready to begin the exercise.

Surface disinfest entire leaves by dipping them for 5–10 sec in 70% ethanol followed by 10% commercial bleach (0.5% sodium hypochlorite) for 15 min. After the bleach treatment, rinse the leaves three times with sterile deionized water and allow the leaves to remain in the last rinse until they are ready to be cultured. Use only sterile forceps or scalpels to hold plant material after disinfestation. To increase the success rate, limit the number of leaves to three or four in a 200-ml beaker containing the bleach solution. To hasten results and/or to eliminate the previous steps, leaves may be used from previously established aseptic cultures.

Place a single disinfested leaf on a sterile surface such as a sterile $100 \times 15$ mm petri dish or an autoclaved paper towel. Use a #10 scalpel blade to remove the leaf margins if they are damaged by the bleach and discard. Cut leaves into $0.5 \times 0.5$ cm pieces and place on the appropriate medium with the abaxial surface (underside) in contact with the medium. Each student should culture at least four explants per medium. If contamination is a possibility, have the students wrap culture vessels with PVC film or Parafilm™. Culture vessels should be labeled with the date and individual's initials.

Incubate cultures at 21–23 C under a 16-h photoperiod provided by cool white fluorescent lamps at 50–60 $\mu mol\cdot m^{-2}\cdot sec^{-1}$. Observe cultures on at least a weekly basis and compare the explant development on each medium. Look for the initiation of shoots, leaves and callus.

The three media needed for this experiment have the same basal formula that includes Murashige and Skoog (1962) salts and vitamins (see chapter two for composition), 88 mM (30 g/L) sucrose, 7 g/L agar and adjusted to pH 5.7. The callus induction medium and the direct organogenesis medium will be needed at the beginning of this exercise. The indirect organogenesis medium is not needed for 4–6 weeks after the initiation of the experiment unless the instructor wants to demonstrate all three types of organogenesis simultaneously. If this latter scenario is desirable, callus will have to be prepared 1–2 months prior to this exercise and supplied to the student at the onset of the experiment.

The callus induction medium includes 0.5 mg/L (2.3 µM) 2, 4-dichlorophenoxyacetic acid (2,4-D), and the direct organogenesis medium includes 0.25 mg/L (1.1 µM ) 6-benzylaminopurine (BA) and 0.05 mg/L (0.25 µM) indole-3-butyric acid (IBA). The indirect organogenesis medium includes 2 mg/L (8.8 μM) BA plus 0.5 mg/L (2.5 μM) IBA. Dispense 20–30 ml of media into each 25 × 150 mm borosilicate glass culture tube, cap and autoclave. If limited space and media are problems, sterile disposable 100 × 15 mm petri dishes can be used for the first 4 weeks instead of culture tubes.

# Exercises

## Experiment 1. Direct organogenesis from leaves of T. fournieri

Plant regeneration directly from cultured explants can be accomplished by using leaves, bulbs, stems, corms, rhizomes, tubers or other plant parts. Although direct organogenesis occurs much less frequently than indirect organogenesis, explants of some species can be placed on nutrient media containing cytokinins and low levels of auxins and initiate shoots directly. Multiplication is achieved through subdivision of the shoot-producing clumps.

The objective of this experiment is to demonstrate direct organogenesis from leaf sections of torenia. The experiment will require approximately 4–6 weeks until shoots are visible.

### Materials

The following items are needed for each student:

* Four culture tubes or 60 × 15 mm petri dishes containing direct organogenesis medium described above
* One healthy, young torenia plant
* 10% commercial bleach

### Procedure

After the leaves have been surface disinfested and dissected, culture one leaf section in each of the culture vessels with direct organogenesis medium (Figure 14.1). In order to statistically analyze the results at the end of the exercise, place the culture vessels in the growth chamber in a randomized complete block design. There should be four blocks, each containing one of the culture vessels from each student.

After 4–6 weeks of growth, the number of shoots per culture vessel should be counted. Students should give their results to the instructor to collate and summarize the data for the students. Each student will be responsible for analyzing the data with ANOVA etc. (see chapter six).

If the student would eventually like to obtain a torenia plant ex vitro, individual shoots should be aseptically separated as they are counted. The shoots can be placed on an MS medium without plant growth regulators (PGRs) for further growth and root development (Stage III of micropropagation). After 3–5 weeks, the plantlets can be removed from the culture vessel and the agar gently washed from the roots with running tap water. The plantlets may then be transferred to a soilless medium in small pots or plug trays and placed under intermittent mist or inside a plastic bag to acclimatize (Stage IV of micropropagation). The plants should be acclimatized in 2–3 weeks and can be grown as "normal" plants.

### Anticipated results

Within 3–7 days after the initial culture, contamination may be visible, and these cultures should be discarded. Depending on the vigor and age of the original source plant, there may be some bleach damage on the leaves and if it is not extensive, the experiment will work. Newly formed shoots should start to be visible in 2–3 weeks. Many shoots will form on the explant except for any of the damaged areas.

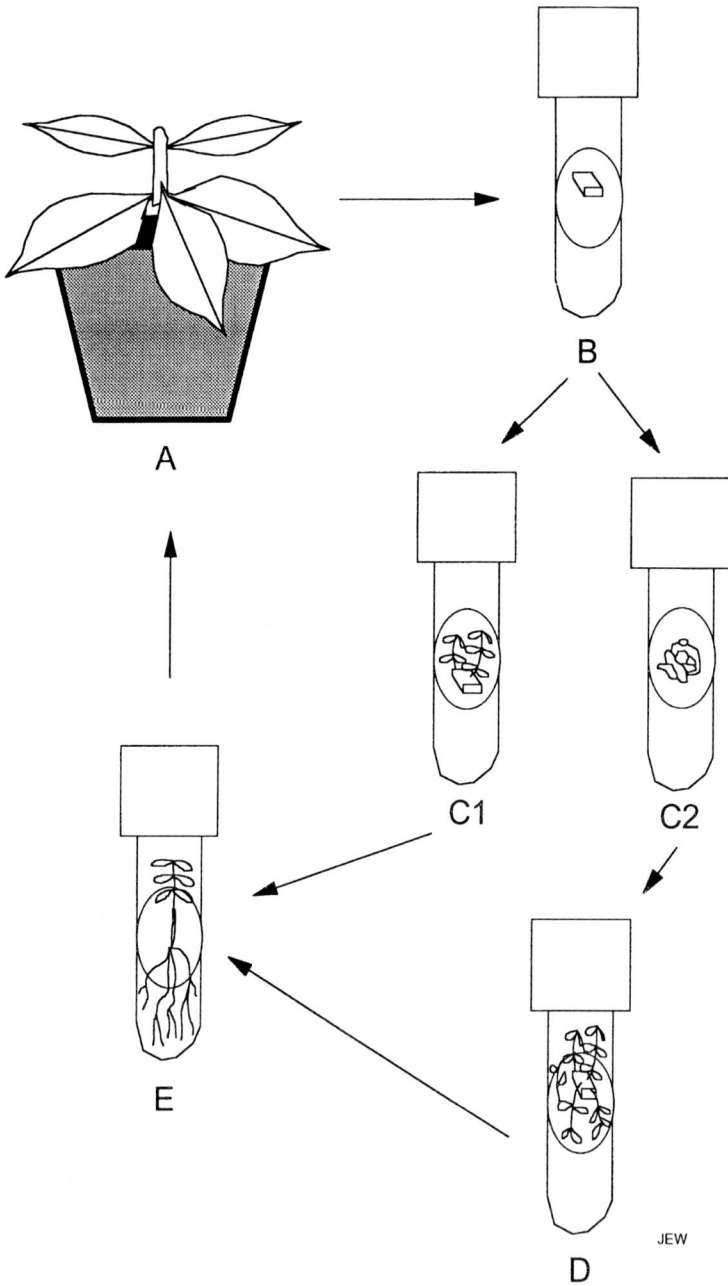

**Figure 14.1**   Plant regeneration in *T. fournieri*. Torenia plants grown in vivo serve as the source of leaf explants (A). Leaf explant sections are established on nutrient medium (B). Explants cultured directly on a cytokinin-containing medium give rise to shoots directly (C1), and callus is derived from explants growing on an auxin medium (C2). Transfer of callus to medium containing cytokinin results in shoot production (D). Shoots derived through either method are rooted on a medium free of PGRs (E).

*Questions*

* Would the culture of explants on their abaxial surface affect organogenesis differ-
  ently than if they were cultured on their adaxial surface?
* What effect does the size of the explant have on organogenesis?
* How does the auxin to cytokinin ratio in a medium determine the type of organo-
  genesis?
* What is the difference between adventitious shoots and axillary shoots?
* Why would adventitious shoot proliferation be a poor way to propagate chimeral
  plants?

## Experiment 2. Callus formation and indirect organogenesis of T. fournieri

Indirect organogenesis is the production of shoots or roots from callus — an amorphous
mass of loosely arranged thin-walled, differentiated parenchyma cells. Types of explants
that produce callus include stems, leaves, flower parts, petioles, roots, or the more mitoti-
cally active shoot tip or its meristematic region. The cells of the explant divide in vitro and
give rise to the organogenic callus (Dodds and Roberts, 1985).

Organogenic callus has meristematically active cells that are the precursors of roots
and shoots. The ploidy levels of these cells may vary tremendously because of the fast rate
of division. This genetic instability is one of serious problems associated with the use of
callus cultures for micropropagation.

The objective of this exercise is to demonstrate indirect organogenesis from torenia
callus. It will take approximately 5–6 weeks to obtain large clumps of callus from leaf
sections and an additional 4–6 weeks for shoot production from callus. Students can
initiate callus at the same time that experiment 1 on direct organogenesis is begun.
Experiment 2 may take a major part of the term or semester. To decrease the time for this
exercise, the instructor can initiate callus for the students 5–6 weeks before the exercise
begins. Refer to Figure 14.1 for a diagrammatic representation of this exercise.

### Materials

In addition to the materials listed for experiment 1, the following items are needed for each
student:

* Four culture tubes or $60 \times 15$ mm petri dishes containing callus induction medium
  or four cultures of torenia callus
* Four culture tubes or $60 \times 15$ mm petri dishes containing indirect organogenesis
  medium described above

### Procedure

After the leaves have been surface disinfested and dissected, culture one leaf section in
each of the culture vessels with callus induction medium. Although callus growth will not
be statistically analyzed, the culture vessels should be placed in the growth chamber in a
randomized complete block design to distribute variability among the cultures. As with
experiment 1, there should be four blocks, each containing one of the culture vessels from
each student. For more precise measurements of callus production, fresh weight values can
be taken aseptically on a weekly basis and plotted over time at the end of the exercise.

After 3–4 weeks, the explants should be subcultured onto fresh callus induction
medium. After a total of 5–6 weeks on the induction medium, callus can be aseptically

removed and cultured on the indirect organogenesis medium. After an additional 4–6 weeks of growth on the organogenesis medium, the number of shoots per culture vessel should be counted. Students should give their results to the instructor to be collated and summarized. The summarized data will be returned to the students for analysis with ANOVA, etc. (see chapter six). If students would like to obtain a torenia plant ex vitro, the procedures that were described in experiment 1 should be followed.

### Anticipated results

Contamination should not be a problem with the indirect organogenesis since cultures will have been established and growing in vitro. After 4–6 weeks on the indirect organogenesis medium, the callus should respond by producing multiple shoots.

### Questions

* Where does callus originate on the leaf explants?
* From which locations on the callus do shoots arise?
* Why is indirect organogenesis from callus a poor way to propagate torenia clonally?

## Literature cited

Bajaj, Y.P.S. 1972. Effect of some growth regulators on bud formation by excised leaves of *Torenia fournieri*. *Z. Pflanzenphysiol.* 66:284–287.

Brand, A.J. and M.P. Bridgen. 1989. 'UConn White' a white-flowered *Torenia fournieri*. *HortScience.* 244:714–715.

Bridgen, M.P., M.Z. Hadi and M. Spencer-Barreto. 1994. A laboratory exercise to demonstrate direct and indirect shoot organogenesis from leaves of *Torenia fournieri*. *HortTechnology.* 4:320–322.

Dodds, J.H. and L.W. Roberts. 1985. *Experiments in Plant Tissue Culture*, 2nd Ed. Cambridge University Press, Cambridge, England.

Hadi, M.Z. 1986. In vitro techniques for the development of pest resistant plants. University of Connecticut M.S. Thesis. 99 pp.

Hadi, M. and M.P. Bridgen. 1990. Screening of somaclonal variants in vitro to produce insect resistant plants. *HortScience.* 25:112.

Murashige, T. and F. Skoog. 1962. A revised medium for rapid growth and bioassays with tobacco tissue cultures. *Physiol. Plant.* 15:473–497.

Tanimoto, S. and H. Harada. 1984. Roles of auxin and cytokinin in organogenesis in *Torenia* stem segments cultured in vitro. *J. Plant Physiol.* 115:11–18.

*chapter fifteen*

# Shoot organogenesis from petunia leaves

*John E. Preece*

*Petunia × hybrida* Hort. Vilm.-Andr. is the common garden petunia. Genetically, it has a complex background consisting of hybrids of three or four *Petunia* species. It is considered a cultigen because its complex genetics clearly indicate that this hybrid species originated in domestication, not in the wild. Petunia is among the most popular bedding plants and is mainly propagated by seeds; however, some hybrids will not come true from seeds and are propagated by rooting stem cuttings (Liberty Hyde Bailey Hortorium, 1976).

Petunia is a member of the Solanaceae (tobacco) family. There have been many in vitro studies on members of this family because they are not difficult to regenerate in tissue culture. Within this family, petunia is particularly responsive to tissue culture techniques and can be regenerated from anthers, protoplasts, seedling tips, stems, leaves and other tissues and organs. Petunia is an ideal plant for a tissue culture laboratory exercise because it is responsive in vitro; seeds are easy to obtain, store and germinate; seedlings are easy to grow; and it can be scheduled year-round. The following exercises use leaves for the initial explant and are designed to familiarize the student with the effects of cytokinin and auxin on the regeneration of shoots and roots and the formation of callus. Additionally, the laboratories will demonstrate both direct and indirect shoot organogenesis. Direct shoot organogenesis occurs when shoots form directly from the original explant without an intervening callus. In contrast, indirect shoot organogenesis happens when callus first forms from the explant, then shoots arise from the callus tissue.

With both direct and indirect shoot organogenesis, cells must return to the meristematic state of growth. Meristems are the least specialized (least differentiated) tissues in the plant. Daughter cells that form in meristems gradually become more specialized as they differentiate into the cells that make up a plant. The balance of hormones influences the direction that differentiation takes. For specialized or differentiated cells to become meristematic requires that they become less specialized. This process is the reverse of differentiation and is called dedifferentiation. Therefore, dedifferentiation is necessary before adventitious shoots can form from the initial explant or from callus tissues.

## General considerations

### Growth of plants

Various petunia cultivars and genotypes will respond in a somewhat dissimilar manner in vitro and will have different optimum requirements for plant growth regulators (PGRs). However, virtually all *Petunia × hybrida* cultivars will respond by regenerating shoots in vitro. In the past, leaf explants from cultivars 'Sugar Daddy', 'Sugar Plum', 'White Pearls' and 'Supermagic Pink' have responded well in vitro. A problem with making recommendations for which cultivar(s) to use in these experiments is that seed companies are continually changing their offerings. Seeds can be purchased from garden centers or directly from seed companies via catalog. Cultivars should be selected from the Grandiflora types, and red-flowering cultivars should be avoided because several of these have responded poorly in vitro (personal communication with Paul E. Read). For the exercises, two cultivars with a different flower color should be selected. Mixed cultivars should not be selected because they will yield variable results. One of many sources for petunia is W. Atlee Burpee & Co., Warminster, PA 18974 (1-800-888-1447).

About 8 weeks prior to the time of the laboratory exercise, seeds can be sown in any new or disinfested greenhouse soilless medium, but a fine textured peatlite medium is preferable. Petunia seeds are tiny and must be sown close to the surface of the medium. A 1-cm-deep furrow should be made, then seeds sown at a density of about two to three per centimeter of row. Be careful not to sow seeds too thickly because it will result in weakened, stretched seedlings that are difficult to separate and transplant. After sowing, the seeds can be covered with a light dusting of medium. Care must be taken during watering because it is easy to wash the seeds out of the furrow. Seeds can be germinated in a growth chamber or in a greenhouse. Because of the varying environmental conditions, extra care must be taken in the greenhouse. The temperature must be maintained at 21 – 24 C.

Fertilization with 100 ppm nitrogen from a water-soluble fertilizer can begin after seedlings emerge. Seedlings, about 1 cm or $^1/_2$ in. tall, can be transplanted into any well-aerated, pest-free soilless medium in 10-cm-diam. pots. The seedlings should be watered immediately after transplanting. There should be one or two plants of each cultivar per student or team of students.

Seedlings should be grown at about 24 C. Leaf explants from seedlings grown at high temperatures >30 C will not respond in vitro. It is best if the seedlings can be subirrigated with an ebb-and-flood system or by use of a capillary mat. Overhead watering can lead to fungal and bacterial contamination in vitro. Pests, especially white flies, must be controlled, or there will be problems with contamination in vitro. Weekly spraying with appropriate pesticides is recommended; use protective clothing and follow label instructions exactly.

Seedlings are sufficiently mature for explants when the leaves touch the sides of the pots, but should still be growing as rosettes. Leaves from young plants that are flowering can be used; however, contamination can be a problem. Avoid using leaves from plants growing outdoors because of contamination and difficulty in disinfestation.

The nutrient medium is Murashige and Skoog (1962) (MS) salts and organics (see chapter two for composition). The medium should contain 30 g/L (88 mM) sucrose and the pH should be adjusted to 5.8 prior to adding 7 g/L of a high quality agar (Difco Bacto agar or Sigma agar) per liter. After autoclaving, and before the agar gels, the medium should be poured into 100 × 15 mm diam. sterile plastic petri dishes. Enough medium should be added to completely cover the bottom of the petri dish (10–15 ml). Two leaf explants will be placed right side up (adaxial side up) in each petri dish. At least four petri dishes of each PGR treatment should be used per cultivar per student or team of students.

## Exercises

### Experiment 1. The effects of cytokinin and auxin on petunia leaf explants

This exercise is designed to familiarize students with the roles PGRs added to the tissue culture medium play in organogenesis. Although two of the treatments in this exercise consider the effects of cytokinin or auxin used alone, the concept of the effect of cytokinin to auxin ratio is illustrated. The cytokinin(s) and auxin(s) that are incorporated into tissue culture media simply add to endogenous PGRs (hormones) that are already present. When a cytokinin or auxin is added singly to a tissue culture medium, the ratio within a plant tissue shifts to a high amount of the PGR that is added and a low amount of the one that is not added (consider that explants will contain both endogenous auxin and cytokinin). It is the changing ratio of cytokinin to auxin that influences whether or not a cell will become determined to undergo a coordinated series of cell divisions followed by differentiation to form shoots, roots or unorganized callus.

Cultures can be incubated on an open laboratory bench with or without cool white fluorescent lamps or in a lighted growth chamber. Although light is required, the quantity of light and photoperiod are not critical. Results will be evident within 2 weeks, and the experiment will be completed in 4–8 weeks.

### Materials

Items needed for each student or group of students:

* One or two stock plants of each cultivar
* 10% bleach solution with Tween 20
* Four 100 × 15 mm petri dishes per petunia cultivar containing medium supplemented with 0.5 mg/L (2.5 μM) benzyladenine (BA, a cytokinin)
* Four 100 × 15 mm petri dishes per petunia cultivar containing medium supplemented with 1.5 mg/L (7.5 μM) indolebutyric acid (IBA, an auxin).
* Four 100 × 15 mm petri dishes per petunia cultivar containing medium supplemented with 0.5 mg/L (2.5 μM) BA and 1.0 mg/L (5 μM) IBA

### Procedure

Remove five or six of the largest, fully expanded leaves from a stock plant of each cultivar and place into an empty petri dish. Fill the dishes with the bleach/Tween 20 solution and cover with the lids; swirl occasionally for 10 min to surface disinfest. Transfer leaves to sterile petri dishes that contain sterile deionized or distilled water. Cover and swirl occasionally for 5 min to remove the bleach solution. Repeat this rinsing procedure once.

Dissect and discard the edges and tip of a leaf (Figure 15.1) to produce a strip that is 1–1.5 cm wide including the midvein. Cut across the leaf and midvein to yield uniform 0.5-cm explants. Be careful not to injure explants with the tips of the forceps. The midvein of the leaf should be positioned across the width of the explant, near the middle. Place two explants, right side up (adaxial side up), on the surface of medium and wrap each petri dish with parafilm or cling-type plastic food wrap. Label every petri dish and place on the top of a laboratory bench or in a growth chamber. Cultures should be incubated in the light, with cool-white fluorescent lamps being best. Light quantity and photoperiod should be uniform for all treatments. This experiment can be arranged as a completely random design with four replications of two samples (explants) per replication.

Collect data on root, shoot and callus formation. Count the number of roots and measure the length of each. Be careful not to confuse root hairs (looks like white fuzz) with fungal contamination. Shoots (stems with leaves) will first appear as bumps (buds) on the edges of the leaf explants near the midveins. They may form directly from the explant tissue or from callus. When leaves first begin to enlarge, distinguishing between stems and leaves can be difficult. Wait until the stems elongate and count the number of shoots that are at least 1 cm long. Callus formation is probably best evaluated using a relative rating scale (i.e., 0 = no callus; 5 = most callus). Look along the veins for the initial formation of callus.

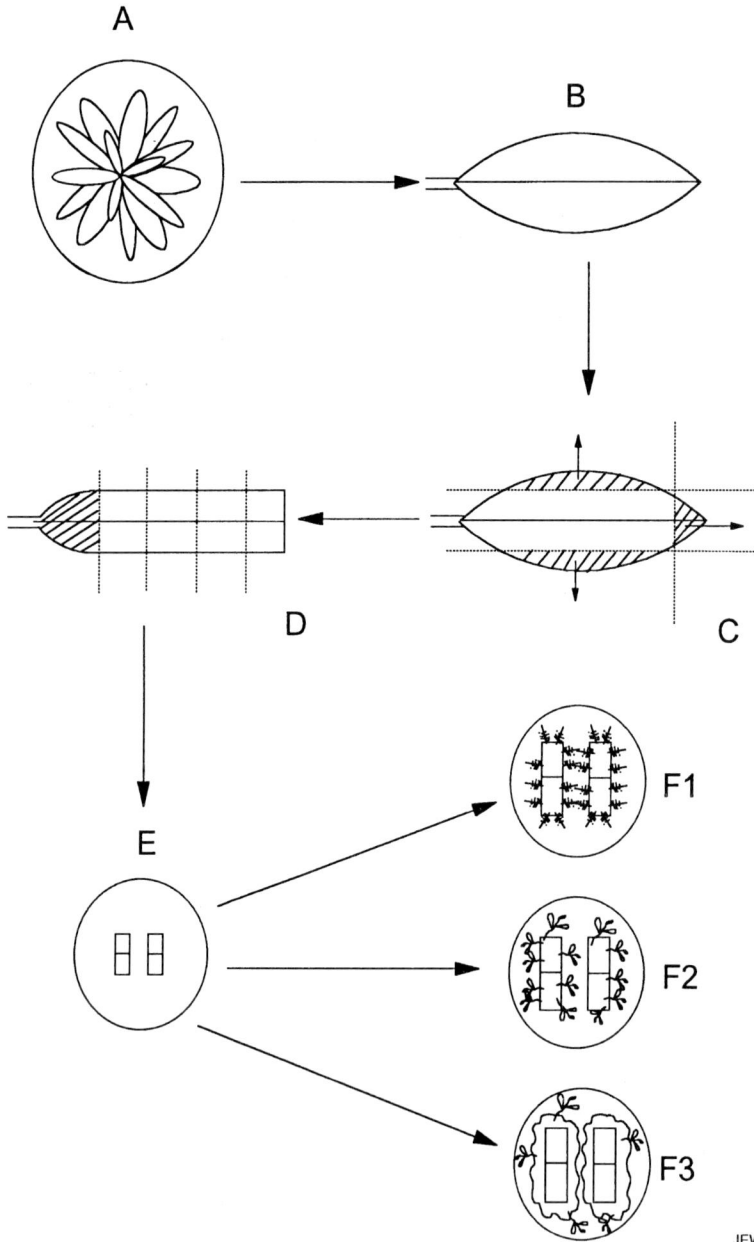

**Figure 15.1** Diagrammatic representation of explant preparation and expected results for *Petunia* experiments. A. Stock plant that is sufficiently mature for collection of leaves for explants. B. Excised leaf. C. Cut away and discard edges and tip of leaf. D. Cut explants so that they are 0.5 cm wide and 1.5 cm long including the midvein going across the width. E. Place two explants of each cultivar into each petri dish. F1. When the medium contains only auxin, only roots should form; F2, when the medium contains only BA, only shoots should form; F3, when the medium contains both BA and IBA, the most callus should form. Some shoots may form from the callus.

### Anticipated results

Most contamination will have appeared after about 1 week of culture. At about this time the first roots will become visible from leaves cultured on medium containing 7.5 μM IBA only. After about 3 weeks, the roots will be well developed and data collection on roots can be terminated. Also after 3 weeks, buds and perhaps small shoots will be visible, especially when the medium contains 2.5 μM BA only. Callus will become visible 7–10 days after culture initiation and will continue to grow regardless of the treatment. The most callus and a few adventitious shoots will form when the medium contains both BA and IBA. Some of this callus can be excised and used in the next experiment. Some roots may develop from adventitious shoots because the shoot tips will be producing hormonal auxin.

### Questions

* What are the roles of auxin and cytokinin in the formation of plant organs?
* In the context of this experiment, why do plants develop roots and shoots instead of simply growing as large masses of callus?
* How can information gained from this exercise be extrapolated to the process of wound healing in plants?
* Why is it important to have major veins in the leaf explants?

## Experiment 2. Regeneration of petunia from callus and monitoring somaclonal variation

This exercise is designed for the student to compare direct and indirect organogenesis and to monitor the performance of the resulting plants. Petunia plants that have been regenerated from callus have demonstrated somaclonal variation (Lewis-Smith et al., 1990). This exercise will take 8–16 weeks to complete depending on how long the plants are monitored in the greenhouse.

### Materials

Items needed for each student or group of students:

* All material from experiment 1 will be needed except for the treatment media
* Four 100 × 15 mm petri dishes per petunia cultivar containing MS medium supplemented with 2.5 μM BA, 1.1 mg/L (5.0 μM) NAA and 0.5 mg/L (0.5 μM) 2,4-dichlorophenoxyacetic acid (2,4-D) (after Binding and Krumbegel-Schroeren, 1984)
* Eight 100 × 15 mm petri dishes per petunia cultivar containing MS medium supplemented with 2.5 μM BA

### Procedure

Prepare explants as described under experiment 1 and place on MS medium supplemented with 2.5 μM BA, 5.0 μM NAA and 0.5 μM 2,4-D. Alternatively, callus tissue can be excised after 3 weeks from the experiment 1 explants growing on the medium containing both BA and IBA. Regardless of the source, after 3–4 weeks excise pieces of callus that are approximately 4 mm$^3$ and place on MS medium with 2.5 μM BA. At this time prepare new leaf explants and place two per petri dish on MS medium containing 2.5 μM BA. After an additional 4 weeks, excise shoots that formed from callus and from leaf explants, but keep shoots from the treatments separate. Shoots can be rooted in vitro on MS without PGRs or under intermittent mist in vermiculite in a greenhouse. After the shoots are rooted, transplant plantlets as described for the original seedlings. Gradually acclimatize plantlets to the greenhouse environment. Compare the growth of plantlets to that of the original stock plants. After a time, the plants will be sufficiently large to be transplanted into 15-cm- (6-in.)-tall pots containing the same soilless medium as before.

This experiment can be arranged as a completely randomized design with four replications, each consisting of two explants, in the laboratory. In the greenhouse, a randomized complete block design is best. The number of replications will depend on the number of plants that are acclimatized. Blocking should be done with consideration of potential temperature gradients within the greenhouse.

Collect data each week on number and length of shoots that form from callus and leaf explants. Measure initial size of acclimatized plantlets in the greenhouse including width and length of fully expanded leaves. Also note if leaves appear normal or abnormal in any way. Measure differences in flower corolla length, width across the tops of mature flowers, color, number of stamens, and note any abnormalities. Determine plant height and number and length of branches.

### Anticipated results

Callus should grow from the original leaf explants placed on the medium with BA, NAA and 2,4-D. Shoots should form directly from leaf explants and indirectly from the callus when they are cultured on medium containing only BA. Acclimatized plants should be obtained from both the leaf explants and the callus. There is a much greater chance of variability among the plants regenerated from callus compared to the plants regenerated directly from the leaf explants or the original stock plants.

### Questions

* What are the disadvantages and/or opportunities that might be presented to the breeder if shoots are regenerated indirectly?
* If people are interested in clonal micropropagation, how might they avoid somaclonal variation?
* Might somaclonal variation be a problem for genetic transformation?
* How might the variation be increased among the plants regenerated from tissue cultures?

## Literature cited

Binding, H. and G. Krumbegel-Schroeren. 1984. Protoplast regeneration. In: *Petunia*. K.C. Sink, Ed., pp. 123–132. Springer-Verlag, Berlin.

Lewis-Smith, A.C., M. Chamberlain and S.M. Smith. 1990. Genetic and chromosomal variation in *Petunia hybrida* plants regenerated from protoplast and callus cultures. *Biol. Plant.* 32:247–255.

Liberty Hyde Bailey Hortorium. 1976. *Hortus Third: A Concise Dictionary of Plants Cultivated in the United States and Canada.* 3rd ed. Macmillan, New York.

Murashige, T. and F. Skoog. 1962. A revised medium for rapid growth and bioassays with tobacco tissue cultures. *Physiol. Plant.* 15:473–497.

*chapter sixteen*

# Nonzygotic embryogenesis

*Dennis J. Gray*

An embryo can be defined as the earliest recognizable multicellular stage of an individual that occurs before it has developed the structures or organs characteristic of a given species. In most organisms, the embryo is a morphologically distinct entity that functions as an intermediate stage in the transition between the gametophytic-to-sporophytic life cycle (Figure 16.1). For example, in higher plants, we are most familiar with embryos that develop within seeds; such embryos usually arise from gametic fusion products (zygotes) following sexual reproduction and are termed zygotic embryos, although seed-borne embryos also can develop apomictically (i.e., without sexual reproduction). However, plants are unique in that morphologically and functionally correct nonzygotic embryos also can arise from widely disparate cell and tissue types at a number of different points from both the gametophytic and sporophytic phases of the life cycle (Figure 16.2).

The first demonstration that plants could produce nonzygotic embryos in vitro was published in 1958 by Steward et al. Subsequently, Reinert (1959) observed bipolar embryos to differentiate in a culture of carrot roots after transfer from one medium to another. While carrot was the first species in which in vitro nonzygotic embryogenesis was reported, in subsequent years many species of angiosperms and gymnosperms have been added to the list of successes. In fact, demonstrations of nonzygotic embryogenesis are so widespread that it may be regarded to be a universal capability of higher plants.

A plethora of terminology has arisen to designate nonzygotic embryos. Such embryos originally were termed *embryoids* to denote perceived significant differences from zygotic embryos, and, unfortunately, this term persists in some literature. However, differences in embryogenic cell origins notwithstanding, distinctions between zygotic and nonzygotic embryos become blurred as our understanding of embryo development increases. Nonzygotic embryos are now shown to be functionally equivalent to zygotic embryos, and the suffix "oid" should be dropped. Other terms for nonzygotic embryos are based primarily upon differences in their specific sites of origin and often are interchangeable, leading to some inconsistencies in the literature. For example, nonzygotic embryos can arise from plant vegetative cells, reproductive tissues, zygotic embryos or callus cells derived from any of these. Thus, "somatic embryos" grow from somatic cells; "haploid or pollen embryos" are derived from pollen grains or microspore mother cells; "nucellar embryos" are formed from nonzygotic nucellar seed tissue; "direct secondary embryos" develop from previously formed embryos; etc.

Most of these embryo types share the common trait of being able to be manipulated via in vitro culture. Such embryogenic culture systems form the basis for many biotechnological approaches to plant improvement, since they allow not only clonal plant propagation, but also specific and directed changes to be introduced into desirable, elite individuals by genetic engineering of somatic cells. Modified individual cells and/or embryos then

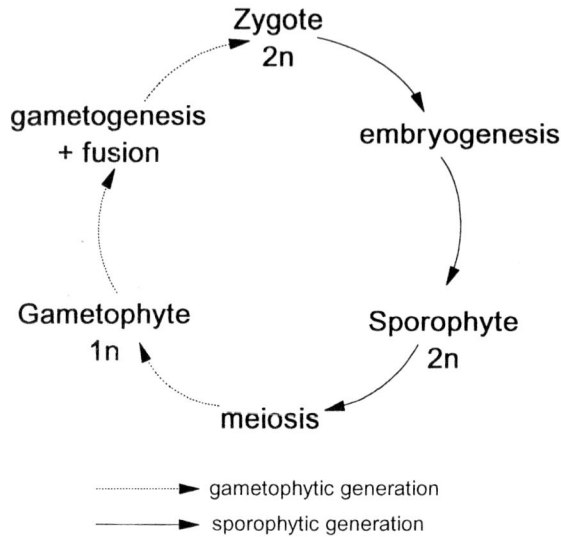

**Figure 16.1**  Typical angiosperm life cycle showing the natural role of embryogenesis in sporophyte development.

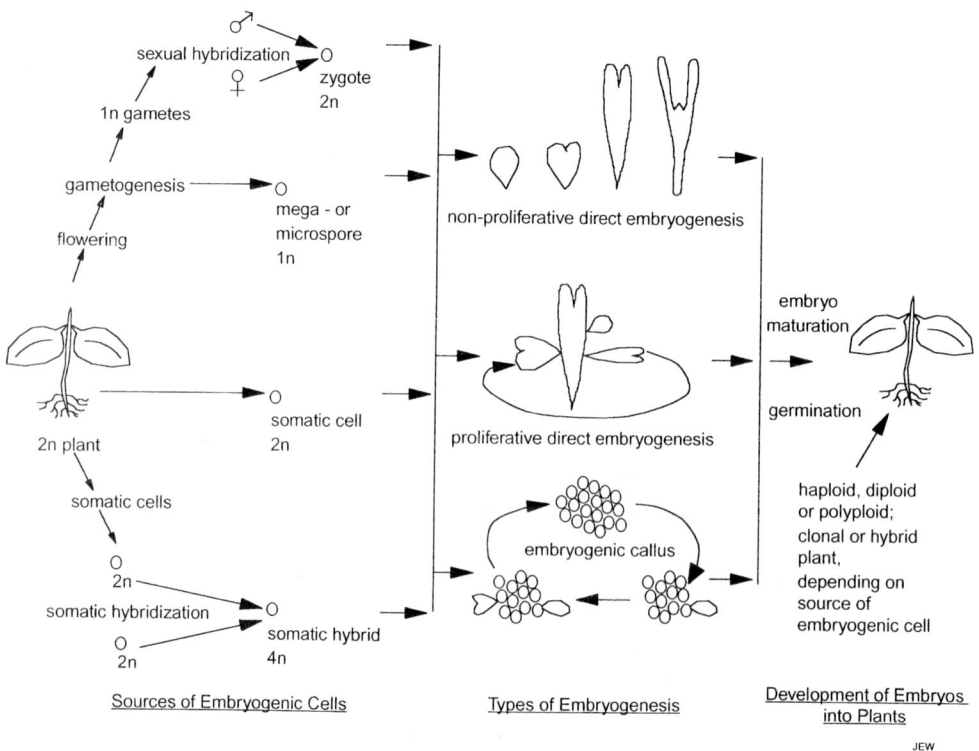

**Figure 16.2**  Sources of embryogenic cells, types of embryogenesis and plants obtained via in vitro culture of higher plants.

A.            B.

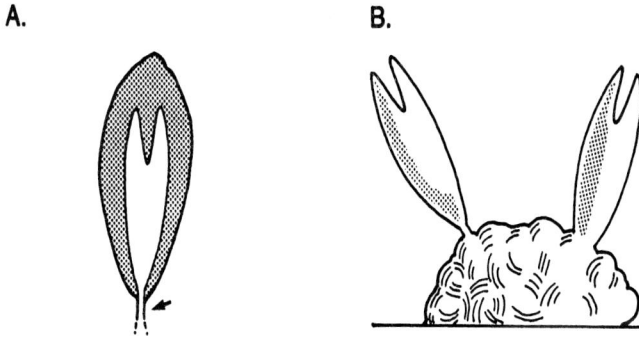

**Figure 16.3** Comparison of typical zygotic embryo development in seed (A) with nonzygotic embryo development from callus (B). Zygotic embryos typically develop within nutritive seed tissues (shaded area) and are connected to the mother plant by a suspensor (arrow). In contrast, nonzygotic embryos often develop perched above subtending tissue; the only source of nutrition being the narrowed suspensor. (From Gray, D.J. and A. Purohit, *Crit. Rev. Plant Sci.* 10:33, 1991.

can be efficiently multiplied in vitro to very high numbers prior to plant development. This approach to genetic improvement bypasses the unwanted consequences of sexual reproduction (mass genetic recombination and required cycles of selection) inherent in conventional breeding technology.

Because of the many commonalities exhibited by these embryo types as well as the various methods available to study and manipulate them, this chapter will, for convenience, only utilize the term *nonzygotic embryo* to designate the embryos that develop through in vitro culture, regardless of origin. Thus, the developmental processes and experimental procedures described will tend to be applicable to many of the embryo types mentioned above.

## In vivo vs. in vitro growth conditions

In seeds, nutritive tissues directly encase the developing embryo. Early in embryogenesis, nutritive substances enter the zygotic embryo either through the suspensor or the embryo body (Figure 16.3A). The relative reliance of the developing embryo on obtaining nutrition via the suspensor vs. endosperm differs greatly depending on the species. In contrast, nonzygotic embryos develop naked, not encased in seed endosperm and, as such, are not subjected to specialized and highly regulated nutritional regimes (Gray and Purohit, 1991). Often, a suspensor is the only link between the embryo and growth medium (Figure 16.3B). This demonstrates that the suspensor can serve as the pathway for all needed nutrition and that endosperm is not absolutely necessary for embryogenesis and germination to occur.

## Common attributes of embryogenic culture protocols

Given the range of different explants, culture conditions and media used to initiate and maintain embryogenic cultures, certain generalities can be made regarding basic methodology. The use of specific plant growth regulators (PGRs), which are required in most instances, will be discussed below with regard to embryogenic cell initiation.

The choice of genotype and explant often is crucial in obtaining an embryogenic response (see Gray, 1990; Gray and Meredith, 1992 for reviews). For example, with corn, only a limited range of genotypes are capable of producing embryogenic cultures, and, with few exceptions, only immature embryo explants can be used. With corn, the age of immature embryos also is important, since embryos that are too young do not survive culture and those that are too old do not produce embryogenic callus. Interestingly, most species tested to date can be induced to produce embryogenic cultures from at least some genotypes and tissues. Additionally, the embryogenic response is heritable, such that it can be bred from embryogenic lines into nonembryogenic lines, albeit with a certain degree of difficulty and transfer of other, often undesirable, traits via sexual hybridization.

Environmental requirements for optimum culture growth often are quite specific, but not unusual. Cultures may require growth in either dark or light or a combination of both over time, the optimization of which can be experimentally quantified. Culture in dark may be necessary in order to prevent the triggering effect of light on many plant biological processes that may adversely affect the growth of embryogenic cell populations. Culture in dark suppresses unwanted tissue differentiation in explant tissue, for example, by limiting the development of plastids into chloroplasts. Similarly, dark conditions may inhibit precocious germination of young embryos. Temperature requirements also are specific, but tend to be in the range of "room temperature" (i.e., 23–27 C). Despite the similarity in culture requirements, improvements in nonzygotic embryo development and maturation increasingly are being obtained by optimization of medium composition.

## Origin of nonzygotic embryos

It is generally accepted that, like zygotic embryos, nonzygotic embryos arise from a single cell, in contrast to budding from a cell mass. This distinction is important in considering efficient genetic engineering, since modification of a single embryogenic cell might eventually result in a modified plant, compared to genetic modification of a cell within a bud, which would result in a chimeric plant.

Nonzygotic embryos growing from isolated cells, such as microspores or protoplasts, clearly develop from single cells. However, the origin of nonzygotic embryos that develop from complex intact primary explant tissue or callus is more difficult to resolve, since the action of microscopic single cells cannot be readily followed. Nonzygotic embryos often develop with a well-defined suspensor apparatus identical to that of zygotic embryos, which somewhat suggests a single cell origin, whereas others can develop with a broad basal attachment, suggesting a multicellular budding phenomenon (Figure 16.4).

## Initiation of embryogenic cells

In complex explants, nonzygotic embryos typically can be initiated only from the more juvenile or meristematic tissues. For example, immature zygotic embryos or zygotic embryo cotyledons and hypocotyls dissected from ungerminated seeds are commonly used explants. The young leaves, shoot tips or even roots of established plants sometimes are used to initiate embryogenic cultures. However, the explant response is highly genotype dependent, so that, for any given species, only a certain type or range of explant can be used to initiate embryogenic cultures.

There are several pathways by which nonzygotic plant cells become embryo initials (Christensen, 1985). In instances where the explant consists of undifferentiated embryonic tissue, such as an immature zygotic embryo, initiation and maintenance of an embryogenic callus is akin to culturing and propagating a preexisting proembryonal complex. Thus, the embryogenic cells present in explant tissue prior to culture initiation are simply propagated and otherwise manipulated in vitro. However, in many instances, embryogenic cells are induced from nonembryogenic cells; this represents a dramatic change in their presumptive fate. This shift in developmental pattern involves a dedifferentiation away from the "normal" fate of the cell followed by redetermination toward an embryogenic cell type.

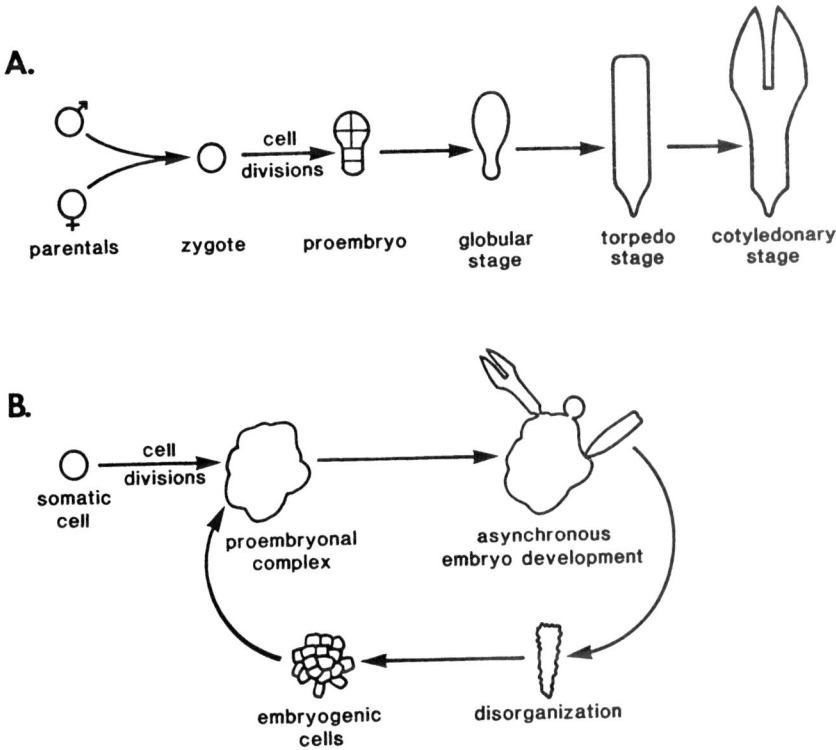

**Figure 16.4**  Comparison of typical zygotic embryogenesis occurring within seeds (A) with proliferative nonzygotic embryogenesis occurring in vitro (B). Zygotic embryogenesis is characterized as
being very regulated, such that embryos synchronously pass through distinct developmental stages,
whereas nonzygotic embryogenesis often is non-uniform, with many stages present at a given time.
Nonzygotic embryos may bypass maturation and become disorganized, adding to the proembryonic
tissue mass. (From Gray, D.J. and A. Purohit, *Crit. Rev. Plant Sci.* 10:33, 1991.)

For example, cells in leaf explants, which normally would develop into constituents of
relatively short-lived parenchymatous tissue, instead become embryogenic under certain
conditions. This is a pivotal change in development, since cells that normally would be
capable of only a few divisions, at most, before senescence, instead become redirected to
become totipotent and capable of possibly unlimited divisions. Such embryogenic cells
become immortal in the sense that they reinstitute the germ line by being capable of
developing into mature, reproductive individuals.

The fact that isolated somatic cells can develop normally into embryos irrevocably
demonstrates that the developmental program for embryogenesis is contained within and
controlled by the cell itself and not by external factors. However, the exact nature of the
triggering mechanism(s) of embryogenesis, whether it be a physical, biochemical and/or
genetic event(s), is unknown. Early attempts to identify genes that became activated as a
direct consequence of embryogenic cell induction failed because of inadequate understanding of the basic induction phenomenon, which resulted in faulty experimental designs. Since the cell cultures used were already induced, any genes critical to the induction
step already had been activated. Hence, the genes that were observed in such studies
actually were those activated later in embryogenesis, such that they were controlling
downstream developmental processes and were not related to the triggering event. Despite the technical difficulty of identifying critical induction-controlling genes, a number
of other genes and their products have been identified during nonzygotic embryo devel-

opment (e.g., Zimmerman, 1993). In some instances, the genetic mechanisms resolved for nonzygotic embryos can be related directly to that of the corresponding zygotic embryo.

## Inductive PGRs

In practice, the initiation of embryogenic cells requires in vitro culture of the appropriate explant on or in a medium that contains specific PGRs. In fact, a preponderance of reports of embryogenic culture initiation employed a very narrow range of PGRs that typically are added to culture medium. Synthetic auxins, notably 2,4-dichlorophenoxyacetic acid (2,4-D), are added to medium in the predominance of reported protocols. Similar auxins used include dicamba, indolebutyric acid (IBA), naphthoxyacetic acid (NOA), picloram and others. In addition, several weaker auxins, such as indoleacetic acid (IAA), a natural plant hormone, and napthyleneacetic acid (NAA) have been utilized in a few culture systems. Auxins serve to induce the formation of embryogenic cells, possibly by initiating differential gene activation, as noted above, and also appear to promote increase of embryogenic cell populations through repetitive cell division, while simultaneously suppressing cell differentiation and growth into embryos. However, an auxin often is not required in instances where the explant consists of preexisting embryogenic cells, as described above, because a discrete induction step then is not required.

In addition to auxinlike PGRs, cytokinins also are required to induce embryogenesis in many dicotyledonous species. In a few instances, only a cytokinin is required to cause embryogenic cultures to develop. The most commonly used cytokinin is benzyladenine (BA), but others, such as thidiazuron (TDZ) and kinetin, and the natural cytokinin, zeatin, also are utilized.

Actual PGR concentration is important for an optimum response, since concentrations that are too low may not trigger the inductive event and concentrations that are too high, particularly when considering phenoxy-auxins, may become toxic. Typically, following the induction period, resulting culture material is transferred to medium lacking PGRs, which removes the auxin-induced suppression of embryo development and allows embryogenesis to occur; however, not all culture systems require this two-step procedure. For example, as mentioned above, certain species do not require any induction step at all, and others, particularly poaceous monocots, become induced and undergo complete embryogenesis in the continuous presence of auxin. Examples of different PGR regimes used in culture systems are illustrated via the laboratory exercises provided after this chapter.

## Embryo development

The physical, observable transition from a nonembryogenic to an embryogenic cell appears to occur when the progenitor cell undergoes an unequal division, resulting in a larger vacuolate cell and a smaller, densely cytoplasmic (embryogenic) cell. Embryogenic cells are readily distinguished by their small size, isodiametric shape and densely cytoplasmic appearance (Figure 16.5). This type of unequal cell division is identical to that observed in zygotes and may be an early indication of developmental polarity. The embryogenic cell then either continues to divide irregularly to form a proembryonal complex or divides in a highly organized manner to form a somatic embryo (Figs. 16.2 and 16.4). However, an often undesirable difference exhibited by nonzygotic embryos is that they frequently deviate from the normal pattern of development, by producing callus, undergoing direct secondary embryogenesis and/or germinating precociously. This tendency toward erratic development clearly is due to environmental factors as discussed below.

**Figure 16.5** Typical embryogenic cells of grape. Note the three-cell stage embryo in the upper center part of the cell mass. (From Gray, D.J., *Somatic Embryogenesis in Woody Plants*, Kluwer, 1995. Reprinted by permission of Kluwer Academic Publishers.)

**Figure 16.6** Somatic embryos of grape at heart and early cotyledonary stages, illustrating morphologies highly faithful to those of zygotic embryos.

Zygotic and nonzygotic embryos share the same gross pattern of development, with both typically passing through globular, scutellar and coleoptilar stages for monocots or globular, heart, torpedo and cotyledonary stages for dicots and conifers. Generally, the anatomy and morphology of well-developed nonzygotic embryos is faithful to the corresponding zygotic embryo type such that they easily can be identified by eye or with the aid of a stereomicroscope (Figure 16.6). For example, nonzygotic embryos of grass and cereal species typically possess a scutellum, coleoptile and embryo axis, which are distinctive embryonic organs of monocots (Figure 16.7A). Embryos of dicotyledonous species have a distinct hypocotyl and cotyledons (usually two) (Figure 16.7B); those of conifers (see chapter twenty-two) also exhibit a hypocotyl and numerous cotyledons (Figure 16.7C), the number of which is similar to that of the given species.

Embryo development occurs through an exceptionally organized sequence of cell division, enlargement and differentiation. During early development the embryo assumes

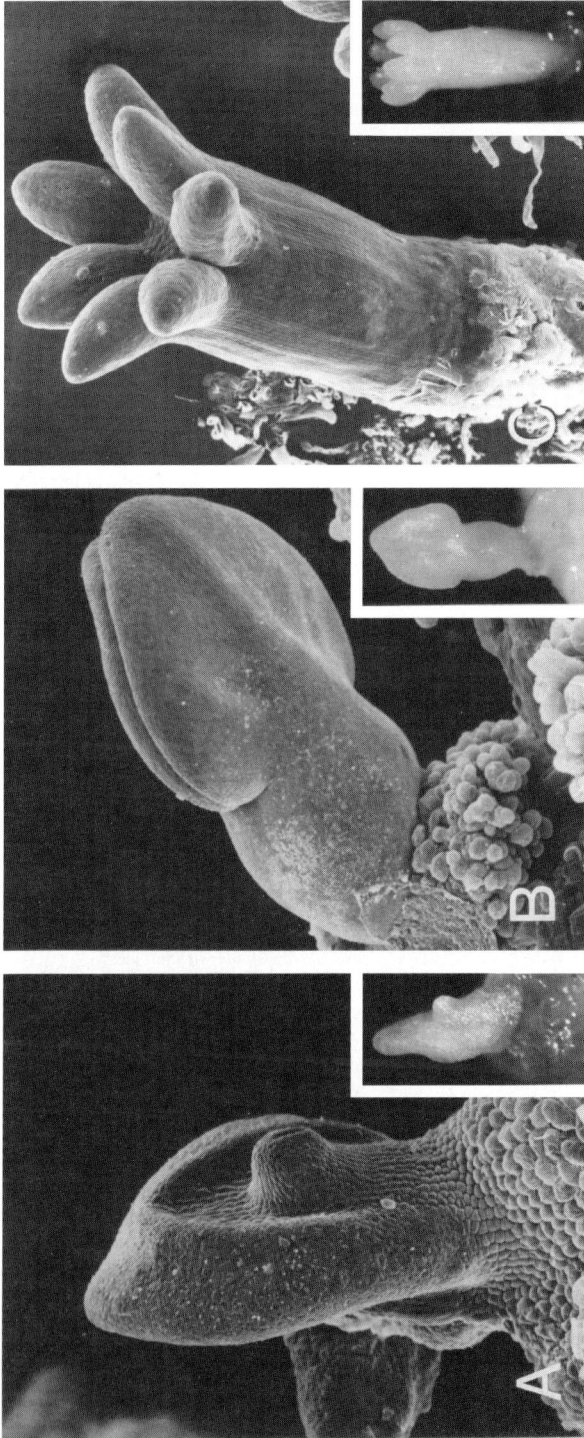

**Figure 16.7** Comparison of monocot (A), dicot (B) and gymnosperm (C) nonzygotic embryos. Scanning electron micrographs, with corresponding stereomicrographs (inset). A. Somatic embryo of orchardgrass growing from an embryogenic callus. Note the protruding coleoptile and notch, through which the first leaves will emerge after germination. The prominent large flattened body of the embryo is the scutellum. (From Gray, D.J. et al., *Protoplasma.* 122:196, 1984. With permission.) B. Somatic embryo of grape growing from an embryogenic callus. Note two flattened cotyledons and subtending hypocotyl (From Gray, D.J., *Somatic Embryogenesis in Woody Plants,* Kluwer, 1995. Reprinted by permission of Kluwer Academic Publishers.) C. Somatic embryo of Norway spruce. Note multiple cotyledons and elongated hypocotyl. (From Fowke, L.C. et al., *Plant Cell Rep.* 13:612, 1994. With permission.)

**Figure 16.8** Longitudinal section through a grape somatic embryo showing typical vascular system. (From Gray, D.J., *Somatic Embryogenesis in Woody Plants*, Kluwer, 1995. Reprinted by permission of Kluwer Academic Publishers.)

a clavate, globular shape and remains, essentially, an undifferentiated, but organized mass of dividing cells with a well-defined epidermis. Subsequent heart through early torpedo stages are characterized by cell differentiation and polarized growth, notably elongation and initiation of rudimentary cotyledons in dicots (Figure 16.6) and development of the scutellum with initiation of the coleoptilar notch in poaceous monocots. At the same time, obvious tissue differentiation begins with the development of embryonic vasculature (Figure 16.8) and accumulation of intracellular storage substances. Final stages of development toward maturation are distinguished by overall enlargement, increase in cotyledon size in dicots (Figure 16.9) and coleoptilar enlargement in monocots. At the same time, the embryonic axis becomes increasingly developed. In dicots, the root apical meristem becomes well established, embedded in tissue located above the suspensor apparatus and at the base of the hypocotyl, whereas the shoot apical meristem develops externally between the cotyledons. In monocots, the embryo axis develops laterally and parallel to the scutellum. The root apical meristem is embedded, and the shoot apical meristem develops externally but is protected by the coleoptile. All of the events described above occur in concert with each other in a manner that is essentially identical to that of zygotic embryos. But, for a number of reasons, as described below, nonzygotic embryos often differ somewhat from their zygotic counterparts in morphology and/or performance.

An obvious difference in gross morphology between nonzygotic embryos growing in vitro and zygotic embryos in seeds is simply caused by the physical constraint on zygotic embryos imposed by the developing seed coat, often causing them to become compressed and/or flattened into a shape and size distinct for a given species or variety. This is made apparent by comparing the morphology of seed-borne zygotic embryos with corresponding nonzygotic embryos of a given species (Figure 16.10). Zygotic embryos excised from seed typically exhibit a highly compressed shape because the embryos become highly flattened during development. In contrast, nonzygotic embryos tend to be larger and have wider hypocotyls and fleshier cotyledons. It is possible that pressure exerted by the seed coat contributes to other aspects of embryo development that are lacking during nonzygotic embryogenesis (Gray and Purohit, 1991).

**Figure 16.9**  Cotyledonary-stage somatic embryos of cantaloupe growing from a cultured cotyledon. Note that the fine suspensor is the only connection to explant. (From Gray, D.J. et al., *J. Am. Soc. Hort. Sci.* 118:425, 1993).

**Figure 16.10**  Comparison of grape zygotic embryo, compressed and flattened by development within a seed (left), with grape somatic embryo, which is not flattened (right). (From Gray, D.J. and A. Purohit, *Crit. Rev. Plant Sci.* 10:33, 1991.)

In addition to differences in development related to simple physical constraints, for several reasons significantly more instances of abnormal development are known to occur during nonzygotic embryogenesis when compared to zygotic embryogenesis. For example, as mentioned above, species that normally produce zygotic embryos with suspensors often produce clusters of nonzygotic embryos from a proembryonal cell complex (Figure 16.4). This basic change in developmental pattern is likely due to differences between the seed and in vitro environments, since immature zygotic embryos dissected from seeds and cultured also often develop abnormally (Gray and Purohit, 1991).

**Figure 16.11** Asynchronous and abnormal somatic embryo development from embryogenic callus of grape. Increasing levels of development (clockwise from lower right globular embryo) are evident in a single group of embryos. In addition, the normal cotyledon number of two (lower center embryo) contrasts to supernumerary cotyledon development of three and four on the two upper embryos. (From Gray, D.J. and J.A. Mortensen, *Plant Cell Tissue Organ Cult.* 9:73, 1987. Reprinted by permission of Kluwer Academic Publishers.)

Nonzygotic embryos growing in mass from proembryonal complexes tend to develop asynchronously so that several stages are present in cultures at any given time (Figures 16.2 and 16.4). Nonzygotic embryos initiated over time are subjected to different nutrient regimes as medium becomes depleted then replenished between and during subcultures. This leads to differences in development even among embryos from a single culture. With such variable and unregulated environmental conditions, nonzygotic embryos often bypass maturation altogether, becoming disorganized, forming new embryogenic cells and contributing to asynchrony (Figure 16.4). Nonzygotic embryos also often exhibit structural anomalies such as extra cotyledons (Figure 16.11) and poorly developed apical meristems.

## Embryo maturation

Maturation is the terminal event of embryogenesis and is characterized by the attainment of mature embryo morphology, accumulation of storage carbohydrates, lipids and proteins (see prior section regarding in vivo vs. in vitro growth conditions), reduction in water content and, often, a gradual decline or cessation of metabolism. Nonzygotic embryos typically do not mature properly when compared to zygotic embryos. In fact, in the preponderance of instances, rapid growth continues to occur, leading to precocious germination.

Although complete maturation is not absolutely necessary in order to obtain plants from nonzygotic embryos, it is required to achieve high rates of plant recovery. As such, factors that influence and/or enhance nonzygotic embryo maturation have been explored.

A number of culture medium components have been shown to promote maturation. In particular, high-sucrose levels (9–12%), timed pulses of abscisic acid (ABA), a naturally occurring plant hormone, and polyethylene glycol (PEG), an osmotically active compound, in conjunction with certain amino acids, notably glutamine, have been remarkably successful. For example, an 8-week treatment with ABA and PEG caused white spruce nonzygotic embryos to accumulate more lipids and to better resemble zygotic embryos morphologically (Attree and Fowke, 1993). Similarly, a short pulse of ABA in conjunction with glutamine resulted in alfalfa nonzygotic embryos that could withstand dehydration as mentioned below (Senaratna et al., 1989)

## Quiescence and dormancy

Perhaps the most obvious developmental difference between zygotic and nonzygotic embryos is that the latter lacks a quiescent resting phase. By comparison, zygotic embryos of many species begin a resting period during seed maturation. Zygotic embryo quiescence in conjunction with the protective and nutritive tissues that comprise a seed are the major factors allowing seeds to be conveniently stored. Thus, the ability to withstand dehydration appears to be a normal step in embryo development.

However, nonzygotic embryos tend either to grow and germinate without normal maturation, become disorganized into embryogenic tissue, or die. They rarely enter a resting stage. Although quiescence and dormancy have only recently been documented in somatic embryos, it is possible that dormancy occurs in instances where plants cannot be obtained from well-developed embryos (see Gray, 1986; Gray and Purohit, 1991 for reviews).

## Embryo germination and plant development

Obtaining plants from nonzygotic embryos often is more difficult than would be expected. The early literature concerning first reports of nonzygotic embryogenesis for a given species or cultivar often did not include information on plant recovery. When plants were obtained, the recovery rate was very low or not reported, suggesting that a majority of nonzygotic embryos were too abnormal to germinate. Typically, plant recovery from nonzygotic embryos ranges from 0–50%. This is very low compared with zygotic embryos in commercial seed, for which germination and plant development typically exceeds 90% in soil. In all but a few instances, nonzygotic embryos develop very poorly if at all when planted in soil like seed. Only in recent years has research been conducted to raise plant recovery rates.

Recent progress in culture methodology has resulted in better development and germination of nonzygotic embryos. In particular, pulse treatments with various amino acids, osmotica and PGRs, particularly ABA, have resulted in nonzygotic embryos with better maturation, including the ability to be dehydrated and stored like seeds, as well as improved germination characteristics (Attree and Fowke, 1993). This demonstrates that careful attention to culture conditions and nutrition, especially with regard to timed pulses of certain factors, results in plant recovery rates from nonzygotic embryos equivalent to those of zygotic embryos. It can be concluded that conditions favoring embryo maturation also favor the recovery of plants.

After the event of germination, the embryo begins development into a plant. Typically, the storage reserves present (lipids, proteins and/or starch — depending on the species) become depleted concomitant with the start of increased mitotic activity in shoot and root meristems. Eventually, a young photosynthetically competent plant develops (Figure 16.12), which then can be gradually acclimated to ambient conditions.

**Figure 16.12**   Plant development from a germinated melon somatic embryo. Note shoot and root development. (From Gray, D.J. et al., *J. Am. Soc. Hort. Sci.* 118:425, 1993).

## Uses for embryogenic cultures

Embryogenic cultures systems are used for a number of purposes:

1.  They constitute an important tool for the study of plant development, both due to the unique convenience of in vitro culture over en planta growth and the contrasts that can be drawn from differences between nonzygotic embryogenesis in vitro and zygotic embryogenesis in seeds.
2.  Embryogenic cultures are an efficient vehicle for genetic engineering, since embryogenic cultures often produce many embryos per volume of cell mass, and isolated genes integrated into single embryogenic cells can become incorporated into the genome of the plant that ultimately develops.
3.  Many nonzygotic embryos can be produced from a single desirable plant for efficient clonal propagation.

Certain potential uses of embryogenic systems as research tools have been described or suggested throughout this chapter and an in-depth discussion of genetic engineering applications is presented in chapter twenty-nine. The use of nonzygotic embryogenesis as a cloning vehicle is discussed in more detail below.

### Synthetic seed technology

An active research front has emerged over the last decade with the goal of developing nonzygotic embryogenesis into a commercially useful method of plant propagation. The technology that has emerged is termed *synthetic* (or *artificial*) *seed technology*. A synthetic seed is defined as a somatic embryo that is engineered to be of practical use in commercial plant production (Gray and Purohit, 1991). Applications for synthetic seed vary depending on the relative sophistication of existing production systems for a given crop and the opportunities for improvement. Whether or not a cost advantage results from synthetic seed will ultimately determine its commercial use. For seed-propagated agronomic crops, quiescent, nonzygotic embryos produced in bioreactors and encapsulated in synthetic seed coats will be necessary. For example, certain vegetable crops with expensive-to-produce seeds, such as seedless watermelon, are attractive candidates for synthetic seed technology, since the cost per plant might be reduced.

Similarly, conifers, which are difficult to improve by breeding because of a long life cycle, would benefit from application of synthetic seed technology if elite individuals could be cloned and planted with seed efficiency (Farnum et al., 1983). For vegetatively propagated crops, particularly those with a high per plant value, naked, hand-manipulated, nonquiescent embryos may be cost-effective. For example, many ornamental crops are painstakingly commercially micropropagated in tissue culture via adventitious bud proliferation with the major expense being the labor needed for multiple culture and rooting steps. Substitution of embryogenic culture systems for such crops would greatly reduce labor costs since mass-produced somatic embryos easily could be hand selected and placed directly into planting flats and rooted (Gray and Purohit, 1991).

Ongoing efforts to develop synthetic seed technology constitute one of the most active areas of nonzygotic embryogenesis research. The potential benefits of using "clonal seeds" in agricultural production is great enough to stimulate continued investigation of factors regulating nonzygotic embryo maturation.

## Conclusion

In vitro embryogenesis from nonzygotic cells must be regarded as a universal, albeit unexpected, property of higher plants. When induced from isolated cells, nonzygotic embryogenesis conclusively demonstrates not only cellular totipotency, but the presence of a highly conserved developmental program that is innate to a wide range of cell types. For the student, nonzygotic embryogenesis represents an opportunity to conveniently study in a culture vessel one of the more developmentally complex aspects of the plant life cycle.

Embryogenic cell culture allows the entire genome of a plant to be manipulated with microbiological techniques. Once the cells are proliferated, genetically engineered and/or otherwise managed, intact plants easily can be reconstituted. From such simple manipulation comes numerous possibilities for better understanding and for ultimately improving our use of plants.

## Acknowledgment

This is Florida Agricultural Experiment Station Journal Series No. R-04837.

## Literature cited

Attree, S.M. and L.C. Fowke. 1993. Embryogeny of gymnosperms: advances in synthetic seed technology of conifers. *Plant Cell Tissue Organ Cult.* 35:1–35.

Christensen, M.L. 1985. An embryogenic culture of soybean: towards a general theory of embryogenesis. In: *Tissue Culture in Forestry and Agriculture*, R.R. Henke, K.W. Hughes, M.J. Constantine and A. Hollaender, Eds., pp. 83-103. Plenum Press, New York.

Farnum, P., R. Timmis and J.L. Kulp. 1983. Biotechnology of forest yield. *Science.* 219:694–702.

Fowke, L.C., S.M. Attree and P.J. Rennie. 1994. Scanning electron microscopy of hydrated and desiccated mature somatic embryos and zygotic embryos of white spruce (*Picea glauca* (Moench) Voss.). *Plant Cell Rep.* 13:612–618.

Gray, D.J. 1986. Quiescence in monocotyledonous and dicotyledonous somatic embryos induced by dehydration. In: Proc. Symp. Synthetic Seed Technology for the Mass Cloning of Crop Plants: Problems and Perspectives. *HortScience.* 22:810–814.

Gray, D.J. 1990. Somatic embryogenesis and cell culture in the Poaceae. In: *Biotechnology in Tall Fescue Improvement*, M.J. Kasperbauer, Ed., pp. 25–57. CRC Press, Boca Raton, FL.

Gray, D.J. 1995. Somatic embryogenesis in grape. In: *Somatic Embryogenesis in Woody Plants*, Vol. 2, S.M. Jain, P.K. Gupta and R.J. Newton, Eds. Kluwer Academic Publishers, pp. 191–217.

Gray, D.J., B.V. Conger and G.E. Hanning, 1984. Somatic embryogenesis in suspension and suspension-derived callus cultures of *Dactylis glomerata*. *Protoplasma.* 122:196–202.

Gray, D.J., D.W. McColley and M.E. Compton. 1993. High-frequency somatic embryogenesis from quiescent seed cotyledons of *Cucumis melo* cultivars. *J. Am. Soc. Hort. Sci.* 118:425–432.

Gray, D.J. and C.P. Meredith. 1992. Grape. In: *Biotechnology of Perennial Fruit Crops,* F.A. Hammerschlag and R.E. Litz, Eds. CAB International, Wallingford, UK.

Gray, D.J. and J.A. Mortensen. 1987. Initiation and maintenance of long-term somatic embryogenesis from anthers and ovaries of *Vitis longii* 'Microsperma'. *Plant Cell Tissue Organ Cult.* 9:73–80.

Gray, D.J. and A. Purohit. 1991. Somatic embryogenesis and the development of synthetic seed technology. *Crit. Rev. Plant Sci.* 10:33–61.

Reinert, J. 1959. Ueber die kontrolle der morphogenese und die induktion von adventiveembryonen an Gewebekulturen aus karotten. *Planta.* 53:318–333.

Senaratna, T., B.D. McKersie and S.R. Bowley. 1989. Desiccation tolerance of alfalfa (*Medicago sativa* L.) somatic embryos — influence of abscisic acid, stress pretreatments and drying rates. *Plant Sci.* 65:253–259.

Steward, F.C., M.O. Mapes and K. Mears. 1958. Growth and organized development of cultured cells. II. Organization in cultures grown from freely suspended cells. *Am. J. Bot.* 45:705–708.

Zimmerman, L.J. 1993. Somatic embryogenesis: a model for early development in higher plants. *Plant Cell.* 5:1411–1423.

*chapter seventeen*

# Embryogenic callus and suspension cultures from leaves of orchardgrass

*Dennis J. Gray, Robert N. Trigiano and Bob V. Conger*

Members of the family Poaceae constitute the single most economically important group of plants. This family includes the nutritious cereal grain crops such as corn (*Zea mays* L.), oats (*Avena sativa* L.), rice (*Oryza sativa* L.) and wheat (*Triticum aestivum* L.), as well as the forage grasses such as annual ryegrass (*Lolium multiflorum* Lam.), tall fescue (*Festuca arundinacea* Schreb.) and the wheatgrasses (*Agropyron* spp.). The cereals are important sources of complex carbohydrates, oils and proteins, whereas the forages are important in dairy and meat production. Sugarcane (*Saccharum officinarum* L.), another member of this family, is an important source of sucrose.

Much research emphasis in the Poaceae has gone toward developing biotechnological approaches to crop improvement. One area of biotechnology that received early attention by researchers was that of in vitro plant regeneration. The grasses and cereals were initially regarded by researchers to be difficult or impossible to regenerate from tissue and cell cultures. However, continuous effort since the early 1980s led to significant advancements, such that plant regeneration has been obtained for all grass and cereal species that have been attempted (Gray, 1990).

Somatic embryogenesis is the most common mode of regeneration for species in the Poaceae. Embryogenic culture systems are useful in the classroom, not only for illustrating plant regeneration, but also for studying embryo development and morphology. Since the Poaceae is contained within the monocotyledoneae, a taxonomic division based on embryo morphology, their zygotic embryos are distinctly different from dicotyledonous embryos. In many poaceous plants, the induction of somatic embryogenesis is not simple and reliable enough to be used for classroom exercises. Many species require use of immature zygotic embryos as explants, which can only be obtained by careful cultivation and pollination of source plants. Other species tend to produce somatic embryos that have relatively abnormal morphologies when compared to zygotic embryos. However, a few poaceous species, notably orchardgrass (*Dactylis glomerata* L.), are much easier to manipulate.

Orchardgrass is a perennial cool-season forage species that is grown in temperate regions of the world to produce high-quality hay. It is genetically self-incompatible, which makes breeding of new varieties difficult and time-consuming. The potential of using somatic embryos as synthetic seed to efficiently clone outstanding parental lines holds promise in the development of improved varieties (see Gray et al., 1992 for details). The following laboratory exercises cover the methods used to produce somatic embryos from cultures of orchardgrass. Selected clones of this species are highly embryogenic and, as a perennial, can be maintained in the greenhouse as potted plants. These factors account for its convenient use. Leaf tissues are used as explant material for initiating cultures from which morphologically correct somatic embryos are produced. Orchardgrass has "clasping-type" leaves, which are easier to dissect than "whorled-type" leaves. Thus, the orchardgrass system is ideal for demonstrating monocotyledonous somatic embryogenesis.

The following experiments represent a greatly expanded version of exercises that were originally published in *HortTechnology* (Gray et al., 1994) and are arranged to lead students through successful culture establishment, observations of embryogenic callus formation and somatic embryo development, including the gradient embryogenic response that is typical of this culture system, culture maintenance, plant regeneration, manipulation of liquid suspension cultures and the production of model "synthetic seeds".

## General considerations

### Growth of plants

A special clone of orchardgrass, 'Embryogen-P' (Conger and Hanning, 1991), is the most convenient source of leaf explant tissue and is available from Dr. B. V. Conger, Department of Plant and Soil Science, The University of Tennessee, Knoxville, TN 37901. The clone should be maintained in greenhouse pots in a high-quality potting mix, with regular fertilizer applications, and kept in vigorous growing condition. Before becoming root bound, plant number can be increased by splitting entire plants, including the root ball and transplanting into two or three new pots. Insect and fungal pests should be controlled, however, culture response may be decreased for up to 6 weeks after chemical spraying, especially with systemic fungicides. Therefore, stock plant health and vigor are the most important factors in successful embryogenic culture initiation of orchardgrass.

### Culture media

The culture medium used is Schenk and Hildebrandt (SH) basal salt mixture (see chapter two for composition) and vitamin powder with 6.6 mg/L (30 µM) dicamba (synthetic auxin plant growth regulator [PGR]), 30 g/L (88 mM) sucrose and 0.7% tissue culture–tested agar. This medium is hereafter referred to as SH30, whereas medium used for embryo germination studies and lacking dicamba is designated as SH0. The pH is adjusted to 5.4 before autoclaving. The cooling medium then is poured, 25 ml, into each $100 \times 15$ mm sterile petri dish. SH30-C is a liquid medium (i.e., without agar) that contains 3 g/L casein acid hydrolysate type 1. The casein hydrolysate is dissolved in 25 ml of distilled water, loaded into a 30-ml syringe and then dispensed through a sterile 0.2-µm syringe filter into cooled autoclaved SH30 liquid medium. The resulting SH30-C then is sterilely pipetted (25 ml) into each 125 ml erlenmeyer flask. Generally, at least four petri dishes or flasks of the appropriate medium are required for each treatment and/or transfer for each student or student team.

## Exercises

### Experiment 1. Culture establishment

Orchardgrass is a tillering-type forage grass in which the plant body consists of dense clumps of tillers. New tillers arise by adventitious budding from the base of previously developing tillers. Each tiller is actually a complete plant in itself, with a root–shoot axis

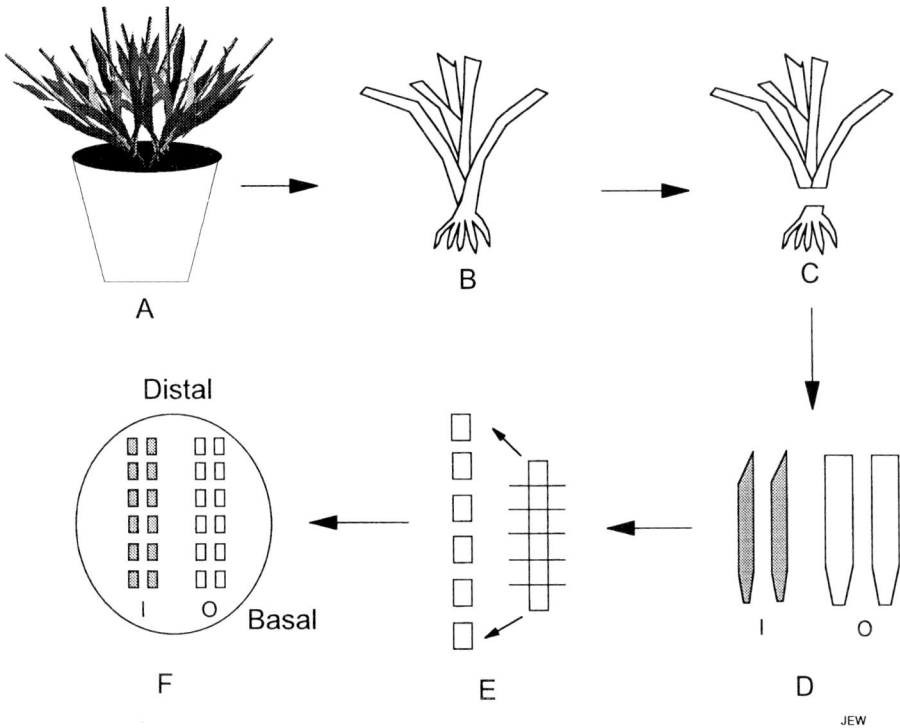

**Figure 17.1** Diagrammatic representation of orchardgrass leaf explant preparation for embryogenic culture initiation. A. Potted plant of orchardgrass variety 'Embryogen-P'. B. An isolated tiller. C. Remove and discard basal stem and roots, leaving only the clasped leaves. D. Carefully separate leaves and retain the innermost and next leaf out. E. Carefully split each leaf longitudinally into two halves. Cut halves of innermost leaf at an angle to retain its identity, then surface disinfest. F. Cut each leaf half into sections and plate sections of all four halves on a single petri dish containing SH30 medium. Be careful to retain the identity of each leaf and to keep the sections within each leaf in order of most basal to distal section.

and several flattened, interfolded clasping leaves. Grass leaves grow from a pronounced basal (intercalary) meristem, such that the youngest part of a given leaf is the portion nearest to the shoot apical meristem. Youngest leaves are those nearest to the center. Thus, the youngest, most meristematically active leaf tissues are those nearest to the bases of the innermost leaves. Figure 17.1 summarizes the steps of this experiment.

*Materials*
The following items are needed for each student or team:

* 100 ml of 50% commercial bleach with a drop of surfactant (such as Triton-X100™, Tween 20™ or Ivory Liquid™ soap)
* 12 (approximately) $100 \times 15$ mm petri dishes containing SH30 medium
* Stereomicroscope with light source

*Procedure*
To initiate cultures, remove individual tillers from a rapidly growing plant. Dissect individual leaves carefully so as to prevent damage and identify the innermost two leaves. Remove the basal-most 3 cm of these leaves and separate the two halves of each

longitudinally. Each leaf base should yield two strips of blade tissue approximately 3 cm long by 4 mm wide. Place bleach solution into a sterile beaker and prepare three additional beakers with sterile distilled water for rinsing. Surface disinfest the leaf halves by gently agitating in the bleach solution for 2 min. Sterile forceps may be used to accomplish agitation. Gently rinse the leaf halves three times in sterile water, then remove one leaf half at a time and cut on a hard sterile surface (such as a petri dish) with a fresh, sharp scalpel blade. Cut the basal-most part of each leaf half crosswise into six approximately square sections (approximate size = 5 × 4 mm) and discard the remaining portion of each leaf half. Immediately plate the sections onto SH30 medium. All of the dissection steps should be accomplished as rapidly as possible.

Arrange the sections from each leaf half in separate rows such that basal sections are at one end of a given row and the most distal sections are at the other end. Sections from four leaf halves should be placed in each petri dish. Seal each petri dish securely with several layers of parafilm.

Incubate cultures at 23 C in dark and evaluate daily for bacterial and fungal contamination. Transfer noncontaminated sections to fresh medium as needed. Forceps should be sterilized between moving of each section when transferring noncontaminated sections in order to avoid spreading of inconspicuous contamination to clean sections. Determine the number of sections from each leaf (i.e., innermost vs. outermost) that became contaminated weekly. Chart the relative contamination rate as a function of leaf position in the tiller.

### Anticipated results
A measure of bacterial and fungal contamination growing on the medium or from cultured leaf sections is expected. While up to 50% of sections may become contaminated, a more normal contamination rate is 10%. It is important to follow directions for flaming instruments and manipulating leaf sections during transfer in order to eliminate, rather than spread, contamination.

### Questions

* Discuss the importance of proper "transfer etiquette" in the establishment of cultures.
* Are the same procedures required in the maintenance of established cultures? Why?
* Did younger leaves become more or less contaminated than older leaves over time? Why?

## Experiment 2. The gradient embryogenic response

The embryogenic response in orchardgrass is related to leaf position within a tiller (i.e., the innermost or youngest leaf vs. the next leaf out or older leaf) as well as the position of cultured leaf sections relative to the shoot apical meristem (i.e., proximal or distal). The following two distinct embryogenic responses are possible (Conger et al., 1983): either (1) an embryogenic callus develops or (2) somatic embryos grow directly from the leaf. Embryogenic callus is white to yellowish in color and friable to wet in consistency. Often, adventitious roots develop from the leaf sections, but never lead to plant regeneration. Somatic embryos, which are described in more detail below, appear as small (approximately 1-mm), white, organized structures, either embedded in callus or growing directly from the leaf section surface.

### Materials
As this experiment utilizes the cultures developed previously, additional materials are not required.

**Figure 17.2** Gradient embryogenic response exhibited by the younger innermost (upper row of sections) and older next out (lower row of sections) leaves of orchardgrass. Note that the basal-most leaf sections (left) produce embryogenic callus, whereas the more distal leaf sections (right) produce embryos directly. Also, more of the basal-most leaf sections produce embryogenic callus than the older leaf sections.

### Procedure

Plate leaf sections on SH30 medium as described above such that the sections from each leaf half are kept together in a single row and arranged in sequential order from the most proximal to the most distal section. Place the sections from four leaf halves in each dish (i.e., four rows of sections/dish — see Figure 17.1). Mark each leaf from each tiller so that the identity of the innermost vs. the outermost of two leaves per tiller is maintained. Culture and screen for contamination as described above.

Determine the number of sections from each location within each leaf half that produces embryogenic callus and/or direct embryos at weekly intervals for 8 weeks. Chart the occurrence of embryogenic callus and somatic embryos as a function of time and position of the section in the original tiller.

### Anticipated results

The gradient embryogenic response is demonstrated by correlating the relative position in a leaf from which a given section is obtained. A typical gradient response is shown in Figure 17.2. After 4–6 weeks, basal sections produce embryogenic callus. The callusing response progressively diminishes in sections from increasingly more distal positions of the leaf, being replaced by a direct embryogenesis response and, eventually, no embryogenic response at all. The differential embryogenic response is related to meristematic activity in the explant tissue. Leaf sections with abundant undifferentiated meristematic cells tend to produce embryogenic callus. As meristematic activity is decreased, embryogenic response is reduced such that only direct somatic embryos develop on sections with limited meristematic activity. Sections in which meristematic activity is too low or has ceased do not exhibit an embryogenic response.

### Questions

* Do sections from younger or older leaves or sections from proximal or distal locations within a leaf produce more callus or direct embryos?
* Which sections respond most rapidly?

* How do observed embryogenic responses of individual sections correlate with the meristematic zones in intact orchardgrass leaves?

## *Experiment 3. Maintenance of embryogenic callus*

Embryogenic callus of orchardgrass can be maintained by monthly transfer of callus and/ or somatic embryos to fresh SH30 medium. Embryogenic callus is very heterogeneous and continuously sectors into embryogenic and nonembryogenic portions. For maintenance of long-term embryogenic cultures, it is important to recognize and transfer only embryogenic sectors (Gray et al., 1984).

### *Materials*
The following items are needed for each student or team:

* Ten 100 × 15 mm petri dishes containing SH30

### *Procedure*
Isolate embryogenic callus and somatic embryos from leaf cultures. Dissect callus utilizing a stereomicroscope to assist in identifying various types of callus and tissue. Separate dry, friable-type callus from wet, mucilaginous-type callus. Identify and separate sectors with numerous root primordia, as well as clumps of large somatic embryos and masses of small somatic embryos embedded in a watery matrix (Gray et al., 1984). Separate the calluses and tissues by type and plate small clumps of each, five clumps per petri dish, onto SH30 medium (produce five cultures). Be sure to also plate single, isolated somatic embryos (five per petri dish) into five petri dishes in order to test their response. Incubate in the dark for 4 weeks as before.

Determine and describe the different callus types obtained based on their morphologies. Determine which of these callus types and/or tissues gave rise to embryogenic callus.

### *Anticipated results*
A typical embryogenic callus is shown in Figure 17.3. Embryogenic calluses or individual embryos are transferred to fresh SH30 medium in order to maintain and increase the embryogenic line. When transferring callus, it is important to isolate only sectors that are producing somatic embryos. The watery matrix material that contains masses of small somatic embryos frequently produces the most vigorous and prolific embryogenic callus. Embryogenic callus continues to sector into embryogenic and nonembryogenic callus types, primarily due to the recallusing of existing somatic embryos. The somatic embryos are anatomically complex and produce a number of different callus types, including embryogenic and rhizogenic (root-forming) calluses, when recultured on SH30.

### *Questions*

* Which of these callus types produced the most embryogenic callus and somatic embryos?
* Where did embryogenic callus arise from plated somatic embryos?
* What type of tissue isolated from embryogenic callus would be best to use in culture maintenance?
* Why does embryogenic callus continually sector?

**Figure 17.3**   Embryogenic callus of orchardgrass.

## *Experiment 4. Observations on monocotyledonous somatic embryos*

Orchardgrass embryogenic cultures produce monocotyledonous somatic embryos that are of the poaceous type. The typical embryo possesses a scutellum and a coleoptile. While the somatic embryos are small, it is possible to see morphological details using a stereomicroscope. Abnormal embryos are also common. For reference, see Figure 16.6, Conger et al. (1983); Gray et al. (1984); Gray and Conger (1985b) and Trigiano et al. (1989).

### *Materials*
The following items are needed for each student or team:

   * Leaf and callus cultures developed through previous experiments

### *Procedure*
Observe somatic embryos developing from callus and directly from leaf cultures. Correlate specific embryos observed with those illustrated in referenced publications. Identify the scutellum and coleoptile. Observe and describe the most commonly occurring types of abnormal embryos.

### *Anticipated results*
When viewed with a stereomicroscope, small white somatic embryos that are morphologically similar to zygotic embryos either appear in embryogenic callus or emerge directly from uncallused leaf sections. Embryos at all developmental stages can be recovered from the cultures (Figure 17.4). The embryos possess a distinct scutellum and a notch, from which a coleoptile develops and enlarges during germination (see Figure 16.6). Direct embryos are frequently attached to the leaf by a distinct suspensor.

### *Questions*

   * What types of abnormalities are most common?
   * How might such abnormalities have occurred during embryo development?

**Figure 17.4** Somatic embryos isolated from embryogenic callus of orchardgrass. The youngest recognizable embryonic stages are shown on the left. Well-developed embryos (right) are white and opaque, possess a large, flattened scutellum and a narrowed embryo axis.

## Experiment 5. Somatic embryo germination and plant recovery

Orchardgrass somatic embryos germinate readily and in a manner similar to that of zygotic embryos when placed on SH0 medium and incubated in light (cool white fluorescent) at 25 C. Small rooted plants are removed from petri dishes, quickly transferred (to avoid dehydration) to moist potting medium in small pots, which are enclosed in plastic bags and cultured in a lighted incubator as above. When plants begin to grow vigorously, often within a week, they are transferred to greenhouse pots and maintained in the same manner as stock plants.

### Materials
The following items are needed for each student or team:

  * Source of embryos from previous experiments
  * Five 100 × 15 mm dishes of SH0

### Procedure
Transfer well-developed somatic embryos, five per petri dish, to SH0 medium and incubate as above. Alternatively, transfer small clumps of embryogenic callus to the same medium. Observe each for signs of germination

Count the number of shoots and roots present after 2, 4, 6 and 8 days to determine whether shoots or roots emerge first. Identify the developing red-pigmented coleoptiles. Count the number of shoots per callus; this provides an indication of the relative number of well-developed somatic embryos per callus. Determine the percentage of somatic embryos that produce acclimated plants.

### Anticipated results
Individual embryos are induced to germinate into plants by transfer to medium lacking dicamba. Relative germination responses can easily be determined by counting the number of roots and shoots produced per culture. Resulting plants are easily acclimatized to greenhouse conditions.

### Questions

  * Why does transfer of embryos to medium lacking dicamba cause germination to occur?
  * What is the first sign of germination, and when does it occur?

## Experiment 6. Initiation and manipulation of embryogenic suspension cultures

Embryogenic suspension cultures are produced by culturing embryogenic callus in liquid SH30 medium (i.e., lacking agar). The development of somatic embryos can be either stimulated or stopped by adding or removing various organic sources of nitrogen (Gray et al., 1984; Gray and Conger, 1985a; Trigiano and Conger, 1987).

### Materials

The following items are needed for each student or team:

* Rotary shaker
* Four 125-ml Erlenmeyer flasks containing 10 ml liquid SH30
* Twelve 125-ml Erlenmeyer flasks containing 20 ml liquid SH30
* Four 125-ml Erlenmeyer flasks containing 20 ml liquid SH30-C
* Five or more 25-ml pipettes

### Procedure

Initiate embryogenic suspension cultures by placing approximately 2 g of rapidly growing embryogenic callus into each 125-ml flask containing 10 ml of SH30 medium. Start with four flasks. Cover the neck of each flask with several layers of sterile aluminum foil and seal the edges of the foil to the flask with parafilm. Rotate the cultures at approximately 75 rpm in the dark. After 2 weeks, add 20 ml of SH30 medium to each flask and rotate at 100 rpm for an additional 2 weeks. Observe the increase of culture mass in rapidly-growing cultures. Culture increase and maintenance is accomplished by pouring half of each rapidly growing culture into new flasks (total of eight flasks) containing 20 ml of SH30 medium at 2-week intervals. Observe the remaining culture material with a stereomicroscope and produce fresh-mounted microscope slides to observe cells with a compound microscope. Inspect for somatic embryos or embryogenic cells, which are characterized as being small and densely cytoplasmic (see chapter sixteen).

To induce embryo development, transfer half of each established, proliferating culture into flasks containing fresh SH30 medium and the other half into flasks containing SH30-C medium. Maintain cultures as before and observe twice weekly for the development of small white somatic embryos, which can be readily observed collecting around the rim of the flask.

After 4-weeks growth in SH30 and SH30-C media, decant the liquid from flasks and spill the remaining culture mass into an empty, sterile petri dish. Using a spatula and/or tweezers, plate small amounts of material (clumps about 3 mm in diam.) from each medium treatment, five clumps per petri dish, onto either solidified SH30 or SH0 medium and incubate in the light. Observe the response weekly for 4 weeks.

### Anticipated results

A typical suspension culture is shown in Figure 17.5. Suspension cultures will proliferate and can be maintained in liquid SH30 medium but will not produce recognizable somatic embryos unless an organic nitrogen source, such as casein hydrolysate, which contains a complex array of amino acids, is present. This indicates that the nitrogen source is an important controlling factor in somatic embryogenesis. In fact, somatic embryo development can be repeatedly started or stopped by adding or deleting casein hydrolysate in medium during transfer. When culture material from SH30 and SH30-C medium is placed on solidified SH30 medium, both produce embryogenic callus cultures. But only the tissue from liquid SH30-C medium produces embryos that readily germinate into plants on SH0

**Figure 17.5**   An embryogenic suspension culture of orchardgrass.

medium. This indicates that, although both SH30 and SH30-C liquid culture media support the proliferation of embryogenic cells, only medium with casein hydrolysate promotes development of the cells into somatic embryos.

*Questions*

* What accounts for the effect of casein hydrolysate on embryo development?
* What do the observed differences between the responses of culture material from the two liquid media indicate when transferred to the two types of solidified media?

## Experiment 7. Encapsulation of somatic embryos to produce synthetic seeds

Somatic embryogenesis is an ideal route for vegetative reproduction, since somatic embryos arise from the cells of one "parent". In fact, somatic embryogenesis represents the most efficient vegetative propagation system that can be envisioned, because of the rapid scale-up potential and prolific plant production exhibited by certain systems. In addition, since somatic embryos are nearly identical to zygotic embryos, research has attempted to add desirable seedlike qualities to somatic embryos, including a protective coating and ability to become quiescent. Such synthetic seeds would revolutionize certain aspects of agriculture, allowing genetically uniform synthetic seed to be produced indoors at will. The following experiment illustrates an aspect of synthetic seed research, the addition of a protective coating. See Gray and Purohit (1991) for a complete treatment of the subject.

### Materials

The following items are needed for each student or team:

* 100 ml of 2% sterile suspension of the sodium salt of alginic acid (Sigma Chemical Co.)
* 100 ml of 25 mM sterile aqueous solution of $CaCl_2$
* 100 ml of liquid SH0 medium
* Ten $100 \times 15$ mm petri dishes, each containing 25 ml of SH0 medium
* Three 150-ml sterile beakers with magnetic stir bars
* Magnetic stir plate

JEW

**Figure 17.6**   Encapsulation of orchardgrass somatic embryos in calcium alginate beads. A. Somatic embryos harvested from 5-week-old suspension, leaf or callus culture. B. Embryos suspended in 2% alginic acid solution. C. Embryos dropped singly into a beaker containing swirling 25 mM $CaCl_2$ solution. D. Alginic acid is converted to calcium alginate, forming bead around embryos. E. Synthetic seeds transferred to liquid SH0 medium to remove excess $CaCl_2$. F. Plants emerging from synthetic seeds on solidified SH0 medium.

* Sterile spoonulas
* Sterile 1000-μm nylon screen

*Procedure*

The steps in this procedure are illustrated in Figure 17.6. Harvest somatic embryos from suspension cultures by sieving the contents of a flask through a 1000-μm screen. Individual and small clumps of embryos, cells and cell aggregates will pass through the screen. Discard the material retained by the screen. Using a stereomicroscope, select individual, morphologically mature embryos and place them into the gently stirring 2% alginic acid solution. Alternatively, leaf or callus cultures with mature somatic embryos may be used in place of suspension cultures by plucking embryos directly from the cultures and placing into alginic acid. For a control treatment, place some embryos directly on solidified SH0 medium without alginic acid treatment.

Using a wide bore pipette, withdraw a single embryo at a time from the alginic acid solution. Express enough of the embryo/alginic acid suspension to form a small drop at the tip of the pipette, such that the drop contains one embryo. Allow this drop to fall into the gently stirring 25 mM $CaCl_2$ solution. Repeat this step numerous times, trying to perfect the art of forming a uniform drop with embryo.

Transfer the resulting calcium alginate/embryo beads to a swirling liquid SH0 solution for 15 min and then plate the beads, five per petri dish to solidified SH0 medium and incubate at 25 C in the light. Compare germination responses with control embryos twice weekly for 2 weeks. Record the number of root and shoot emergences at each time for each treatment.

*Anticipated results*

Immediately after dropping the alginic acid/embryo complex into $CaCl_2$, a cloudy-appearing bead should form around the embryo. This is the salt of alginic acid formed with the divalent cation $Ca^{+2}$, which is relatively insoluble in water. It may take several minutes for the calcium alginate drop to solidify completely to a soft gellike consistency. Typically, plants are obtained from encapsulated embryos, but the plant recovery rate often is lower than that of controls.

Encapsulated somatic embryos constitute one type of synthetic seed. The calcium alginate can be regarded as a synthetic seed coat, which, theoretically, offers protection as well as a matrix to hold nutrients, pesticides, etc. See Gray and Purohit (1991) for a detailed discussion.

*Questions*

* Why did encapsulated embryos differ from controls in plant recovery rate?
* What are some potential advantages of coating somatic embryos?

# Acknowledgment

This is Florida Agricultural Experiment Station Journal Series No. R-04836.

# Literature cited

Conger, B.V. and G.E. Hanning. 1991. Registration of 'Embryogen-P' orchardgrass germplasm with a high capacity for somatic embryogenesis from in vitro cultures. *Crop Sci.* 31:855.

Conger, B.V., G.E. Hanning, D.J. Gray and J.K. McDaniel. 1983. Direct embryogenesis from mesophyll cells of orchardgrass. *Science.* 221:850–851.

Gray, D.J. 1990. Somatic cell culture and embryogenesis in the Poaceae. In: *Biotechnology in Tall Fescue Improvement.* M.J. Kasperbauer, Ed., pp. 25–57. CRC Press, Boca Raton, FL.

Gray, D.J. and B.V. Conger. 1985a. Influence of dicamba and casein hydrolysate on somatic embryo number and quality on cell suspensions of *Dactylis glomerata* (Gramineae). *Plant Cell Tissue Organ Cult.* 4:123–133.

Gray, D.J. and B.V. Conger. 1985b. Time-lapse light photomicrography and scanning electron microscopy of somatic embryo ontogeny from cultured leaves of *Dactylis glomerata* (Gramineae). *Trans. Am. Microsc. Soc.* 104:395–399.

Gray, D.J., B.V. Conger and G.E. Hanning. 1984. Somatic embryogenesis in suspension and suspension-derived callus cultures of *Dactylis glomerata*. *Protoplasma.* 122:196–202.

Gray, D.J. and A. Purohit. 1991. Somatic embryogenesis and development of synthetic seed technology. *Crit. Rev. Plant Sci.* 10:33–61.

Gray, D.J., R.N. Trigiano and B.V. Conger. 1992. Liquid culture of orchardgrass somatic embryos and their use in synthetic cultivar development. In: *SynSeeds: Applications of Synthetic Seeds to Crop Improvement*, K. Redenbaugh, Ed., pp. 351–366. CRC Press, Boca Raton, FL.

Gray, D. J., R. N. Trigiano and B. V. Conger. 1994. Classroom exercises in the study of orchardgrass somatic embryogenesis. *HortTechnology.* 4:322–324.

Trigiano, R.N. and B.V. Conger. 1987. Regulation of growth and somatic embryogenesis by proline and serine in suspension cultures of *Dactylis glomerata*. *J. Plant Physiol.* 130:49–55.

Trigiano, R.N., D.J. Gray, B.V. Conger and J.K. McDaniel. 1989. Origin of direct embryos from cultured leaf segments of *Dactylis glomerata*. *Bot. Gaz.* 150:72–77.

*chapter eighteen*

# Somatic embryogenesis from seeds of melon

*Dennis J. Gray*

Melon, also known as cantaloupe or muskmelon (*Cucumis melo* L.), has been diversified horticulturally through breeding and selection into hundreds of cultivars. It belongs to the family Cucurbitaceae, the cucurbits, which includes several other important species, notably, cucumber, *Cucumis sativa* L., squash and pumpkin, *Cucurbita pepo* L., and watermelon, *Citrullus lanatus* (Thunb.) Matsum. and Nakai. Collectively, cucurbits are one of the most important horticultural crops in the world, as can be seen by their abundance year-round in nearly all produce markets.

Melon is subject to a number of bacterial, fungal and viral diseases that severely limit yield and for which adequate levels of native resistance are not available. A solution to this problem may be to genetically engineer resistance into existing germplasm (see chapter thirty-two for transformation methodologies). Adaptation of this technology to commercial varieties and breeding lines of *Cucumis melo* requires a dependable, high-frequency cell regeneration system. Although plant regeneration through organogenesis has been described for melon (Chee, 1991) and an organogenic system is demonstrated for watermelon (Compton and Gray, 1993), somatic embryogenesis is more efficient (Gray et al., 1993). Melon is an ideal subject for demonstrating somatic embryogenesis, since explants can be obtained from commercially available seeds and the culture system is simple and rapid.

The purpose of this exercise is to demonstrate the development of somatic embryos of a dicotyledonous species and to investigate the effects of differences in explants and genotypes on the frequency of somatic embryogenesis. This exercise will illustrate concepts of nonzygotic embryogenesis discussed in chapter sixteen. The referenced publication, Gray et al. (1993), is required to illustrate some of the experiments.

## General considerations

### Seed sources

Purchase 1 lb each of the following seeds: (1) 'Rocky Ford Green Flesh', also known as 'Eden Gem', from D. V. Burrell Seed Growers Co., Rocky Ford, CO, (2) 'Super Market' hybrid from D. V. Burrell Seed Co. and (3) 'Top Mark' also from D. V. Burrell Seed Co. One pound of seed is a sufficient quantity to repeat the described experiments many times. Dry seeds stored at approximately 4 C will retain viability for several years.

## Preparation of explants

To facilitate removal of seed integuments and dissection, seeds first are allowed to imbibe water. Rinse seeds briefly in a small kitchen colander or similar screen under running tap water. Place 100 seeds per 100 ml of sterile distilled water in 150-ml beakers for 1 h. Remove the seed coat and underlying papery integument to reveal a white embryo. Place the embryos immediately into fresh sterile water until all seeds have been processed. Surface disinfest embryos by agitation for 5 min in 100 ml of an aqueous solution of 25% commercial bleach containing a drop of Triton X-100 surfactant, then rinse twice and place for immediate use in sterile distilled water.

## Culture media

Chemicals and supplies are available from Sigma Chemical Company, St. Louis, MO. The following two culture media are required for the experiment: embryo initiation (EI) and embryo development (ED) media. The EI medium contains Murashige and Skoog's basal salts (see chapter two for composition) and vitamin powder with 30 g/L (88 mM) sucrose, 0.7% tissue culture–tested agar, 5 mg/L (22.5 μM) of 2,4-D and 0.1 mg/L (0.45 μM) of thidiazuron. The latter two substances are a synthetic auxin and cytokinin, respectively. ED medium is identical to EI except that 2,4-D and thidiazuron are omitted. The pH of all media is adjusted to 5.4 before adding agar and then autoclaved. Cooling medium is dispensed, 30 ml, into each $100 \times 15$ mm sterile petri dish.

# Exercises

## Experiment 1. Effect of explant type on somatic embryogenesis

In melon, the predominance of embryogenesis occurs only from certain regions of specific embryonic organs. In this exercise, the seed-derived embryos will be dissected into well-defined explant types, cultured and examined for embryogenesis to determine the site(s) of embryogenically competent cells.

### Materials
The following items are needed for each student or student team:

* Twenty surface-disinfested zygotic embryos of 'Rocky Ford Green Flesh' (use only this variety for this experiment)
* Twenty $100 \times 15$ mm petri dishes containing EI medium
* Fourty $100 \times 15$ mm petri dishes containing ED medium

### Procedure
Using fresh, sharp scalpel blades, aseptically dissect zygotic embryos as shown in Figure 18.1. Each embryo should yield one embryo axis explant plus four basal cotyledonary explants and four distal cotyledonary explants. Basal cotyledonary explants will appear similar to those shown in Figures 18.2 and 32.3. Place all nine explants into a single petri dish containing EI medium and arrange in a manner such that the explant type can be identified after callusing occurs. Maintenance of cultures in complete darkness (as in a dark incubator or in a cardboard box covered with aluminum foil) and at 25 C is an absolute requirement for the desired responses to occur. Check cultures every other day under dim light for contamination and move uncontaminated explants away from contaminated explants to fresh medium (see chapter seventeen — laboratory exercise concerning orchardgrass somatic embryogenesis for procedures for managing contamination).

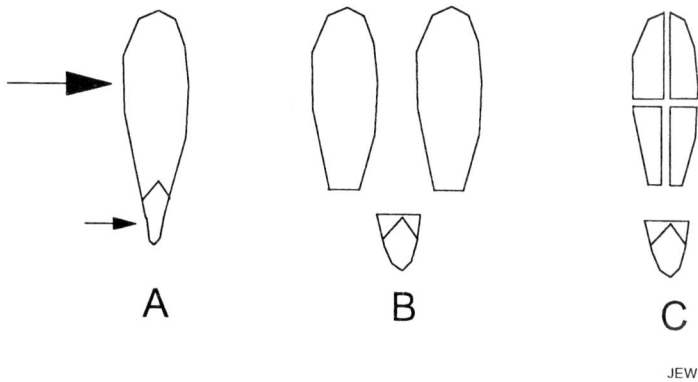

**Figure 18.1** Dissection of explants from imbibed melon seed. A. Embryo from which seed coat and papery integument have been removed. Cotyledons (large arrow) are easily distinguishable from embryo axis (small arrow). B. Remove embryo axis and separate cotyledons. Be careful to make clean, sharp cuts and to not damage the tissue. C. Cut each cotyledon into two basal and two distal explants.

After 2 weeks on EI medium, transfer explants to ED medium, but place the embryo axis explants on a separate petri dish to provide more space for growth. Be sure to maintain the identity of each explant type. Incubate these cultures under light (16 h cool white fluorescent light [60 $\mu$mol·m$^{-2}$·sec$^{-1}$]/8 h dark cycle).

After 3 weeks and, again, after five weeks on ED medium, determine the total number of each explant type that produced at least one somatic embryo, recognizable through a stereomicroscope. In addition, at 3 weeks estimate the number of somatic embryos on each responding explant without disturbing the cultures and, at 5 weeks, open the petri dishes and count the exact number of embryos. Calculate the standard error of response per explant type for number of responding explants and number of somatic embryos per explant and determine which explant type was most responsive.

*Anticipated results*

During incubation on EI medium, the explants will enlarge to several times their original size and begin to produce callus. This response is illustrated in Figure 18.2. Explants normally should remain pale yellow in color. Change in pigmentation to a green color on EI medium indicates that adequate dark conditions were not maintained and suggests that a reduced embryogenic response subsequently will occur. Microscopic examination of explants after 2 weeks on EI medium often will reveal small hyaline (clear) clavate (club-shaped) structures, some of which are the earliest visible stages of somatic embryogenesis.

After transfer to ED medium in the light, explants continue to enlarge and turn green in color (Figure 18.2). White to yellow to green-colored callus also will develop. Somatic embryos in various stages (see the last experiment in this series) will begin to appear. Younger embryos initially are white to pale yellow in color, but soon become green. Although some somatic embryos often can be found on all explant types, the greater abundance of embryos will develop from the basal cotyledonary explants.

*Questions*

* What is the rationale behind initiating cultures on medium containing auxin and cytokinin, followed by transfer to basal medium?
* Why is an initial incubation period in dark so important?
* Why do certain explant types respond better than others?

**Figure 18.2**  Development of basal cotyledon explant on EI and ED media. Newly plated explant (left) is white in coloration. After 2 weeks on EI medium in the dark (center), the explant enlarges, begins to callus along its edges, but remains white to pale yellow in color. Two weeks after transfer to ED medium in the light (right), further enlargement and callusing occurs and the explant becomes green in color. It is at or after this stage that well-developed somatic embryos typically appear.

## Experiment 2. Effect of genotype on somatic embryogenesis

Embryogenesis in melon is genotype dependent. Relatively high or low embryogenic responses occur depending on the genotype used. In this experiment, three genotypes that differ in embryogenic capacity are compared.

### Materials
The following items are needed for each student or student team:

* Fifteen surface-disinfested embryos each of 'Rocky Ford Green Flesh', 'Super Market' and 'Top Mark'
* Thirty 100 × 15 mm petri dishes containing EI medium
* Thirty 100 × 15 petri dishes containing ED medium

### Procedure
Dissect embryos as described in experiment 1, but culture only the basal cotyledon explants. Place the four explants from each seed on a single petri dish containing EI medium. Prepare ten petri dishes for each explant. Incubate in the dark and manage as described in experiment 1. Determine the relative difference in embryogenesis between the three genotypes at 3 and 5 weeks after transfer to ED medium, and process the data as in experiment 1.

### Anticipated results
Dramatic differences in embryogenic response will occur among the three genotypes tested. 'Rocky Ford Green Flesh' will exhibit the greatest embryogenic response, whereas 'Super Market' will be intermediate in response and 'Top Mark' will show the least. Varying rates of contamination may also occur between the genotypes.

### Questions

* What is the apparent relationship, if any, between contamination rate and genotype in terms of embryogenic response?

* Speculate as to precise reasons why genetic differences among genotypes may influence their embryogenic response.

## Experiment 3. Embryogenic culture maintenance

### Materials

The following items are needed for each student or student team:

* Embryogenic explants and callus-bearing somatic embryos obtained from the previous experiments
* Fifteen $100 \times 15$ mm petri dishes of EI medium
* Fifteen $100 \times 15$ mm petri dishes of ED medium

### Procedure

Place five pieces of explants, small pieces of callus (about 5 mm diam.) and single somatic embryos into five petri dishes each of EI and ED medium. After 2 weeks, observe differences between culture growth and embryogenesis, then transfer cultures to fresh medium. Repeat this cycle one more time (i.e., transfer for three cycles and utilize a total of 15 petri dishes of each medium). At the end of the experiment, compare the embryogenic responses for each type of inoculum on each medium and with the responses that were obtained initially from primary explants.

### Anticipated results

It is not possible to maintain high-frequency somatic embryogenesis of melon under the conditions described. Cultures placed on ED medium either produce a few plants from germinating embryos or decline and die. Cultures placed on EI medium continue to proliferate as callus, but the number of embryos produced declines rapidly. Additionally, the embryos that are produced become increasingly abnormal in morphology until they barely can be identified. Although the exact reasons for this decline are not known, a relatively high proportion of polyploid plants are known to occur from somatic embryos of melon (Oridate et al., 1992).

### Questions

* Why does embryogenesis decline so rapidly in cultures of melon?
* Considering theories relating to induction of embryogenic cells, how can the response of recultured somatic embryos be explained?

## Experiment 4. Observation and categorization of embryonic stages

Because of the rapidity and high frequency of the embryogenic response exhibited by melon, it provides a perfect system for studying somatic embryo development. In addition, this system allows all stages of embryogenesis to be followed dynamically, which is not possible with zygotic embryogenesis because of the presence of obscuring seed integuments. Often, all stages of embryogenesis, including instances of abnormal development, can be seen in a single explant. In this experiment, cultured explants are examined and various embryogenic stages are categorized.

*Materials*

The following items are needed for each student or student team:

* Approximately ten 100 × 15 mm petri dishes containing ED medium and each containing five cultured cotyledonary bases of 'Rocky Ford Green Flesh' that already have been induced on EI medium and transferred to ED medium as described in experiment 1
* High-quality stereomicroscope and light source

*Procedure*

After transfer to ED medium, number each culture and observe at least two times each week. Keep a log of explant response. Identify a few of the explants that begin to produce the most somatic embryos and follow their development. On these explants, identify several young embryo stages and sketch their relative location on the explant, as well as their morphology. Draw these embryos weekly and determine the changes in morphology. Identify globular, heart-shaped, torpedo, and cotyledonary embryonic stages (see Figures 5.2 and 19.9, as well as Gray et al., 1993 for examples of some stages). Attempt to identify cotyledonary-stage embryos with pronounced shoot apical domes and follow the stages of early shoot development. Identify instances of abnormal embryo development and precocious germination and categorize different types (see Gray et al., 1993 for examples). Identify somatic embryos that have developed into plants (see Figure 19.13 for an example). Determine which embryonic stages result most often in plant development.

*Anticipated results*

The process of embryogenesis in this system is very rapid. Cultures must be observed twice a week. Once highly responsive explants are identified, the development of a select few embryos will illustrate both normal and abnormal embryogenesis. All typical stages of embryogenesis can be identified and followed.

*Questions*

* Suggest reasons for the development of fused somatic embryos or those with missing or extra cotyledons.
* When in the development of the embryos must certain abnormalities occur?
* Do normal or precociously germinating embryos ultimately produce more plants?

# Acknowledgment

This is Florida Agricultural Experiment Station Journal Series No. R-04839.

# Literature cited

Chee, P.P. 1991. Plant regeneration from cotyledons of *Cucumis melo* 'Topmark'. *HortScience*. 26:908–910.

Compton, M.E. and D.J. Gray. 1993. Shoot organogenesis and plant regeneration from cotyledons of diploid, triploid and tetraploid watermelon. *J. Am. Soc. Hort. Sci.* 118:151–157.

Gray, D.J., D.W. McColley and M.E. Compton. 1993. High-frequency somatic embryogenesis from quiescent seed cotyledons of *Cucumis melo* cultivars. *J. Am. Soc. Hort. Sci.* 118:425–432.

Oridate, T., H. Atsumi, S. Ito and H. Araki. 1992. Genetic differences in somatic embryogenesis from seeds in melon (*Cucumis melo* L.). *Plant Cell Tissue Organ Cult.* 29:27–30.

# chapter nineteen

# Somatic embryogenesis from mature peanut seed

*Hazel Y. Wetzstein and Charleen Baker*

The cultivated peanut or groundnut (*Arachis hypogaea* L.) is a crop of great economic importance. In the United States, primary uses of peanut seeds are for confectionery purposes or making peanut butter. However, worldwide it is an important source of oil, protein, food and feed. A major objective of peanut improvement programs is the production of cultivars that are genetically resistant to pests and diseases. The use of genetic transformation technologies has created the potential to improve agronomic characters and food quality properties. Some valuable traits that could potentially be introduced into peanut include resistance to insects, viruses and fungi, herbicide tolerance, increased adaptability to different environments and improved oil quality.

Before biotechnological methods, such as genetic transformation, can be utilized for improving the agronomic and culinary characteristics of peanut, a reliable regeneration system had to be developed. Somatic embryogenesis is a promising method, because potentially high numbers of regenerates can be obtained, each theoretically originating from a single or few cells. High regeneration rates and unicellular origin of somatic embryos increase the likelihood of successfully introducing desirable genes and recovering stably transformed plants.

The following laboratory exercises will illustrate a direct somatic embryogenic system using mature peanut embryo axes as the explant. Students will be acquainted with the following concepts: (1) the effect that explant part (i.e., hypocotyl vs. epicotyl) has on embryogenesis, (2) how auxin concentration in the primary and secondary media affect embryo induction and development, (3) how environmental factors, such as light, can modify embryo form and (4) germination of embryos and acclimatization of plants. These exercises are adapted from publications by Baker et al. (1995) and Wetzstein and Baker (1993).

## General considerations

### Explant material and medium preparation

Most of the media used in the following experiments were adapted for peanut seed cultivars 'GK7' and 'AT127' and frequently produce embryogenic cultures. However, embryogenic cultures have also been obtained using a number of other cultivars including 'Florunner', 'Arkansas Valencia', 'NC7', 'Georgia Runner', 'GK17' and 'VC-1', and these are suitable for substitution if 'GK7' and 'AT127' are not available.

**Figure 19.1**  A. Dissected peanut seed showing cotyledons and embryo axis (point). B. Whole, isolated embryo axes. Note epicotyl (e) and radicle (r) ends.

Shelled peanut seed should have a reddish brown seed coat and well-filled cotyledons. Dissect out the small embryo axes on clean, but not necessarily sterile, paper towels. This can be easily accomplished by using a scalpel blade to separate the two cotyledons, after which the embryo axis can be popped out using the tip of the blade. See Figure 19.1 of an opened peanut seed illustrating cotyledons and an embryo axis. As axes are removed, store them in a petri dish lined with moist filter paper to prevent desiccation. To obtain satisfactory and reproducible results, discard any damaged or malformed axes (i.e., injured during dissection or by pests). For each experiment, some extra axes should be collected since some may be lost or damaged during disinfestation or subsequent cutting.

Under a flow hood, transfer the axes to a 250-ml beaker and surface disinfest in 10% commercial bleach (0.26% NaOCl) with 2–3 drops of Tween 20 (a detergent) with occasional agitation for 5 min. Rinse the axes three times in sterile deionized water.

The basal culture medium is composed of Murashige and Skoog (MS) (1962) salts supplemented with $B_5$ vitamins (Gamborg et al., 1968) (see chapter two for composition), 30 g/L (88 mM) sucrose and 0.4% Gel-gro (ICN Biochem., Irvine, CA). Plant growth regulators (PGRs) will be used as described in the different experiments. Adjust the pH of the culture media to 5.8 before adding Gel-gro and autoclave. Dispense 25 ml of media into each 100 × 15 mm petri dish.

## Exercises

### Experiment 1. Effect of explant tissue on somatic embryo development

The type of explant tissue used is critical for obtaining success in many embryogenic systems. Particularly in species that are difficult to work with, actively dividing meristematic tissues often give much better results. The mature cotyledons of peanuts, which are well-differentiated tissues with abundant storage reserves, typically give a poor embryogenic response. In contrast, the embryo axis, which has young, meristematic regions, usually gives a good response. Even parts within the embryo axis react or respond differentially. Having high-frequency embryogenesis and understanding where embryos emerge can be critical if an objective is to apply genetic transformation techniques to an embryogenic system. In this exercise, the whole embryo axis, epicotyl portion and radicle portion will be used as explants. Two 2,4-dichlorophenoxyacetic acid (2,4-D) concentrations (15 mg/L [68 $\mu$M] and 30 mg/L [135 $\mu$M]) will be used. Sites of embryo development will be noted. It is recommended that this exercise be completed in groups of two to three students, with four petri dishes used for each explant type and auxin concentration (six treatments in all). This experiment will require about 8 weeks for completion. The experiment may be considered as a completely randomized design with four replicates of each treatment. Data should be analyzed, e.g., standard deviation or ANOVA etc. (see chapter six).

### Materials

The following items are needed for each student or team of students:

* About 150 mature peanut seeds
* Petri dish lined with moistened filter paper
* 100 ml of 10% commercial bleach with 2–3 drops of Tween 20
* Eight $100 \times 15$ mm petri dishes per explant type: four containing MS + 15 mg/L 2,4-D and four containing MS + 30/L mg 2,4-D
* Twenty-four $100 \times 15$ mm petri dishes containing MS without 2,4-D
* Incubator providing darkness at 26 C

### Procedure

Excise and surface disinfest embryo axes as described in the general considerations section. Retain one third of the explants as whole or entire embryo axes. Using a sharp scalpel, cut the remaining two thirds of the axes into two pieces each; the cut should be made at about one third of the distance from the shoot apex. The apical portion will contain the upper third of the axis, whereas the radicle portion will comprise the lower two thirds of the axis. Place each of the three explant types (whole axis, apical portion and radicle portion) onto media with either 15 or 30 mg/L 2,4-D. Refer to Figure 19.2 for a diagrammatic representation of this procedure. Incubate cultures in the dark at 26 C. After 30 days, transfer all explants to fresh basal MS medium without 2,4-D and return to the darkened incubator. Observe cultures every 2 weeks from initiation and note from which tissues somatic embryos are emerging. Record callus production and somatic embryo numbers after 4 weeks on basal medium.

### Anticipated results

Explants will exhibit some swelling and enlargement on induction medium containing 2,4-D. Somatic embryos will begin to form after transfer to basal MS medium without 2,4-D. Embryos will originate from apical leaves with little or no embryo formation on the radical portion of the whole axes or radicle portion of cut explants. See Figure 19.3 for a representation of developing somatic embryos. Radicle portions may exhibit some browning and callus growth.

### Questions

* Why is it necessary to transfer explants after 4 weeks from induction medium to a medium without PGRs?
* Which explant type would be preferable to use in genetic transformation studies and why?

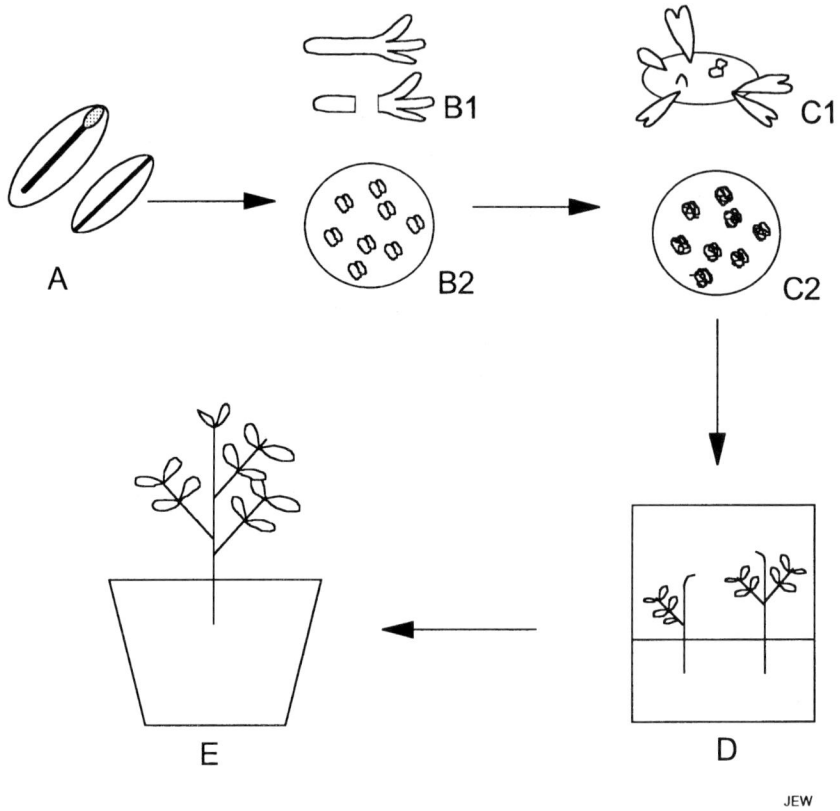

**Figure 19.2** Diagrammatic representation of the somatic embryogenic system for peanut. A. Mature seed showing cotyledons and embryo axis. B. Embryo axis explants (whole axis, epicotyl portion, radicle portion) placed on induction medium with 2,4-D. C. Explants after transfer to secondary medium with no or low 2,4-D concentration. Somatic embryos are evident within 4 weeks. D. Somatic embryos on germination/conversion medium. E. Plants transferred to pots.

**Figure 19.3** Embryogenic culture from epicotyl explants showing numerous somatic embryos about 4 weeks after transfer to basal secondary medium.

## Experiment 2. Effects of auxin concentration and photoperiod on somatic embryo development and morphology

Somatic embryos are usually induced by incorporating an auxin into a primary medium. An important first step in developing any embryogenic protocol is to establish the most appropriate auxin type and concentration for induction of somatic embryogenesis. Also, the proper environmental conditions, such as temperature and light, for cultures should be determined. In this experiment, different concentrations of 2,4-D in the induction medium will be evaluated as well as the effects of growing cultures in dark vs. light conditions.

Students should work in teams of two or three with four petri dishes used for each auxin concentration and light/dark treatment (eight treatments in all). If materials or time is limited, omit the light culture treatments, since cultures grown in the dark will produce better embryos. Alternatively, complete all of the dark treatments and include only the 15 and 30 mg/L 2,4-D treatments in the light. This experiment will require about 8 weeks to complete. The experiment may be considered as a completely randomized design with four replications of each treatment. Data should be statistically analyzed, e.g., standard deviation or ANOVA, etc. (see chapter six).

### Materials
The following are needed for each student or team of students:

* About 200 peanut seeds
* Four $100 \times 15$ mm petri dishes containing MS + without 2,4-D per photoperiod
* Four $100 \times 15$ mm petri dishes containing MS + 15 mg/L 2,4-D per photoperiod
* Four $100 \times 15$ mm petri dishes containing MS + 30 mg/L 2,4-D per photoperiod
* Four $100 \times 15$ mm petri dishes containing MS + 60 mg/L 2,4-D per photoperiod
* Thirty-two $100 \times 15$ mm petri dishes containing MS without 2,4-D
* Incubators or growth room providing dark or a 16-h photoperiod at 26 C

### Procedure
Dissect out and surface disinfect embryo axes as described previously. Cut the axes in two, using only the apical portions as explants (see experiment 1 for details). Prepare several modifications to the basal culture medium by adding 2,4-D at 0, 15, 30 or 60 mg/L prior to autoclaving. Place five explants in each petri dish and culture in a 26 C incubator in the dark. If an incubator is not available, merely place the cultures within a box in the growth room. Repeat the procedure for the PGR concentrations to be incubated in the light under a 16-h photoperiod at about 130 $\mu mol \cdot m^{-2} \cdot sec^{-1}$.

After 30 days on induction medium, transfer all explants to fresh basal MS medium without PGRs. Return the cultures to the same respective light or dark conditions. Observe cultures every 2 weeks from culture initiation and record somatic embryo numbers after 4 weeks on basal medium.

### Anticipated results
Somatic embryos will begin to develop after transfer to the secondary medium in cultures grown under both light and dark conditions. Embryos in dark-grown cultures will be white, well formed and easily separated for germination. In cultures grown in the light, embryos will be green, less succulent and more difficult to separate. A white nonembryogenic callus may be prevalent in cultures grown in the light. Embryos will not form in cultures that were initiated on medium without 2,4-D. Most explants induced on 15 or 30 mg/L 2,4-D will form embryos. On induction media with 60 mg/L 2,4-D, explants will exhibit browning and a relatively lower rate of embryogenesis compared with other treatments in which embryos were formed.

*Questions*

* What 2,4-D concentration for induction and photoperiod would you recommend using for embryogenesis and why?
* Do you think that embryo morphology will influence embryo germination and development? How?

## Experiment 3. Secondary medium effects on repetitive embryogenesis and somatic embryo development

In peanut and many somatic embryogenic systems, explant material is placed sequentially on two types of media. A primary medium, which has an auxin, is used to induce somatic embryos. This is followed by transfer to a secondary medium without or with a lower concentration of auxin. Auxin is proposed to be necessary for somatic embryo induction, but is inhibitory to somatic embryo development. In this experiment, secondary media with varying concentrations of 2,4-D will be used. Some cultures will be retained on induction medium with high 2,4-D (30 mg/L), whereas others will be transferred to low (10 mg/L) or medium without 2,4-D. Observe the effects that secondary medium has on repetitive embryogenesis (that is, the continued proliferation of somatic embryos from preexisting somatic embryos) and embryo development. This exercise should be completed by teams of two or three students with four petri dishes used per secondary medium (three treatments). The experiment will require about 12 weeks for completion. The experiment may be considered as a completely randomized design with four replicates of each treatment and data should be analyzed, e.g., standard deviation or ANOVA, etc. (see chapter six).

### Materials
The following items are needed for each student or team of students:

* About 100 mature peanut seeds
* Twelve $100 \times 15$ mm petri dishes containing MS + 30 mg/L 2,4-D
* The following media for subculture after 4 weeks and again after 8 weeks:
* Four $100 \times 15$ mm petri dishes containing MS without 2,4-D
* Four $100 \times 15$ mm petri dishes containing MS + 10 mg/L 2,4-D
* Four $100 \times 15$ mm petri dishes containing MS + 30 mg/L 2,4-D

### Procedure
Dissect out and surface disinfest the embryo axes as previously described. Cut the axes in two, using only the apical portions as explants (see experiment 1 for details). Place five explants in each petri dish, all of which contain MS medium with 30 mg/L of 2,4-D. Incubate dishes in a 26 C incubator in the dark. After 30 days, transfer the explants to one of three secondary media (0, 10 or 30 mg/L 2,4-D). After 4 weeks, subculture onto fresh petri dishes containing the same secondary medium. Observe cultures after transfer to secondary media and record embryo numbers 8 and 12 weeks after initiation.

## Anticipated results

Somatic embryos will begin to form when cultures are transferred to secondary media. Some of the embryos on secondary medium without PGRs will exhibit precocious or premature germination in culture. Explants on secondary medium with 10 mg/L 2,4-D will produce secondary embryos. Some callus production will occur on explants on 30 mg/L 2,4-D, with few repetitive embryos formed.

### Questions

* * What is repetitive or secondary embryogenesis and of what significance is it?
* * What are the advantages and disadvantages of somatic embryos germinating in culture?
* * What level of 2,4-D would you choose for the secondary medium and why?

## Experiment 4. Germinating, planting and acclimatization of somatic embryos

Plants can be obtained from the somatic embryos produced in previous experiments. Germination is enhanced by transferring the embryos to Magenta vessels in the light. Peanut somatic embryos can have very divergent forms, including those with an elongated axis, trumpet or horn-shaped embryos and fan-shaped embryos (Wetzstein and Baker, 1993). In this experiment, students will be able to observe germination and shoot growth of embryos and acclimatization of plants out of culture. The experiment will illustrate how embryo morphology can affect germination and ability to produce plants. This experiment will take about 5–8 weeks beyond the production of somatic embryos.

### Materials

For germination:

* * Two to five GA-7 Magenta vessels containing 50 ml basal MS medium (or baby food jars or similar vessels)

For planting in soil and acclimatization:

* * 4-in. plastic pots
* * Equal parts of soilless potting mix÷perlite÷vermiculite
* * MS salts for watering
* * Plastic bags

### Procedure

Under sterile conditions, carefully separate individual somatic embryos from the cultures. Insert the radicle end of about ten embryos into each Magenta vessel containing basal MS medium. Observe the different types (i.e., funnel-shaped, well-defined axes and cotyledons, etc.) of somatic embryos and note the treatment conditions from which embryos were obtained. Incubate the boxes under a 16-h photoperiod at about 28 C.

Periodically observe the embryos for root emergence and shoot growth. Transfer germinated embryos to pots filled with a mixture composed of equal parts of soilless potting mix, perlite and vermiculite that has been wetted with water. Gently wash agar from roots, place roots in planting mix and wet each pot with about 100 ml MS salts solution. Cover the plant in the pot with a plastic bag and grow under light conditions. After 2–3 days, slowly acclimatize the plants over the next week by cutting progressively larger slits in the plastic bag. Remove the bags after this time. To maintain plants, water when dry and fertilize with MS salts every 2 weeks.

*Anticipated results*

Some somatic embryos should exhibit root and shoot growth within 1 month of transfer into Magenta vessels. Embryo form will affect the germination and growth into plants. Horn or trumpet-shaped embryos will be poor producers of plants. Those with a well-defined axis and cotyledons or foliar structures will produce plants in higher frequencies.

*Questions*

* Why is it important to wash agar from the roots prior to transfer to pots?
* Why is acclimatization of plants necessary?

## Literature cited

Baker, C.M., R.E. Durham, J.A. Burns, W.A. Parrott and H.Y. Wetzstein. 1995. High frequency somatic embryogenesis in peanut (*Arachis hypogaea* L.) using mature, dry seed. *Plant Cell Rep.* (in press).

Gamborg, O., R. Miller and K. Ojima. 1968. Nutrient requirements of suspension cultures of soybean root cells. *Exp. Cell Res.* 50:150–158.

Murashige, T. and F. Skoog. 1962. A revised medium for rapid growth and bioassays with tobacco tissue cultures. *Physiol. Plant.* 15:473–497.

Wetzstein, H.Y. and C.M. Baker. 1993. The relationship between somatic embryo morphology and conversion in peanut (*Arachis hypogaea* L.). *Plant Sci.* 92:81–89.

*chapter twenty*

# Direct somatic embryogenesis from leaves and flower receptacles of cineraria

**Robert N. Trigiano, Mary Catherine Scott and Kathleen R. Malueg**

The florists' cineraria, *Senecio × hybridus* Hyl., is a daisylike flowering pot plant and a species belonging to the family Asteraceae. The most notable horticultural attributes of the plant are its flowers that are produced in a wide range of colors including white, pink, red, purple, maroon, magenta and several striking hues of blue. Cineraria is a perennial, but is usually grown as an annual, and is very well suited to low-light intensities typical of winter production in the Northern Hemisphere. Cineraria is usually considered to be a minor crop of little economic importance and is underutilized as a floricultural crop. However, cineraria is an excellent subject to demonstrate direct somatic embryogenesis. The following two exercises that are depicted in Figure 20.1 will illustrate the effects of medium composition on the direct formation of somatic embryos from either leaves or flower receptacles of cineraria.

## General considerations

### Growth of plants

'Hansa' or 'Cindy' mix cineraria seeds (Park Seed Co., Greenwood, SC) are sown sparingly, about four to five per inch, in very shallow trenches of prewetted vermiculite or other suitable seedling mix and then lightly covered with medium. We have found that an 11 × 21 in. plastic flat of seedlings will provide enough material for a class of 10–15 students. Cover the trays with plastic wrap and incubate at 21 C with about 50 $\mu mol \cdot m^{-2} \cdot sec^{-1}$ of light provided by cool fluorescent tubes. If a growth chamber is unavailable, a bench top in a cool laboratory will work well. The plastic wrap should be removed after the seedlings emerge in about 10 days to 2 weeks. The first true leaves will be formed in about 2 weeks and are suitable for explants. Flower formation requires about 26–28 weeks of growth, including some time in a cooler. Five weeks after sowing, seedlings are transplanted into bedding plant cell packs. After 4 weeks, plants are moved up to 4-in. pots (any soilless medium will do) and, after an additional 4 weeks of growth, plants are placed in a cooler at 15 C. The plants are moved after 6 weeks to a cool greenhouse (about 18 C) and should flower between 8 and 10 weeks later (see Larsen, 1985 for scheduling). For receptacle explants, choose flowers buds that are showing slight color, but are more or less tightly closed.

**Figure 20.1** Diagrammatic representation of exercises for direct somatic embryogenesis from leaves and receptacles of cineraria. A. Leaf and receptacle tissue before surface disinfestation. $B_1$. Tissue cut into equal halves and placed on SH medium "A" and MS medium "B" (experiment 1). $B_2$. Tissue cut into equal halves and both placed on the MS medium containing PGRs (experiment 2). C. Tissue cultured on "A" transferred to charcoal-containing medium while "B" remains on the original medium. D. Tissue cultured on "A" transferred to medium without PGRs, whereas tissue on "B" is transferred to fresh MS medium containing PGRs. E. Somatic embryo formed on explants in "A" germinated on medium without PGRs in a Magenta vessel. F. Plantlet grown in a peat pellet and acclimatized to ambient growing conditions. G. Plantlet potted in soilless medium and transferred to a greenhouse or cooler.

## Explant preparation and basal culture media

The first true leaves, including the petiole, may be harvested for explants about 2 weeks after the seedlings emerge. Surface disinfest the leaves first by immersing in 70% ethanol for 30 sec, followed by gently agitating in 10% bleach solution for 10 min and finally rinsing three times with sterile distilled water. Excise and discard the petiole and bisect the leaves (lamina) longitudinally (lengthwise) into two equal halves (Figure 20.1). If receptacle tissue is to be used as explants, flower buds with about 2 cm of the subtending pedicle (stem) should be immersed in 70% ethanol for 1 min and then the alcohol ignited by quickly passing the bud through a flame. The buds are soaked in 20% Clorox for 10 min followed by three rinses in sterile distilled water. Using aseptic technique, remove the subtending pedicle and strip the calyx (green portion) and all the florets from the receptacle. Cut the receptacle into two equal halves. An optional disinfecting treatment is to immerse the naked receptacle in 10% Clorox solution for 5 min followed by three rinses with sterile water before bisection.

There are two treatments in both of the experiments included in this chapter; therefore, the exercises may be set up simply as paired variate designs. In this experimental design, we assume that an individual leaf or flower receptacle is composed of relatively homogenous tissue in which any portion (e.g., a leaf half) is capable of responding to any treatment in a similar manner compared with the corresponding sister half. Therefore, if the sister halves are placed on different treatments, variations in responses will be due to treatment and not to differences in the ability of the tissue to respond or to other extraneous factors. This is a very powerful experimental design and eliminates the need for extensive replication. Regardless of the type of tissue used in these experiments, place one half of the explant onto one medium (treatment) and place the corresponding sister half onto the other (this is the "pair" in the paired variate design). Label the first petri dish A-1 and the second B-1, where the letter indicates the treatment and the numeral designates the replicate or explant (see Figure 20.1). Data (e.g., the number of somatic embryos) should be analyzed using a t-test for paired variates (Snedecor and Cochran, 1967 or other comparable statistics textbook).

The culture media in the following two experiments are composed of either Murashige-Skoog (MS) (1962) or Schenk and Hildebrandt (SH) (1972) basal salts (see chapter two for composition) supplemented with 30 g (88 mM) sucrose, 1 g (0.55 mM) myo-inositol, 5 mg (2.9 µM) thiamine HCl, 3 mg (13.5 µM) 2,4-dichlorophenoxyacetic acid (2,4-D), 1 mg (4.5 µM) benzyladenine (BA) and 8 g of agar per liter. The pH of the medium is adjusted to 5.8 before sterilization, and approximately 10 ml of medium is poured into each $60 \times 15$ mm plastic petri dish.

## Exercises

### Experiment 1. The effect of basal medium on initiation of somatic embryos

The composition of the basal medium can have dramatic effects on the response of an explant in culture. This simple experiment is designed to test the ability of cineraria explants to form somatic embryos on two commonly used tissue culture media, SH and MS. The experiment will require about 4 weeks to complete.

#### Materials

The following items are needed for each student or team of students:

* Six seedlings or six flower receptacles
* Twelve $60 \times 15$ mm petri dishes, six containing SH medium (treatment "A") and six containing MS medium (treatment "B") and both media supplemented with 3 mg/L (13.5 µM) 2,4-D and 1 mg/L (4.5 µM BA)
* 10 and 20% Clorox solutions + 0.1% Triton X-100

*Table 20.1*  Example of Data Analysis Using a t-Test for Paired Variates

| Pair number | Treatments "A"[a] | "B"[a] | $D = A - B$ | $D - D_m$[b] | $(D - D_m)^2$ |
|---|---|---|---|---|---|
| 1 | 1 | 9 | −8 | −2 | 4 |
| 2 | 0 | 9 | −9 | −3 | 9 |
| 3 | 1 | 12 | −11 | −5 | 25 |
| 4 | 1 | 6 | −5 | 1 | 1 |
| 5 | 0 | 0 | 0 | 6 | 36 |
| 6 | 2 | 5 | −3 | 3 | 9 |
|   |   |   | −36 | 0 | 84 |

Calculations: 1. $D_{mean} = -36/6 = -6$; 2. standard deviation $(84/5)^{1/2} = (16.8)^{1/2}$; 3. standard error $(16.8)^{1/2}/(6)^{1/2} = (2.8)^{1/2} = 1.65$; 4. $t = (-6 - 0)/1.65 = -3.64$: Note "0" is used to include the null hypothesis of no difference between treatments; 5. For 5 degrees of freedom and $p = 0.05 = -2.57$ (see Snedecor and Cochran, 1967); 6. Since $t = -2.57$ is greater than −3.64, the result is significant — explants cultured on treatment "B" produced significantly more embryos than were produced on treatment "A".

[a] Number of somatic embryos
[b] Mean difference between treatments

*Procedure*

Surface disinfest six leaves or receptacles, cut into halves as described above and place one half of each explant onto each of the experimental media (Figure 20.1). Label each of the dishes with the treatment ("A" or "B") and be absolutely certain that the dishes containing the sister halves are labeled with the same number. Incubate the dishes in the dark at 22–25 C. Examine weekly for embryo development and contamination. Note: if one half of a pair is discarded because of contamination (e.g., A-2), then the other half (B-2) should also be discarded. After 2 and 4 weeks, count the number of somatic embryos formed on explants cultured on each of the treatments (media). We suggest that the class share and analyze the data using a t-test for paired variates (see Table 20.1 for an example).

*Anticipated results*

Both leaf explants should begin to curl and become somewhat contorted after 4–5 days in culture. Late in the second week of culture, explants grown on SH medium or treatment "A" (Figure 20.2a) will produce some crystalline-looking callus and occasionally, a somatic embryo, whereas numerous globular-stage yellow somatic embryos will begin to form directly from explants cultured on MS medium or treatment "B" (Figure 20.2b). Four weeks after culture initiation, leaf portions cultured on SH medium will have produced abundant callus, whereas receptacle explants will be curled with little evidence of callus growth (Figure 20.2c). Explants cultured on MS medium will usually be covered with yellow to green globular to heart-shaped somatic embryos that may be beginning callus (Figure 20.2d). Very few, if any, morphologically mature embryos with well-developed cotyledons will be present.

*Questions*

* Why did MS and not SH medium support the development of somatic embryos? *Hint:* Consider the nitrogen sources and total salt concentration of each medium.
* Why was the ontogeny of somatic embryos on MS medium arrested in the globular or heart-shaped stage of development? What are some culture strategies that would permit continued morphological development?

**Figure 20.2**  Somatic embryogenesis from leaf and flower receptacles tissues of cineraria. a. A leaf explant cultured on SH medium (treatment "A") does not produce somatic embryos. b. The sister leaf half cultured on MS medium (treatment "B") produced numerous somatic embryos (arrows). c. One half of a receptacle cultured on SH medium becomes contorted, but does not produce somatic embryos. d. The other half of the receptacle formed many embryos (arrows), which in this case, are beginning to callus.

## Experiment 2. Maturation of somatic embryos and regeneration of plants

Somatic embryo initiation and development are often affected by nutrition, as demonstrated in the previous exercise, and also by plant growth regulators (PGRs). Sequential media transfer systems are often necessary to promote induction, maturation and germination of somatic embryos (see chapter sixteen). Typically, induction or primary medium for somatic embryogenesis contains relatively high concentrations of PGRs, especially auxin or auxinlike compounds, whereas the levels of PGRs in maturation medium are usually low and, in some, may even be excluded. Germination medium may be the same as maturation medium, may have reduced concentrations of basal salts and sugar or may have incorporated into it low levels of some PGRs. The following exercise illustrates a three-medium scheme to produce somatic embryos that can germinate and form plants and is adapted from a publication by Malueg et al. (1994). The experiment will require about 5 weeks for somatic embryo production, or if acclimatized regenerated plants are desired, about 11 weeks.

### Materials

Each student or team of students will need the following materials:

* Eighteen $60 \times 15$ petri dishes containing MS medium amended with 13.5 μM 2,4-D and 4.5 μM BA

* Six 60 × 15 petri dishes containing MS medium without PGRs, but supplemented with 0.5% (5 g/L) activated charcoal (AC)
* Six 60 × 15 petri dishes containing MS medium without PGRs
* Three GA-7 vessels containing MS medium without PGRs
* Jiffy 7 Peat Pellets (Jiffy Products of America, Inc., Chicago, IL)
* Fafard No. 2 potting mix (Conrad Fafard, Inc., Agawam, MA)
* Small plastic bags

*Procedure*

Surface disinfest and prepare explants for culture as described under general consider-ations. Place one half of a leaf or receptacle on MS with PGRs and label "A-1" and the sister half in another dish containing the same medium and label "B-1". As before, the letters and numerals designate treatments and a specific leaf or receptacle, respectively. Prepare six replications and incubate in the dark at 22 C for 2 weeks. Transfer the explants from the petri dishes labeled "A" to medium containing AC and label with "A" and the explant number. Incubate as before for 3 days and then transfer the same explants to MS medium without PGRs. Transfer the explants from "B" treatment to fresh MS medium with PGRs. Be sure to retain the original labels on all new petri dishes. Incubate the cultures in the previously described conditions for an additional 3 weeks. Count and categorize somatic embryos by developmental stage (i.e., globular, cotyledonary) for each of the treatments. Analyze data by embryo developmental stage and total number of embryos using a t-test for paired variates (see Table 20.1).

Remove only cotyledonary-stage somatic embryos from the explants, transfer to MS medium without PGRs in GA-7 vessels and incubate in the light at about 22 C. When the plants are about 2 cm tall, transfer to moist Jiffy 7 peat pellets and cover with clear plastic bags. Plants are easily acclimatized to ambient conditions by opening the bags for increas-ingly longer intervals over a 2-week period. Plants can then be potted in Fafard No. 2 potting medium and grown as discussed under General considerations, growth of plants.

*Anticipated results*

Somatic embryos should form directly on all explants capable of responding to the treat-ments. However, there should be more developmentally advanced (torpedo and cotyle-donary-stage) embryos on explants that were transferred through the series of three media. Most embryos formed on explants that remained on the initial medium with PGRs should be callused as shown in Figure 20.2d.

About one third of the cotyledonary-stage embryos transferred to MS without PGRs should produce a radicle and then a shoot in less than 10 days; the others will only form a radicle or will never germinate. Most of the plants derived from somatic embryos will have multiple stems, acclimatize easily and produce normal color and shape, but smaller flowers.

*Questions*

* Why do some or most germinating embryos develop multiple shoot apices?
* Why do fewer than half of the embryos placed on germination medium form plants?
* How would you determine if somatic embryos were formed directly or indirectly?

# Literature cited

Larsen, R. 1985. Temperature and flower induction of *Senico × hybridus* Hyl. *Swed. J. Agric. Res.* 15:87–92.

Malueg, K.R., G.L. McDaniel, E.T. Graham and R.N. Trigiano. 1994. A three media transfer system for direct somatic embryogenesis from leaves of *Senico × hybridus* Hyl. (Asteraceae). *Plant Cell Tissue Organ Cult.* 36:249–253.

Murashige, T. and F. Skoog. 1962. A revised medium for rapid growth and bioassays with tobacco tissue cultures. *Physiol. Plant.* 15:473–497.

Schenk, R.U. and A.C. Hildebrandt. 1972. Medium and techniques for induction and growth of monocotyledonous and dicotyledonous plant cell cultures. *Can. J. Bot.* 50:199–204.

Snedecor, G.W. and W.G. Cochran. 1967. *Statistical Methods*, 6th ed., 593 pp. Iowa State University Press, Ames.

*chapter twenty-one*

# Somatic embryogenesis from immature seeds of yellow poplar

*Scott A. Merkle*

*Liriodendron tulipifera* L. (yellow poplar, tulip poplar, tulip tree) is one of the most distinctive and valuable hardwoods in the eastern United States. The tree is characterized by rapid height growth on good sites, reaching up to 40 m in 50 years and by its straight trunk and conical crown. Large volumes of yellow poplar wood are used for furniture, plywood, corestock, millwork, siding and other light construction lumber. It is also used for pulp and for products manufactured from chips or flakes (Russell, 1977). Yellow poplar is known as an excellent source of pollen and nectar for honeybees and as a highly desirable ornamental, for which a number of horticultural cultivars have been described (Santamour and McArdle, 1984).

As with other forest trees, vegetative propagation of yellow poplar is a desirable goal as a means of capturing genetic superiority. Usually less than 10% of open-pollinated yellow poplar seeds are filled (Wilcox and Taft, 1969), making them unreliable for propagation. A number of vegetative propagation methods have been reported for yellow poplar, including rooted stem cuttings, root cuttings and grafting. The ability to root stem cuttings of yellow poplar was very poor until it was discovered that more than 70% success could be obtained by using cuttings taken from stump sprouts or epicormic shoots (McAlpine, 1964; Kormanick and Porterfield, 1966). However, in order to obtain stump sprouts or epicormic shoots, the tree must be either partially girdled or sacrificed entirely. Currently, the most successful method reported for in vitro propagation of yellow poplar is via somatic embryogenesis (Merkle and Sommer, 1986; Merkle et al., 1990). As with most embryogenic systems reported for forest trees, the major drawback of the yellow poplar system with regard to application to tree improvement is that initiation of the cultures relies on the use of immature zygotic embryos, which are of unknown genetic value. However, even with this limitation, the cultures have proved to be valuable tools for such applications as protoplast culture (Merkle and Sommer, 1987) and microprojectile-mediated gene transfer (Wilde et al., 1992).

Traditionally, woody plants in general and forest trees in particular have been challenging subjects for in vitro propagation compared with agronomic and other herbaceous species, because of a number of properties intrinsic to their biology. The following laboratory exercises will illustrate some of these problems and how they can be handled. In addition, yellow poplar embryogenic cultures have the rare property among forest tree embryogenic cultures of proliferating as proembryogenic masses (PEMs) (Halperin, 1966). PEMs may be thought of as very early stage proembryos caught in a repetitive cycle of proliferation. This property is one feature that makes yellow poplar cultures excellent material for mass production of somatic embryos and microprojectile-mediated gene transfer. The following laboratory exercises will illustrate the handling of PEMs for some of these applications. The objectives of the exercises are to give students experience with the following techniques: (1) initiation of repetitive embryogenic cultures from immature seeds and (2) production of somatic embryos from PEMs.

## Exercises

### Experiment 1. Effect of seed developmental stage on embryogenic culture initiation

As stated in the introduction, most embryogenic cultures of forest trees are initiated from seed tissues or seedlings. One factor that has been repeatedly demonstrated to be critical for the successful initiation of embryogenic cultures of many species, including most woody plants, is the developmental stage of the zygotic embryo or seed when it is placed in culture. In many cases, it has been shown that immature seeds have a much higher potential to produce embryogenic cultures than do mature seeds or seedlings. This experiment was designed to illustrate this concept and will also demonstrate the special challenges that forest trees present with regard to procuring the optimum explant. In fact, it should be noted that precisely because this culture initiation experiment deals with obtaining material that is only available during a certain time of year, it may not fit easily into a class schedule, unless the class is offered during the summer months. If the class schedule does not fit in with the requirements of the experiment, it will be up to the instructor to collect the explant material and initiate the cultures so that the embryogenic material will be available for the other experiment in the chapter. Refer to Figure 21.1A–F, for a diagrammatic representation of this exercise. Experiment 1 will require approximately 15 weeks to complete.

#### Materials
The following materials will be needed for each student or team of students:

* 80 ml of 5% commercial bleach
* 10 ml of Roccal (National Laboratories, Montvale, NJ)
* 80 ml of 0.01 N HCl
* Eight sterile 150-ml beakers
* One grafting knife
* Scissors
* Two spoonulas (Fisher Scientific, Pittsburgh, PA)
* One stand magnifier or dissecting microscope
* Ten $60 \times 15$ mm plastic petri dishes containing yellow poplar induction medium (Table 21.1)
* Scalpels with #11 disposable blades
* Scalpels with fixed blades

#### Procedure

*Collection of plant material.*   Collection of yellow poplar fruit should probably be handled by the instructor, but would be an interesting activity for each student group to conduct, if the timing of the class allows. Yellow poplar flowers in the eastern United States any time between April and June, depending on latitude and elevation. The fruit is a bright yellow green, conical aggregate of samaras that is easy to see in the tree after the flower petals fall off. These aggregates should be collected for the first time approximately 8 weeks following flowering, and, if possible, again 4 weeks or more later. Sufficient aggre-

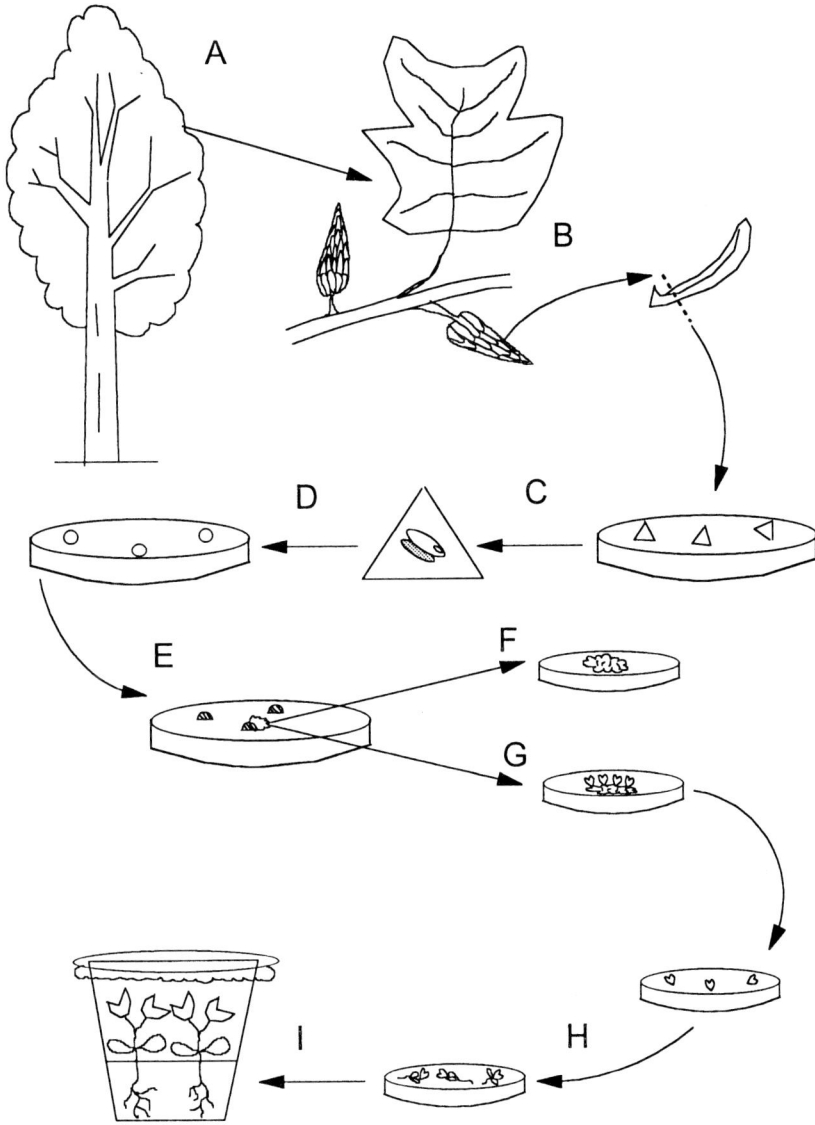

**Figure 21.1** Diagrammatic representation of yellow poplar embryogenic culture initiation and somatic embryo and plantlet production. A. Immature fruit are collected from yellow poplar tree. B. Aggregates are dissected to individual samaras and dewinged. C. Dewinged samaras are sliced open with a scalpel to reveal seed contents (embryo and endosperm). D. Seed contents are scraped out and cultured on dishes of induction medium. E. Explants are transferred to fresh medium after 1 month, and within another month, PEMs proliferate from some of the explants. F. PEMs transferred to basal medium continue to proliferate as PEMs. G. PEMs transferred to basal medium begin to produce somatic embryos within 1 month. H. Cotyledonary-stage somatic embryos transferred to basal medium lacking casein hydrolysate and cultured in the light germinate and produce green leaves within 2 weeks. I. Germinated somatic embryos grow into yellow poplar plantlets following transfer to potting mix.

gates should be collected so that each student or team of students has at least two fruits from each collection date. Often, fruit can be reached from the ground on young, open-grown trees, but for larger trees reaching the fruit may require pruning poles. Fruit should be kept refrigerated in plastic bags until used.

*Table 21.1*   Composition of Yellow Poplar Basal and Induction Media

| Component | Concentration (mg/L) | Reference |
|---|---|---|
| *Major salts* | | |
| $KNO_3$ | 1000 | Witham et al. (1971) |
| $Ca(NO_3)_2$ | 500 | |
| $MgSO_4 \cdot 7H_2O$ | 72 | |
| KCl | 65 | |
| $NH_4NO_3$ | 1000 | |
| *Minor salts* | | |
| $MnSO_4 \cdot 4H_2O$ | 8 | Sommer and Brown (1980) |
| $ZnZO_4 \cdot 7H_2O$ | 4 | |
| $H_3BO_3$ | 2 | |
| KI | 0.6 | |
| $NaMoO_4 \cdot 2H_2O$ | 0.025 | |
| $CuSO_4 \cdot 5H_2O$ | 0.025 | |
| $CoCl_2 \cdot 6H_2O$ | 0.025 | |
| $FeSO_4 \cdot 7H_2O$ | 27.85 | Murashige and Skoog (1962) |
| $Na_2$-EDTA | 41.25 | |
| *Vitamins* | | |
| Thiamine HCl | 1 | Gresshoff and Doy (1972) |
| Nicotinic acid | 0.1 | |
| Pyridoxine HCl | 0.1 | |
| Myo-inositol | 0.1 | |
| *Others* | | |
| Casein hydrolase | 1000 | |
| Sucrose | 40000 | |
| Phytagar | 8000 | |
| 2,4-D[a] | 2 | |
| BA[a] | 0.25 | |

[a] Plant growth regulators are included only in induction medium.

*Seed surface disinfestation, dissection and culturing.*   Use the grafting knife to cut each aggregate in half lengthwise and begin separating the samaras from each other. Before the fruit are ripe, the samaras are fused together tightly, so it will require some effort with the knife and fingers to pry the samaras apart. Once the samaras are separated, cut off the wing from the seed end of each samara with scissors and discard. Be careful not to let the samaras dry out. If they will not be used immediately, store them in a plastic bag in the refrigerator.

Surface disinfest the samaras using eight beakers in the following sequence: 70% ethanol for 20 sec, 10% Roccal (with sterile water) for 3 min, repeat the first two steps using the same solutions over again, 5% bleach for 5 min, sterile water for 3 min, 0.01 N HCl for 3 min, and three more sterile water rinses for 3 min each. Samaras may remain in the last rinse beaker for a few hours without damage. We have found it easiest to move large numbers of samaras between beakers by flipping them out of one beaker and into the next using spoonulas, rather than picking them up with forceps or using cheesecloth sacks.

Lift three or four samaras from the last rinse with sterile forceps and place them on the filter paper in a sterile glass petri dish. Use the scalpel with #11 blade to cut completely through each samara, slicing down through the vertical ridge. Inside each samara are two seeds. Unfortunately, as mentioned earlier, a very low percentage of open-pollinated yellow poplar seeds are filled, so a few hundred will probably have to be opened to find enough filled seeds for the experiment. By cutting through the ridge, the blade should nearly bisect both seeds. It is easy to determine if one or both are filled, since the white

**Figure 21.2** Proliferation of PEMs from explanted yellow poplar embryo and endosperm following 7 weeks of culture on induction medium.

endosperm will be exposed. Using the fixed blade scalpel, scoop as much of the contents out of each seed as possible and place on induction medium. For removing the seed contents, we recommend using a scalpel with a relatively dull blade, so that the contents can actually be pushed out of the cavity in a single clump without being chopped up. Embryos are only about 1 mm long and are very difficult to distinguish from the endosperm, but if the whole endosperm is removed, chances are very good that the embryo is embedded in it. Culture the contents of two or three seeds per petri dish. Seal the petri dishes with Parafilm and incubate in the dark at about 25 C. After 1 month, transfer the explants and any callus that they have produced to fresh petri dishes of induction medium and incubate for another month.

### Anticipated results

At the end of the first month on induction medium, some of the explants may have produced a small amount of callus, which is usually nonmorphogenic. By the end of the second month, however, a yellow, nodular cluster of PEMs should appear at one end of some explants (Figure 21.2). At that time, since each of these PEM-producing explants represents an independent clonal line, each should be transferred to its own petri dish of induction medium. PEM proliferation can be maintained by monthly transfer to fresh medium. If fruit was collected for culture initiation on both of the dates suggested above, most if not all of the cultures producing PEMs will probably be derived from seeds collected on the first date. This is because in yellow poplar, as in a number of other woody species, immature zygotic embryos have a higher potential for production of embryogenic material than do more mature embryos (Sotak et al., 1991).

### Questions

* Why do the immature zygotic embryos usually have a higher potential for production of embryogenic material than the more mature embryos?
* Does an embryogenic culture derived from the seed embryo of a given tree effectively "clone" the genotype of that tree? If not, why not? Is it possible for a seed explant to produce an embryogenic culture that does "clone" the source tree? How?

## Experiment 2. Production of somatic embryos and plants from yellow poplar PEMs

Once proliferating PEM cultures are obtained, it is not difficult to produce somatic embryos and plantlets from them. Like many embryogenic systems, including the model carrot system, yellow poplar PEMs are maintained in a proliferative state by some minimal level of exogenous auxin, and once the concentration of auxin is reduced below this level or removed entirely, clusters of cells on the surface of the PEMs are "released" to develop into somatic embryos. This experiment is designed to demonstrate how PEMs are "released" by removal of 2,4-dichlorophenoxyacetic acid (2,4-D) from the medium, and how germination of these embryos is enhanced by transfer to medium without casein hydrolysate (CH). Refer to Figure 21.1F–J, for a diagrammatic representation of this exercise. Experiment 2 will require about 8 weeks to complete.

### Materials

The following materials will be needed for each student or team of students:

* One dish each of five embryogenic yellow poplar lines producing PEMs (or as many lines as are available). PEMs should have been transferred to fresh induction medium 2–3 weeks previous to the experiment
* Ten $60 \times 15$ mm plastic petri dishes containing yellow poplar induction medium (see recipe — Table 21.1)
* Twenty $60 \times 15$ mm plastic petri dishes containing yellow poplar basal medium (see recipe — Table 21.1)
* Ten $60 \times 15$ mm plastic petri dishes containing yellow poplar basal medium without CH
* Two spoonulas (Fisher Scientific)

### Procedure

For each embryogenic line, transfer about $1/4$ spoonula of PEMs to two petri dishes of basal medium and two petri dishes of induction medium. Use the back of the spoonula to spread the PEMs into a thin layer on the surface of the medium. Seal the dishes with Parafilm and incubate in the dark at about 25 C. Check the petri dishes starting at about 3 weeks after transfer and score the number of somatic embryos that arise from each by the end of 6 weeks.

After 6 weeks, from each of the cultures that produces somatic embryos, use forceps to collect individual cotyledonary-stage embryos and transfer them to either a fresh petri dish of basal medium or basal medium lacking CH. Place the embryos horizontally on the surface of the medium, spacing them at least 0.5 cm apart. Try to harvest enough embryos from each culture to place five to ten embryos on each of two petri dishes of each medium. Seal petri dishes with Parafilm and incubate under cool white fluorescent lights for 16 h per day at about 30 C. After 2 weeks, score each of the petri dishes for the number of embryos that have produced roots and/or have green cotyledons or other leaves.

Germinated yellow poplar somatic embryos can be directly planted in moist potting mix (e.g., Fafard Peatlite) and grown in a chamber with controlled humidity for plantlet production. A 16-h photoperiod provided by cool white fluorescent lights and weekly fertilization with Hoagland's solution or commercial plant food are recommended. The chamber should be maintained at close to 100% relative humidity until two or three new leaves have been produced (usually about 6 weeks); then the humidity may be lowered to ambient levels over the course of about 2 weeks, after which the plants can be transferred to the greenhouse. If a humidifying chamber is not available, plantlets may be grown in

**Figure 21.3**   Developing yellow poplar somatic embryos 6 weeks following transfer of PEMs from induction medium to basal medium.

moistened potting mix in Magenta GA-7 vessels covered with plastic wrap. After a few new leaves have been produced, a few holes can be punched in the wrap; as new leaves appear, more holes can be punched, and, finally, the plantlets can be repotted and moved to the greenhouse.

### Anticipated results
Following 6 weeks of culture, PEMs transferred to basal medium should have produced a number of somatic embryos (Figure 21.3). Globular-stage somatic embryos can be distinguished from PEMs by their smooth, shiny appearance (as opposed to the bumpy surfaces of the PEMs), and later stages are characterized by the emergence of cotyledons and radicle. PEMs transferred to induction medium should continue to proliferate as PEMs. After 2 weeks, some of the somatic embryos transferred to basal medium should have germinated, with elongating radicles and greening cotyledons. A significantly higher percentage of the somatic embryos transferred to basal medium lacking CH should have germinated, and those that have germinated should have longer radicles and more apical development than those cultured on basal medium with CH. Plantlets transferred to potting mix and fertilized weekly should produce at least one new leaf within a few weeks and a branching root system 5–10 cm long should develop within 6 weeks.

### Questions

* Why do PEMs transferred to basal medium begin to produce somatic embryos, whereas those transferred to induction medium continue to produce PEMs?
* What are some possible explanations for why the removal of CH from the basal medium enhances germination of somatic embryos?
* Why is a gradual lowering of relative humidity required in order for the plantlets to survive transfer to greenhouse conditions?

## Literature cited

Gresshoff, P.M. and C.H. Doy. 1972. Development and differentiation of haploid *Lycopersicon esculentum* (tomato). *Planta.* 107:161–170.

Halperin, W. 1966. Alternative morphogenic events in cell suspensions. *Am. J. Bot.* 53:443–453.

Kormanick, P.P. and E.J. Porterfield. 1966. Rooting yellow poplar cuttings. *Forest Farmer.* 26:24, 41–42.

McAlpine, R.G. 1964. A method for producing clones of yellow poplar. *J. For.* 62:115–116.

Merkle, S.A. and H.E. Sommer. 1986. Somatic embryogenesis in tissue cultures of *Liriodendron tulipifera* L. *Can. J. For. Res.* 16:420–422.

Merkle, S.A., and H.E. Sommer. 1987. Regeneration of *Liriodendron tulipifera* (family Magnoliaceae) from protoplast culture. *Am. J. Bot.* 74:1317–1321.

Merkle, S.A., R.J. Sotak, A.T. Wiecko and H.E. Sommer. 1990. Maturation and conversion of *Liriodendron tulipifera* somatic embryos. *In Vitro Cell. Dev. Biol.* 26:1086–1093.

Murashige, T. and F. Skoog. 1962. A revised medium for rapid growth and bioassays with tobacco tissue culture. *Physiol. Plant.* 15:473–497.

Russell, T.E. 1977. Planting yellow poplar — where we stand today. USDA Forest Service Gen. Tech. Rep. SO-17. South. For. Expt. Stn., New Orleans, LA. 8 pp.

Santamour, F.S., Jr. and A.J. McArdle. 1984. Cultivar checklist for *Liquidambar* and *Liriodendron. J. Arboricult.* 10:309–312.

Sommer, H.E. and C.L. Brown. 1980. Embryogenesis in tissue cultures of sweetgum. *For. Sci.* 26:257–260.

Sotak, R.J., H.E. Sommer, and S.A. Merkle. 1991. Relation of the developmental stage of zygotic embryos of yellow poplar to their somatic embryogenic potential. *Plant Cell Rep.* 10:175–178.

Wilcox, J.R. and K.A. Taft, Jr. 1969. Genetics of yellow poplar. USDA Forest Service Research Paper WO-6. 12 pp.

Wilde, H.D., R.B. Meagher and S.A. Merkle. 1992. Expression of foreign genes in transgenic yellow poplar plants. *Plant Physiol.* 98:114–120.

Witham, F.H., D.F. Blaydes and R.M. Devlin. 1971. *Experiments in Plant Physiology.* Van Nostrand-Reinhold Co., New York. 245 pp.

*chapter twenty-two*

# Induction of somatic embryogenesis in conifers

*Stephen M. Attree, Patricia J. Rennie and Larry C. Fowke*

The genus *Picea*, the spruces, is a member of the Pinaceae, the largest family of conifers. There are about 40 species of spruce restricted mostly to the Northern Hemisphere. Due to whiteness of the wood, low resin content, and fiber qualities, the main use of spruce is as pulpwood for the manufacture of paper.

After logging, forests are often left to regenerate naturally, which is slow and unpredictable, as a succession of different plant species will follow. However, increased forest productivity is needed to meet an increasing world demand for wood and wood products. Improved forest productivity can be achieved by more intensively managing forests by, for example, using improved silvicultural practices and replanting with nursery-grown trees following logging. The nursery trees are usually grown from seed collected from the wild; however, forest productivity may be further improved by replanting with genetically superior trees grown from seed obtained from tree breeding programs. Characteristics that can be selected for in tree-breeding programs include faster growth rate, resistance to pests and diseases and tolerance to environmental stresses such as frost or drought. Also important is the lignin content of the wood. Wood with a low lignin content is more suitable for pulping, whereas wood with a high lignin content is more desirable for lumber. Elite seed from breeding programs is expensive and available only in low quantity; therefore, it is used to generate clonal seed-production orchards that mass-produce seed for reforestation. The clonal orchards are established using various conventional means of vegetative propagation, such as rooting cuttings or grafting scions (shootstocks) onto seedling rootstocks. Mature trees in such orchards are pollinated by wind, and the resulting seed is then collected. This improved seed is then grown in nurseries, and the young plants are replanted at the reforestation sites.

Conventional breeding of conifers is a very long-term endeavor. Vegetative propagation would alleviate time constraints, but has a low success rate and is especially difficult with mature trees. However, it is the mature trees that are of most interest for establishing clonal orchards as desirable traits can be readily identified. Because of these difficulties, clonal propagation and breeding procedures are not extensively used to supply improved trees to forest nurseries.

In vitro culture techniques, particularly somatic embryogenesis, provide an alternative means to rapidly clone large numbers of plants with improved characteristics cost-effectively for direct use by forest nurseries. Somatic embryogenesis may be induced from elite seed of known parents from tree improvement programs. Somatic embryogenesis also provides a regeneration system suitable for producing genetically modified plants. The steps to produce improved plants for direct reforestation via conventional means or

Tree breeding program

↓

Genetically superior seed

Grow plants                    Induce somatic
                               embryogenesis

↓

Propagate vegetatively

↓

Grow clonal orchard

↓

Bulk improved seed             Bulk mature
                               somatic embryos

Nursery planting

↓

Reforestation          JEW

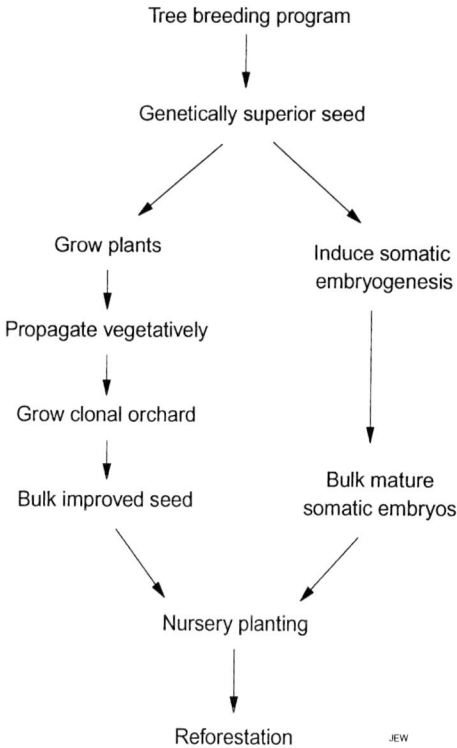

**Figure 22.1**   Steps for improving forest productivity by conventional or in vitro means.

somatic embryogenesis are outlined in Figure 22.1. While somatic embryogenesis in angiosperms has been studied for almost 30 years, somatic embryogenesis in conifers was only reported relatively recently (see chapter sixteen and review by Attree and Fowke, 1993). Since the initial reports of conifer somatic embryogenesis, our understanding of factors controlling induction and development of somatic embryos has made remarkable progress. For conifers, somatic embryogenesis usually involves the culture of zygotic embryos on medium containing auxin and cytokinin. From a proportion of the cultured zygotic embryos arise somatic embryos that undergo continued proliferation in the presence of auxin and cytokinin with rapid increase occurring in liquid suspension culture. Thus, from a single zygotic embryo, unlimited numbers of immature somatic embryos can be derived. These may be induced to undergo synchronous development to mature embryo forms (see Figure 16.7C) by removing the auxin and cytokinin and culturing in bioreactors in the presence of abscisic acid and osmoticum (Attree et al., 1994).

The simple methods provided in this chapter allow students to establish and maintain embryogenic cultures of spruce on gelled medium. The students can then become familiar with the appearance of conifer embryogenic cultures, and the cultures can be compared with embryogenic cultures of angiosperms, as described in other chapters. While methods in this chapter may be applied to a wide variety of conifers, spruces are most suitable for study, as somatic embryos may be induced from immature developing seed and mature stored seed (Tautorus et al., 1990) or cotyledonary explants excised from in vitro–germinated seedlings (Attree et al., 1990). Induction of somatic embryogenesis in pines (*Pinus* spp.) is more difficult than in the spruces, often requiring precotyledonary-stage zygotic embryos to be cocultured with the megagametophyte. Even then, induction occurs at only about a 5% frequency. Also, firs (*Abies* spp.) may require cytokinin alone for induction. Not all conifer species are, at present, capable of embryogenic induction; a recent list is given by Attree and Fowke (1993).

***Table 22.1*** Composition of $^1/_2$ LV Medium

| Components | Concentration (mg/L) |
| --- | --- |
| $NH_4NO_3$ | 825.0 |
| $H_3BO_3$ | 15.5 |
| $CoCl_2 \cdot 6H_2O$ | 0.06 |
| $CuSO_4 \cdot 5H_2O$ | 0.25 |
| $CaCl_2$ | 8.3 |
| $Na_2EDTA \cdot 2H_2O$ | 18.65 |
| $Fe_2SO_4 \cdot 7H_2O$ | 13.9 |
| $MgSO_4$ | 451.5 |
| $MnSO_4 \cdot H_2O$ | 10.5 |
| KI | 2.1 |
| $KNO_3$ | 950 |
| $KH_2PO_4$ | 170 |
| $Na_2MoO_4 \cdot 2H_2O$ | 0.63 |
| $ZnSO_4 \cdot 7H_2O$ | 21.5 |

## Exercises

### Experiment 1. Induction and maintenance of somatic embryos

#### Seed sources and basal medium

Seed may be obtained from a seed orchard; however, the ubiquitous nature of spruces should allow cones containing seed to be collected from local trees. For spruce, mature seed is suitable, but induction frequencies may be higher if immature developing seed is collected every 1–2 weeks in early summer. Seed collected shortly after pollination will contain early stages of zygotic embryos at a stage comparable to the embryos within the embryogenic cultures. A variety of culture media are suitable for culturing conifer somatic embryos; however, half strength Litvay ($^1/_2$ LV) medium (Litvay et al., 1981) has been found to work well for many spruces and is available commercially (Sigma product no. L4272) or Table 22.1. The $^1/_2$ LV medium should also contain 1% sucrose, 2 mg/L (9 µM) 2,4-D, 1 mg/L (4.5 µM) benzyladenine (BA), 500 mg/L casein hydrolysate, 250 mg/L glutamine and 0.6% agar. The plant growth regulators (PGRs) may be added prior to autoclaving at 121 C for 20 min.

#### Materials

For inducing and culturing somatic embryos, the following items are needed for each student or team of students:

* 50–150 seeds, preferably of a spruce
* 50 ml of 15% hydrogen peroxide
* A nonsterile flask capped with muslin cloth secured with a rubber band
* Binocular dissecting microscope
* 100 × 15 mm petri dishes containing culture medium
* Darkened incubator at 25 C

#### Procedure

The procedure for induction and maintenance of somatic embryos is summarized diagrammatically in Figure 22.2. Soak seed overnight in nonsterile tap water to imbibe seeds and soften the seed coat for ease of dissection. Discard the empty nonviable seeds that float.

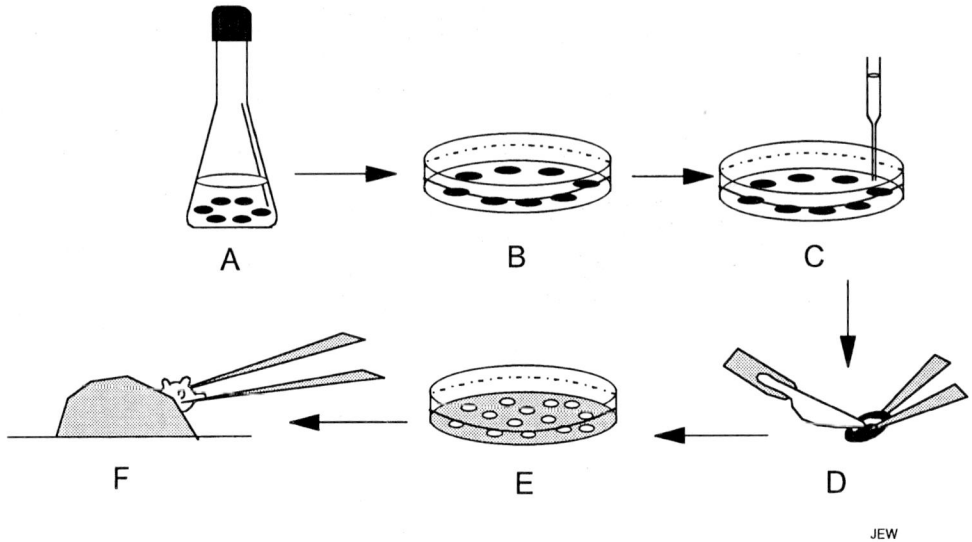

**Figure 22.2** Diagrammatic representation of the methods used for inducing somatic embryos in conifers. A. Soak seeds overnight in water. B. Surface disinfect seeds in 15% hydrogen peroxide. C. Rinse seeds in sterile, distilled water. D. Using a scalpel, excise zygotic embryos from seeds. E. Incubate embryos on culture medium. F. Subculture embryogenic tissues on medium.

Drain the water and transfer the seeds to a sterile $100 \times 15$ mm petri dish and surface disinfect for 15–20 min in about 20 ml of a 15% solution of hydrogen peroxide. Remove the hydrogen peroxide by pipetting and rinse the seeds three times with the sterile distilled water. Using a dissecting binocular microscope and aseptic techniques, excise the zygotic embryos from within the white megagametophyte and place 10–15 embryos on the surface of the culture medium in each petri dish. Excision is best performed by holding the seed in the forceps and gently cutting the seed coat lengthwise, then turning the seed over and cutting the seed coat down the other side. Remove the white megagametophyte, incise lengthwise and pry open with the forceps to reveal the contents. The embryo is situated in the central cavity running lengthwise down the megagametophyte (Figure 22.3a).

Conifers undergo single fertilization to produce the diploid zygote, and the mega-gametophyte is haploid. The megagametophyte carries out a nutritive function similiar to that of the endosperm of angiosperms, which results from double fertilization. Zygotic embryos early in development may be barely detectable as a small, thin hairlike strand of translucent tissue consisting largely of suspensor cells. As zygotic embryo development proceeds, the embryos become easier to distinguish and manipulate as they pass through larger globular and then cotyledonary stages.

Seal the dishes containing the plated zygotic embryo explants and incubate in the dark at 20–25 C. Transfer to fresh medium after 2–4 weeks. Within 3–6 weeks, a translucent white mucilaginous embryogenic tissue, consisting of hairlike immature somatic embryos,

**Figure 22.3** Light micrographs showing zygotic and somatic embryos of conifers. a. Dissected seed of white spruce (*Picea glauca* [Moench.] Voss.) to show mature zygotic embryo (arrow) (bar = 1 mm). b. Translucent embryogenic tissue arising from a cultured zygotic embryo (large arrow), of white spruce. Note the immature somatic embryos (small arrows) protruding from the surface of the embryogenic tissue (bar = 1 mm). c. Single white spruce somatic embryo (arrow) growing from the surface of a cultured white spruce shoot explant (bar = 0.25 mm). (From Attree, S.M. et al., *Can. J. Bot.* 68:30–34, 1990. With permission.) d. Somatic embryo of white spruce showing early stage of embryo head cleavage (large arrow). The suspensor (small arrow) consists of elongate highly vacuolated cells (bar = 0.1 mm). e. Late stage of cleavage showing two somatic embryos (arrows) of white spruce (bar = 0.2 mm). f. Early jack pine (*Pinus banksiana* Lamb.) zygotic embryos undergoing cleavage. One of the embryos matures (large arrow) while the others (small arrows) degenerate (bar = 0.2 mm). (From Attree, S.M. and Fowke, L.C., *Plant Cell Tissue Organ Cult.*, 35:1–35, 1993. With permission.) g. Cleaving white spruce somatic embryo at stage similar to that shown in Figure 22.3d stained with aceto-orcein to reveal the distribution of nuclei (bar = 0.1 mm). h. White spruce somatic embryo head stained with aceto-orcein to show nuclei and mitotic figures. Note the cell at metaphase (arrow) (bar = 50 μm). i. Cells of white spruce somatic embryo head stained to show nuclei and mitotic figures (bar = 20 μm).

should be evident proliferating from a proportion of the explants (Figure 22.3b). Using a pair of fine forceps, separate this tissue from the nonembryogenic tissues and culture in a separate petri dish. Nonembryogenic tissue is readily distinguishable as a white or green undifferentiated hard callus.

### Maintenance of embryogenic cultures

Subculture every 2–4 weeks to fresh maintenance medium. During subculture remove a small portion (about 5 mm diam.) of the white tissue from the edge of the embryogenic tissue and transfer to fresh medium. About seven clumps can be maintained per petri dish, and several cultures should be kept to guard against contamination.

### Anticipated results

The frequency of induction varies according to conifer species and maturity of seed or age of seedling if the latter is used as explant material. However, induction frequencies of 30% can be expected and may go as high as 70% if immature spruce seeds are used. Spruce somatic embryos tend to arise usually within 4–6 weeks of planting; however, they sometimes appear as early as 2 weeks after planting if immature zygotic embryos are used as explants. Also, the embryogenic tissue may arise over a large proportion of the explant when immature zygotic embryos are used (Figure 22.3b). It generally takes longer (e.g., up to 12 weeks) for mature zygotic embryo explants to give rise to embryogenic tissue. Mature embryo explants more often undergo some nonembryogenic callusing prior to the appearance of embryogenic tissue, and somatic embryos may arise from just one point on the explant (Figure 22.3c). A lack of any growth of the explants is probably due to lack of viability of the embryo explant, which is not uncommon.

### Questions

* Why is somatic embryogenesis attractive to the forest industry?
* What physical features are characteristic of embryogenic tissue?

## Experiment 2. Morphology of somatic embryos

Conifer somatic embryos provide excellent material for light microscopic observation. Embryos can be examined in a bright field compound microscope either unstained or using a simple aceto-orcein stain that colors the nuclei and chromosomes red.

### Materials

For microscopic examination of somatic embryos, the following items are needed:

* Aceto-orcein stain (prepare a 45% acetic acid solution by adding 45 ml of glacial acetic acid to 55 ml distilled water; add 0.15 g orcein powdered stain to 15 ml of the 45% acetic acid solution)
* 1.5-ml eppendorf tubes
* Forceps and dissecting needles
* Slides and coverslips
* Alcohol lamp or other heat source
* Paper tissues for blotting
* Compound microscope
* Small beaker or water glass

*Procedure*

Place a small mass of embryogenic tissue in a drop of water on a microscope slide. Tease tissue apart with dissecting needles, add a coverslip and examine using a light microscope. Close the condensor diaphram slightly to increase contrast. Locate individual embryos and try to determine which part of the embryo contains actively dividing cells and whether or not cytoplasmic streaming is occurring in any of the cells.

To stain nuclei and chromosomes, place a small solid clump or mass of cells into a few drops of aceto-orcein stain in an eppendorf tube for 30 min or longer depending on the size of tissue mass (1 h in order for stain to diffuse into large compact embryo heads). Heat the tube in a beaker of water to 60–80 C for 10 min. Remove a small sample of stained tissue from the tube and spread on a slide in a drop of stain. Observe under the microscope, being careful to keep stain solution away from the microscope lenses (acetic acid will damage lens mount) and skin. The nuclei of the embryos should stain an intense red against the background of stain on the slide. To intensify the stain, gently warm the slide plus stain over the alcohol lamp, being careful not to boil the stain. When the nuclei are adequately stained, place a coverslip over the sample and remove residual stain by placing folded tissues at one side of coverslip and adding 45% acetic acid solution to the opposite side. When most of the stain has been replaced with acetic acid, blot slide and observe. Nuclei and chromosomes should remain red, whereas other cellular materials are colorless.

*Anticipated results*

With light microscopy, the proliferating embryogenic tissue consists of differentiated immature somatic embryos, which resemble immature zygotic embryos. The somatic embryos are polarized structures that are organized into a meristematic embryonic head region subtended by elongate suspensor cells (Figure 22.3d and e and Hakman et al., 1987). With unstained living preparations, it is sometimes possible to observe cytoplasmic streaming in the suspensor cells closest to the embryonic region.

Under the influence of auxin and cytokinin, the immature somatic embryos continue to proliferate by splitting into two without further development (Figure 16.7c and Figure 22.3d and e) — a process termed cleavage polyembryogenesis. This type of cleavage occurs naturally within seeds of certain conifer species when the early developing zygotic embryo separates into four independently developing embryos (Figure 22.3f). Competition among these embryos results in only the most aggressive surviving and maturing. Conifer embryogenic tissues differ in appearance from those of angiosperms. The early stages of angiosperm somatic embryogenesis generally consist of proliferating cell clusters, called proembryonic cell complexes (Haccius, 1978), that are disorganized and so do not form recognizable structures unless auxin is reduced to encourage the later stages of embryo development. These differences may reflect a long period of evolutionary separation of these two important plant groups (Attree and Fowke, 1993).

With stained preparations at low magnification, areas of meristematic cells should stain strongly in contrast to suspensor regions that remain unstained (Figure 22.3g). At higher magnification, mitotic figures may be observed in embryo head regions (Figure 22.3h and i). Mitotic frequencies may be calculated by dividing the number of cells with mitotic figures by the total number of cells in a sample.

*Questions*

* Describe the general morphology of a conifer somatic embryo. Is it similar to the corresponding zygotic embryo or angiosperm zygotic and somatic embryos?
* How do conifer somatic embryos proliferate in culture?
* Where do cell divisions occur in the somatic embryos?

## Literature cited

Attree, S.M., S. Budimir and L.C. Fowke. 1990. Somatic embryogenesis and plantlet regeneration from cultured shoots and cotyledons of seedlings germinated from stored seed of black and white spruce (*Picea mariana* and *Picea glauca*). *Can. J. Bot.* 68:30–34.

Attree, S.M. and L.C. Fowke. 1993. Embryogeny of gymnosperms: advances in synthetic seed technology of conifers. *Plant Cell Tissue Organ Cult.* 35:1–35.

Attree, S.M., M.K. Pomeroy and L.C. Fowke. 1994. Production of vigorous, desiccation tolerant white spruce (*Picea glauca* [Moench.] Voss.) synthetic seeds in a bioreactor. *Plant Cell Rpt.* 13: 601–606.

Haccius, B. 1978. Question of unicellular origin of non-zygotic embryos in callus cultures. *Phytomorphology.* 28:74–81.

Hakman, I., P. Rennie and L.C. Fowke. 1987. A light and electron microscope study of *Picea glauca* (white spruce) somatic embryos. *Protoplasma.* 140:100–109.

Litvay, J.D., M.A. Johnson, D. Verma, D. Einspahr and K. Weyrauch. 1981. Conifer suspension culture medium development using analytical data from developing seeds. *Tech. Pap. Ser. Inst. Paper Chem.* 115:1–17.

Tautorus, T.E., S.M. Attree, L.C. Fowke and D.I. Dunstan. 1990. Somatic embryogenesis from immature and mature zygotic embryos, and embryo regeneration from protoplasts in black spruce (*Picea mariana* Mill.). *Plant Sci.* 67:115–124.

*Part IV*

*Crop improvement techniques*

*chapter twenty-three*

# Use of protoplasts for plant improvement

*Michael E. Compton, James A. Saunders and Richard E. Veilleux*

A fundamental difference between plant and animal cells is that certain plant cells are totipotent and can be induced to divide and develop into complete plants (see chapters seven, eleven and sixteen). Another significant difference is the presence of a cell wall in plants. The plant cell wall, although playing a key role in processes involving plant structure and function, is a major impediment in exploiting direct DNA transfer to individual cells and the production of somatic hybrids by cell fusion. Fortunately, the cell wall can be temporarily removed from plant cells without a significant loss of viability. Plant cells from which the cell wall has been removed are termed protoplasts. Protoplasts are somewhat unique in plant cell culture in that they exist as separate cells without cytoplasmic continuity with neighboring cells.

Plant protoplasts were first isolated by Klercker in 1892 by slicing onion bulb scales with a fine knife in a plasmolyzing solution, resulting in the release of protoplasts when cells were cut through the wall (Bhojwani and Razdan, 1983). Protoplast yield was low and the procedure was restricted to highly vacuolated, nonmeristematic cells. Isolation of plant protoplasts became more widely practiced in the 1960s with the extraction and purification of enzymes that could be used to degrade the plant cell wall. Cocking (1960) discovered that intact protoplasts could easily be obtained by incubating plant tissues in a concentrated cellulase solution prepared from the fungus *Myrothecium verrucaria*. By 1968 commercial preparations of purified cell wall–degrading enzymes such as macerozyme and cellulase were available, which allowed easy isolation of protoplasts from plant leaves and nonvacuolated cells.

## Isolation of plant protoplasts

Plant protoplasts can be produced by two procedures: (1) mechanically slicing or chopping of plant tissues, or (2) enzymatically digesting the cell wall using various hydrolytic enzymes.

To isolate protoplasts by mechanical slicing, numerous cuts are made through plant tissue. This eventually allows some of the protoplasts to slip out of tangentially sectioned cells. This procedure produces protoplasts rapidly, but the yields are low and the technique has never gained popularity. When the procedure is used, it works best on thick tissues like potato tubers or apple fruit (Leigh and Branton, 1976).

Protoplast isolation by the more common procedure of enzymatic cell wall digestion involves the use of cellulase, hemicellulase and pectinase, which are extracted from various sources including fungi, snail gut and termite gut. These hydrolytic enzymes are available commercially in differing formulations of varying purity. Digestion by a combination of these three enzymes is generally conducted at a pH of 5.5–5.8 over a period of 3–16 h. No matter which procedure is used, protoplasts can be collected and purified by centrifugation techniques designed to separate broken and damaged cells from intact protoplasts by taking advantage of their differing buoyant densities.

## Uses for plant protoplasts

Plant breeders have used sexual hybridization to improve cultivated crops for centuries. This process is generally limited to plants within a species or to wild species that are closely related to the cultivated crop. When possible, sexual hybrids between distantly related species have been useful for the incorporation of single gene traits such as insect and disease resistance. However, intraspecific and interspecific incompatibility barriers limit the use of sexual hybridization for crop improvement.

Protoplast culture has been used to develop plants with improved agronomic and horticultural characteristics and improved disease resistance through recovery of culture-induced variants, parasexual hybrids and genetically engineered plants. By isolating protoplasts and regenerating plants from them, variants (somaclones) with improved characteristics have been obtained. Parasexual or somatic hybrids can be obtained by fusing protoplasts of unrelated or distantly related species. Through genetic engineering, foreign genes can be transferred to plant cells and improved plants with the new genes regenerated from the transformed cells. Improved plants obtained by protoplast manipulations can be used in breeding programs to develop new cultivars. This chapter discusses the use of protoplasts in plant improvement

## Procedures for regenerating plants from protoplasts

Leaf mesophyll tissue from plants grown in a greenhouse or growth chamber with controlled light (16:8 or 18:6 h light to dark photoperiod of, 60–300 $\mu mol \cdot m^{-2} \cdot sec^{-1}$ light intensity) and temperature (20–25 C) is commonly used as a source of plant protoplasts. Uncontrolled greenhouse environment may hinder the repeatability of experiments. Cells within the leaf mesophyll are loosely packed, which allows penetration of digestive enzymes that facilitate protoplast release. Typically $10^6$ to $5 \times 10^7$ protoplasts can be obtained per gram of leaf tissue. Generally, the youngest, fully expanded leaves from young plants or seedlings are used. Preconditioning, e.g., by placing plants in darkness for 24–72 h before protoplast isolation or giving them a cold treatment (4–10 C), may improve protoplast yield.

Leaves from greenhouse or growth chamber-grown plants require surface disinfestation prior to protoplast isolation. In vitro plantlets are often preferred as a source of protoplasts, because of the advantages conferred by aseptic conditions and a controlled environment. In vitro sources of protoplasts include callus, cotyledons, hypocotyls, embryogenic suspension cultures, leaves, shoots or somatic embryos.

Before protoplast isolation, plant tissues are cut into small pieces and floated in an osmotically adjusted solution at 20–25 C for 1–24 h. During this step water moves out of the cells, which causes their contents to shrink and draw away from the cell wall (plasmolysis). This allows cells to retain their integrity after the cell wall is removed. Sometimes it is beneficial to conduct this step at low (4–10 C) temperature. Next, the pieces are incubated in digestive enzymes for removal of the cell walls and middle lamellae. Incubation is conducted in darkness on a shaker (30–50 rpm) for 4–18 h at 25–30 C. Incubation time and temperature vary with species and tissue. Plasmolysis and enzyme incubation steps may be conducted simultaneously. Following enzyme treatment, the digest is gently

**Figure 23.1** Freshly isolated protoplasts from in vitro plantlets of *Solanum phureja*.

swirled to release protoplasts and filtered (50–100 μm mesh size) to separate protoplasts from large debris. The protoplast suspension is centrifuged at low speed ($50 \times g$) for about 10 min. Protoplasts and debris collect in the pellet at the bottom of the tube. The supernatant is discarded and protoplasts resuspended in a high sucrose (flotation) medium, which is overlaid with 1 ml of rinse medium containing mannitol. During centrifugation, viable protoplasts collect at the interface of the two media while debris concentrates in the pellet. Protoplasts are removed from the interface with a pasteur pipet and washed two to three times before transfer to culture medium. There are several variations for purification of protoplasts that combine or separate the various steps outlined in this general procedure.

After the final wash, protoplasts are suspended in a small volume (about 1 ml) of liquid culture medium. A sample is removed and transferred to a hemacytometer to determine protoplast density (see chapter twenty-five for details). Enough culture medium should be added to adjust the protoplast density to $10^4$–$10^6$ cells per milliliter. Protoplasts may be cultured in a thin layer of liquid medium, embedded in agarose, in liquid medium on top of solid medium or in agarose droplets suspended in liquid. The type of culture method depends on the species and/or purpose of the experiment. Liquid medium is often used because cell density and osmotic pressure can be efficiently adjusted. In addition, cells of some species are unable to divide in agar-solidified medium. The development of individual cells can be observed if protoplasts are embedded in agarose because they remain stationary.

Protoplast viability may be estimated by staining a sample with fluorescein diacetate (FDA) (see chapter twenty-five). The stain is mixed with protoplasts and fluorescence observed under ultraviolet light. Viable protoplasts actively absorb the stain and exhibit a green fluorescence, whereas nonviable protoplasts do not. Although nonviable mesophyll protoplasts may fluoresce red because of the autofluorescence of chlorophyll, they are not metabolically active. Another viability stain, Evans blue, does not require fluorescence microscopy to detect protoplast viability (see chapter twenty-five).

Freshly isolated protoplasts are spherical (Figure 23.1, see also Figures 25.2, 25.4 and 26.2) because they are unrestricted by a cell wall. Under suitable conditions, viable protoplasts regenerate a new cell wall within 48–96 h after isolation. Presence of the cell wall can be determined by a change in shape from spherical to ovoid or by staining cells with Calcofluor White, which detects the presence of cellulosic wall materials. Stained cells or

**Figure 23.2** First mitotic division after protoplast isolation of *Solanum phureja*. Notice that the cells are no longer spherical and are more opaque than the freshly isolated protoplasts. (From Cheng, J. and R.E. Veilleux, *Plant Sci.*, 75:257, 1991. With permission.).

protoplasts with remnant cell wall material or protoplasts that have resynthesized a new cell wall fluoresce bluish white under ultraviolet light. Protoplasts without a wall do not fluoresce (see chapter twenty-five). Protoplasts that fail to regenerate a wall generally will not divide and eventually die.

Not all healthy protoplasts divide. Therefore, the plating efficiency (PE = number of dividing protoplasts/total number of protoplasts) is used to estimate cell vigor. PE is usually calculated 1 week after protoplast isolation and varies from 0.1–80%. However, a PE of 20–30% is most common. The first cell division (Figure 23.2, see also Figure 26.3) often occurs within 2–10 days after isolation. Division of protoplasts is affected by culture medium, osmoticum, plating density, storage conditions and source tissue.

After protoplasts have regenerated a new wall and divided, they are diluted with culture medium that has lower osmoticum. Multicellular colonies (Figure 23.3) develop within 14–21 days and macroscopic calli (Figure 23.4) within 4 weeks. Plant regeneration occurs by shoot regeneration or somatic embryogenesis. The first shoots or embryos may be seen as early as 1 month after protoplast isolation but may require 6 months to 1 year.

## Factors that influence protoplast growth and development

Modifications to normal tissue culture procedures often are required because of the delicate nature of protoplasts. Changes often include adjustment to the inorganic salt concentration, addition of organic components, vitamins, sugars to control osmolarity and plant growth regulators (PGRs) to stimulate cell division.

### Culture media

The best culture medium for protoplasts is often similar to, or slightly modified from, that used for organ regeneration from other explants. Adjustment of ammonium nitrate and

**Figure 23.3** Multicellular colonies seven days after isolation of protoplasts of *Solanum phureja*. (From Cheng, J. and R.E. Veilleux, *Plant Sci.*, 75:257, 1991. With permission.).

**Figure 23.4** Macroscopic p-calli several weeks after isolation of protoplasts of *Solanum phureja*.

calcium is often required to stimulate cell division. Ammonium nitrate is essential to promote cell division, but is toxic to protoplasts at concentrations (20 mM) used in most tissue culture media. To avoid toxicity and promote cell division, the ammonium nitrate concentration is lowered to quarter to half strength. In contrast, the calcium concentration is usually increased for most protoplast procedures. The calcium concentration of most tissue culture media ranges from 0.5–3 mM. At this low concentration, protoplasts typically aggregate and brown rapidly. Raising the calcium concentration (14–40 mM) promotes early cell division, cell synchronization, decreases aggregation and reduces browning of protoplasts during the early stages of culture. Normal ammonium nitrate and calcium levels are restored after protoplasts have regenerated a new wall and divided.

Organic components such as inositol, nicotinic acid, pyridoxine, thiamine, glycine, folic acid and biotin are found in many tissue culture media. In addition, casein hydrolysate, D-Ca-pantothenate, choline chloride, cysteine, malic acid, ascorbic acid, adenine sulfate, riboflavin and glutamine are often added to protoplast media in small amounts (0.01–10 mg/L) to hasten cell wall synthesis and promote cell division. Most of these compounds are not used after protoplasts have regenerated a new cell wall and divided.

Sugars are used in protoplast culture media as osmotic stabilizers and carbon sources. Because of the elimination of wall pressure (i.e., pressure potential), a component of water potential, culture media must be osmotically adjusted to prevent rupture of protoplasts during isolation and early culture. Mannitol, sorbitol and glucose are used as osmotica at concentrations from 0.3–0.7 M. Sucrose and glucose are used as carbon sources at concentrations from 0.2–0.6 M. The sugar concentration is gradually reduced after cells synthesize a new wall and divide. Osmotic sugars are usually eliminated by the time macroscopic colonies are visible.

Exogenous PGRs are necessary to promote cell division. Both auxins and cytokinins typically are used. The type and concentration differ with the species and mode of regeneration. Naphthaleneacetic acid (NAA) and 2,4-dichlorophenoxyacetic acid (2,4-D) at 0.45–10.7 µM are typical auxins. BA and zeatin are usual cytokinins used at concentrations of 2–5 µM. Kinetin and zeatin riboside are sometimes used instead of, or in combination with, BA or zeatin at similar concentrations. Coconut milk has been substituted for cytokinin at concentrations of 20–40 ml/L.

## Culture environment

Following isolation, protoplasts are sensitive to light and should be incubated in darkness or exposed only to filtered light (5–30 µmol·m$^{-2}$·sec$^{-1}$) until they synthesize a new wall and divide. At this point, they can be exposed to elevated light levels (30–300 µmol·m$^{-2}$·sec$^{-1}$). White light is often best as blue or red light alone may inhibit shoot formation. Protoplasts are generally incubated at 20–25 C. Optimum environmental conditions vary and should be tested for each species.

## Somaclonal variation

Spontaneous mutation rates in plants vary depending on the species and DNA sequence. When mutations occur in somatic cells, they generally will not be observed because somatic cells within mature plant tissue do not divide to give rise to mutated sectors of tissue. Tissue culture can be used as a means of recovering somatic mutations because the culture environment encourages the division of individual somatic cells within explanted tissue. In addition, a large number of explants can be cultured in a small space. From leaves or cotyledons placed in culture, four to ten shoots often can be regenerated per explant.

Such shoots originate from single cells or small groups of cells. Under these circumstances, only 4–100 cells may contribute to the shoots regenerated from a single explant, limiting the probability of obtaining somatic variants. In contrast, about $10^4$–$10^6$ protoplasts are cultured per milliliter of medium. A typical $60 \times 15$ mm petri dish with 4–10 ml of medium contains 40,000 to 10 million cells. Given that each cell is viable and has a chance of regenerating into a plant, the possibility of recovering somatic variants is improved.

Somaclonal variants have been recovered from protoplast cultures of various species. Some variations include changes in leaf and flower morphology, pollen viability, variation in secondary product production and improved disease resistance. Protoclones, i.e., pro-toplast-derived plants of an asexually (clonally) propagated crop, have been examined in potato with the hope of obtaining plants with improved horticultural characteristics. Protoclones that exhibited compact growth habit; variation in tuber characteristics such as maturity date, color, russeting, protein content, size and number of tubers per plant; changes in flowering response to photoperiod; improved resistance to early blight (*Alternaria solani* [Ell. and Martin] Sor.) and late blight (*Phytophthora infestans* [Mont.] d By.) were regenerated from protoplasts of potato (*Solanum tuberosum* L. cvs. Russet Burbank [Shepard et al., 1980] and Bintje [Burg et al., 1989]). Plants with improved resistance to fusarium wilt (*Fusarium oxysporum* Schlecht. F. sp. *lycoperici* [Sacc.] Snyd. and Hans. race 2) were obtained by challenging tomato (*Lycopersicon esculentum* Mill. cv. UC 82) protoplasts and protoplast-derived calli (p-calli) with fusaric acid (20–30 µM) or regenerating plants without selection pressure (Shahin and Spivey, 1986). Plants with two- to four-fold higher lanatoside C (a glycoside that yields digoxin, glucose and acetic acid on hydrolysis) content than plants produced from meristem culture or seed, respectively, were regenerated from *Digitalis lanata* J.F. Ehrh. protoplasts (Diettrich et al., 1991).

Somatic variants often constitute only a small (about 10%) portion of the total number of regenerants. However, there have been instances where 50–100% of the regenerants were variants. For example, 75% of tobacco (*Nicotiana tabacum* L. cv. Vielbättriger [2n=4x=48]) plants regenerated from protoplasts underwent chromosome doubling (Vyskot et al., 1991), and only tetraploid (2n=4x=48) or aneuploid (2n=46–96) plants were regenerated from protoplasts of monohaploid (2n=1x=12) and dihaploid (2n=2x=24) potato (*S. tuberosum*) (Ramulu et al., 1989). Approximately 98% of plants regenerated from diploid *S. phureja* Juz. and Buk. (2n=2x=24) were tetraploid (2n=4x=48) or aneuploid (2n=44-47) and less than 2% were diploid. In most cases ploidy changes were caused by endopolyploidization of individual cells during the first cell divisions rather than spontaneous cell fusion, based on the fact that freshly isolated protoplasts had the expected DNA content, whereas nuclei from 8-day-old cells exhibited at least twofold higher DNA content. Lack of regenerants with the original ploidy level may be due to a greater competence of polyploid cells for shoot regeneration.

## Protoplast fusion and somatic hybridization

Once purified protoplasts have been obtained from two different plant or tissue sources, various treatments can be applied to induce them to fuse together to form hybrid cell lines. The simplest way to do this is to use spontaneously fusogenic protoplasts. Such cell lines have been described in detail in carrot and other systems (Boss, 1987). Unfortunately, although fusogenic cell lines exist for some species, they are not common enough for widespread use in plant protoplast research. Generally, chemical agents or electrical manipulation (see chapter twenty-four) are necessary to induce membrane instability that leads to protoplast fusion.

## Protoplast

**Figure 23.5**   Typical protocol for protoplast fusion of potato using PEG. 1. Dilute protoplast mixture to $10^4$ cells per milliliter in 0.5 M glucose, 3 mM $CaCl_2$, 0.7 mM $KH_2PO_4$, pH 5.7. 2. Add three 200-µl volumes of agglutinating solution (0.09 mM PEG 6000, 10 mM $CaCl_2$, 0.7 mM $KH_2PO_4$, pH 5.5). 3. Add five 100-µl volumes of diluting solution (50 mM $CaCl_2$, 0.5 M glucose, 50 mM glycine buffer, pH 10.0). 4. Wash cells with culture medium and culture in hanging drop suspensions.

### Polyethylene glycol-mediated protoplast fusion

A number of aqueous solutions have been used to induce chemical fusion of plant proto-plasts (see Saunders and Bates, 1987 for review). These include various salt solutions (NaCl, KCl, $NaNO_3$, $KNO_3$), dextran sulfate, polyvinyl alcohol, lysolethicin and polyeth-ylene glycol (PEG). Of these fusogenic agents, PEG is used most frequently in conjunction with alkaline pH and high calcium concentrations. Figure 23.5 depicts a typical protocol for the use of PEG, alkaline pH and high calcium for protoplast fusion (see chapter twenty-four).

There are a number of steps in the fusion of plant protoplasts using PEG as a chemical facilitator. Initially, the cell membranes must be brought into close physical contact. As most cell membranes possess a net negative surface charge, adjacent cells with similar charges tend to repel each other. Some type of cell agglutination reaction that overcomes the repellent effect of similar net negative surface charges on the protoplast membranes is often the primary role of a chemical fusogenic facilitator. Some lectins, such as concanava-lin A and immune antisera, promote cell fusion simply by this mechanism. PEG is a potent cell agglutinator, but also functions in an important secondary role as a membrane modi-fier. There is evidence to suggest that, once cell membranes are in close contact, membrane surface proteins migrate to create areas of lipid-rich regions. It is during this period that the dehydrating effect of PEG on the cell membrane, and the ability of PEG to bind to phospholipids within the membrane, becomes important. Continuities between adjacent cells begin to form during this cell adhesion period. Subsequently, successive washing with high concentrations of calcium in a buffer with an alkaline pH are typically done.

## Electrofusion-mediated protoplast fusion

Another type of cell fusion that has emerged in recent years involves the manipulation of cell membranes by electrical currents. The electrofusion of plant protoplasts has several advantages over PEG fusion in that it does not employ reagents that are often toxic to the cells being fused. As with all other procedures, however, conditions for electrofusion must be optimized for specific cell types to achieve maximal effectiveness. Typically, we can achieve a yield of 20% or greater fusion products by electrofusion of protoplasts compared to less than 1% fusion products with PEG.

There are two processes necessary for effective electrofusion of protoplasts. These two steps are similar to those described for PEG fusion. Initially, the protoplasts must be brought closely together (agglutination), and, secondly, the cell membranes must be perturbed by an electrical shock to promote membrane fusion between the two cells. With electrofusion, the most successful process for bringing the protoplasts into close physical contact is a technique called dielectrophoresis. This procedure subjects the protoplasts to a low voltage (50–100 V/cm electrode gap) oscillating sine wave current of approximately 1 MHz frequency. The dielectrophoretic phenomenon polarizes the cells so that each protoplast has a positive and negative area. The positive end of one protoplast is drawn toward the negative area of an adjacent protoplast such that "pearl chains" of protoplasts are formed prior to fusion. The fusion itself may be accomplished by a relatively high voltage (0.5–1.5 kV/cm electrode gap) pulse that destabilizes the cell membranes at specific sites on each protoplast, resulting in the fusion of adjacent cells. The electrical pulse does not affect the organelles inside the cell because the current only passes along the cell membranes that are generally more conductive than the cytoplasm.

It is necessary to supply the voltage of the fusion pulse for some critical duration at a minimal threshold level in order to achieve successful fusions. Pulses that are too brief (less than 10 μsec) or of insufficient voltage will be ineffective. If the minimal threshold voltage of the pulse is greatly exceeded, then the cells will not recover from the treatment.

There are several reports of useful somatic hybrid plants that have been produced by electrofusion. For example, the Colorado potato beetle voraciously eats the leaves of domestic potato. This is the major insect pest in the United States of cultivated potato; however, it will not eat the foliage of a closely related, wild species, *S. chacoense* Bitt., because of the presence of specific alkaloids, leptines, in the leaf tissue. We have successfully fused protoplasts from somatic leaf tissue of *S. tuberosum* with those of *S. chacoense* by electrofusion. The fused cells grew as p-calli for several weeks in tissue culture. These p-calli could be subsequently regenerated into hybrid plants with traits of both "parents". When leaves of the hybrid plants were tested in bioassays with Colorado potato beetles, the insects showed a threefold greater preference for the cultivated potato than for either *S. chacoense* or the somatic hybrid. Figure 23.6 shows a choice-preference bioassay in which adult Colorado potato beetles were presented with a selection of leaves from *S. tuberosum*, *S. chacoense*, and from the hybrid plant. Although the tubers of these hybrids are unsuitably small for potato production, the somatic hybrids can be used in potato-breeding programs to incorporate insect resistance into cultivated lines.

## Selection of hybrid cell lines

An important feature of any somatic hybridization procedure, whether using chemical facilitators such as PEG or electrofusion, is the identification and selection of the fused cells. Typically, an investigator is interested in separating the hybrid cell lines, which are the result of two fused cells from different parents (heterokaryotic fusion), from all other cells in the mixture. Unless there is some identification and selection system incorporated

**Figure 23.6**    An in vitro feeding preference assay using adult Colorado potato beetles feeding on freshly excised leaves of *S. tuberosum* (S.t.), *S. chacoense* (S.c.) or somatic hybrids produced from these two parents. A. (somatic hybrid SH4) and C. (somatic hybrid SH1) depict typical assays of feeding trials of leaves from two somatic hybrids. Pictures were taken 4 h after initiation of the feeding. In all cases, a preference was shown to the leaf of the control parent, *S. tuberosum* cv. Kennebec, which was included as a choice in each assay. (From Cheng, J. et al., *In Vitro Cell. Dev. Biol*, 31:90, 1995. With permission.)

into each of the parental cell lines prior to fusion, this stage of the process can be very difficult. For example, cell lines that result from fusion of two or more cells from the same parent or from cells that did not undergo fusion may greatly outnumber the heterokaryotic fusion products. Furthermore, there is no guarantee that just because a cell fusion occurred, nuclear fusion will follow. Fusion of the two membrane-bound nuclei within a single cell and coordination of mitosis involving the chromosomes from what were formerly two independent nuclei are required to form a stable hybrid. A number of selection systems have been developed based on specific requirements of the prospective parent cell lines. These selective systems include antibiotic-resistant cell lines, herbicide-resistant cell lines and cell lines with the ability to grow on specific amino acid analogues, among others. It is most convenient to use two parental cell lines with differing requirements as selective screens. By invoking these selective screens on the populations of cells subjected to a fusion treatment, only those cells that possess the complementary traits of both parents, i.e., the fused cell lines, will thrive.

Unfortunately, most commercially important cultivars of crop plants that are candidates for cell fusion research do not have such sophisticated selection systems. Prior transform ation of the parental plants to incorporate a selective trait may be required. This is a very time-consuming process and can sometimes be avoided by utilizing internal traits of each parent using a manual selection system. In the case of the insect-resistant potato hybrids cited above, the *S. chacoense* cell line contained a visible red-colored anthocyanin pigment, but grew poorly in tissue culture and did not regenerate shoots (Figure 23.7). The *S. tuberosum* cell line did not contain any visible anthocyanins, but regenerated shoots readily. By looking for regenerated shoots that contained a red pigment, we were able to identify putative hybrid plants efficiently. We also noticed that the callus tissue from hybrid fusions grew much more vigorously than callus tissue from either parent. By selecting the most rapidly growing p-calli that regenerated shoots with anthocyanins, we were able to identify putatively transformed potato hybrids.

Even when such obvious differences are lacking, parental cell lines can be given different properties by the use of vital stains that can be recognized by fluorescence

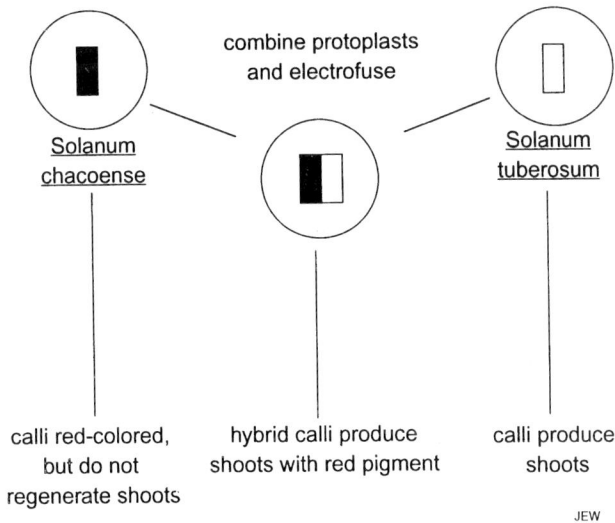

**Figure 23.7** Schematic diagram of selection of calli and shoots produced from electrofused protoplasts from *S. chacoense* and *S. tuberosum*. *S. chacoense* protoplasts regenerate calli with red pigment but do not regenerate shoots from p-calli. *S. tuberosum* protoplasts regenerate calli that produce shoots without red pigmentation. Calli from fused cells contain anthocyanin and produce shoots with red pigment.

detectors. Vital stains are those that do not adversely affect the viability of cells. Two such stains, fluorescein isothiocyanate (FITC) and rhodamine isothiocyanate (RITC) have been used successfully on plant protoplasts. FITC stains cells green, and RITC stains cells red when viewed under a fluorescence microscope. Fusion products that have both FITC and RITC fluoresce yellow. The yellow cells can be isolated mechanically using micromanipulators, and the fused cells can then be cultured individually on an enriched tissue culture medium. Because most protoplast populations do not grow very well at low cell densities, these enriched media are supplemented either with nurse cultures on the opposite side of semipermeable membranes or cell-free culture media taken from an actively growing cell culture source, i.e., "conditioned" medium. To obtain large numbers of fused cells marked with fluorescent dyes, a laser-activated flow cytometer, known as a cell sorter, can be used to isolate cells according to the wavelength of their fluorescent emission. The sorted cells can then be expected to yield a high frequency of somatic hybrids in subsequent culture. Although these procedures have been used successfully, one difficulty is retaining sterile conditions throughout the lengthy process.

Another procedure that has been successfully used to identify putative somatic hybrid plants is to count the number of chloroplasts present in the guard cells of leaves. Although this is a rapid method of obtaining data about the hybrid nature of regenerated plants, occasionally true hybrids lose the characteristic of doubled chloroplasts. Somaclonal variation of this type is not uncommon in cell lines grown in tissue culture for multiple generations, particularly when multiple chromosome ploidy levels are involved.

## Genetic transformation

Genetic transformation of single cells has been accomplished by co-cultivating protoplasts with *Agrobacterium tumefaciens* or by direct DNA transfer using polycationic chemicals, electroporation, liposomes, microinjection or sonication (Sawahel and Cove, 1992; Fisk and Dandekar, 1993). Transformation of plant protoplasts using polycationic chemicals and/ or electroporation has been most often reported.

## Agrobacterium-*mediated transformation*

As discussed in chapters twenty-nine to thirty-one, *A. tumefaciens*, the causative agent of crown gall disease, stimulates gall formation by inserting a portion of the tumor-inducing (Ti) plasmid into plant cells. For *Agrobacterium*-mediated gene transfer, protoplasts are isolated and cultured for 2–3 days prior to co-cultivation with *Agrobacterium* cells. Protoplasts are mixed with *Agrobacterium* at a rate of $10^4$–$10^6$ plant cells to $10^6$–$10^8$ bacterial cells per milliliter and incubated together for 24–30 h. After co-cultivation, antibiotics that inhibit growth of the bacterium are added to the culture medium. Treated plant cells are allowed to grow to the multicellular stage before the addition of selective antibiotics (kanamycin, hygromycin, etc.) that inhibit growth of nontransformed cells. Transformed cells grown on selective medium are able to regenerate plants with the introduced gene. Transformation frequencies usually range from 0.26–15.3%. Use of *Agrobacterium* to transform plant protoplasts has lost favor in recent years because not all species, especially cereals and grasses, are amenable to *Agrobacterium*-mediated transformation and it is sometimes difficult to inhibit growth of *Agrobacterium* following co-cultivation. In addition, use of simpler regenerative culture systems has eliminated some of the need to use protoplasts for transformation (see chapters twenty-nine to thirty-one).

## *Polycationic chemical–mediated transformation*

The most commonly used polycationic chemical for direct DNA transformation is PEG. It has been used to transform protoplasts from a wide variety of plant species. Transformation frequencies vary from $0.8 \times 10^{-6}$ to $6 \times 10^{-3}$ depending on the plant species.

To transform plant cells using PEG, protoplasts are isolated and adjusted to the proper cell density before adding transforming DNA and PEG. The mixture is incubated for 10–30 min before dilution with culture medium. With some species it is beneficial to subject protoplasts to a heat shock treatment (about 45 C) prior to the addition of DNA. It is not completely known how PEG promotes direct DNA uptake. Current thought is that the polycationic charge of PEG interacts with negatively charged DNA to form a positively charged complex that binds to the anionic plasmalemma. Absorption of DNA into cells occurs through endocytosis or active uptake. PEG-mediated transformation is a popular method for transforming plant protoplasts because it has low cell toxicity, promotes high transformation frequencies, increases membrane permeability, protects DNA against degradation by nucleases, is independent of DNA structure (supercoiled, circular or linear), allows integration of DNA into the host genome without a carrier and does not have a limited host range.

Hexademethrine bromide (Polybrene) acts similarly to PEG in that it promotes gene transfer by interacting with the anionic plasmalemma to develop regions of positive charge that react with negatively charged DNA. The polybrene method has been used to transform maize (*Zea mays* L. cv. Black Mexican Sweet [BMS]) protoplasts with genomic DNA from BMS kan$^R$ cell line 28 (Antonelli and Stadler, 1990). Polybrene-mediated direct DNA uptake is a promising method for transforming plant protoplasts because it has low cell toxicity, promotes high transformation frequencies and is not limited by DNA size or structure.

## *Electroporation-mediated transformation*

In electroporation, an electric pulse is used to induce DNA uptake by plant protoplasts. It is thought that DNA uptake occurs through pores in the cell membrane generated in response to the electric pulse. In general, transforming DNA is mixed with plant protoplasts prior to one or more electric pulses of 200–1000 V/cm and 16–1000 µF capacitance.

Short pulses (typically microsecond) are typically less lethal to protoplasts than long (millisecond) pulses. Optimum electroporation parameters vary and should be tested for each new species.

Transformation frequencies and cell viability are generally lower for electroporation than polycationic chemical methods but can be improved by adding PEG (4–8%) to the electroporation medium. Stable transformation using electroporation either with or without PEG has been obtained in protoplasts of many plant species. Genes incorporated into plant protoplasts using electroporation have included antibiotic resistance, GUS A or phosphoenolpyruvate from *Sorghum bicolor* (L.) Moench. Electroporation is a fast and efficient means of delivering DNA to plant cells and protoplasts, is not limited by the size or structure of transforming DNA and does not have a limited host range. However, electroporation is often lethal to protoplasts, resulting in low viability.

### Sonication-, liposome- and microinjection-mediated transformation

Other methods of incorporating foreign DNA into plant protoplasts have been described recently and show considerable promise. During sonication, protoplasts are mixed with DNA and given sonic pulses (500–900 msec at 0.65–1.6 V/cm) that disrupt the cell membrane and allow DNA to enter cells through pores created in response to the electric pulse. Liposomes are hydrogenated phospholipids in which DNA or RNA is encapsulated and delivered directly to plant cells through endocytosis or active uptake (Antonelli and Stadler, 1990). Microinjection uses a fine capillary needle to deliver DNA, chromosomes or nuclei directly into plant cells in a manner that preserves cell integrity and viability.

## Conclusions and future prospects

Plants with improved characteristics have been obtained through somaclonal variation and somatic hybridization. Direct gene transfer utilizing protoplasts can be used to insert beneficial genes into crops such as cereals and grasses, which are not amenable to *Agrobacterium*-mediated transformation. These methods can be used to create plants with improved agronomic and horticultural characteristics as well as increased insect and disease resistance.

## Acknowledgment

This is Florida Agricultural Experiment Station Journal Series No. R-04865.

## Literature cited

Antonelli, N.M. and J. Stadler. 1990. Genomic DNA can be used with cationic methods for highly efficient transformation of maize protoplasts. *Theor. Appl. Genet.* 80:395–401.

Bhojwani, S.S. and M.K. Razdan. 1983. *Plant Tissue Culture: Theory and Practice.* pp. 237–260. Elsevier, Amsterdam.

Boss, W. 1987. Fusion-permissive protoplasts. A plant system for studying cell fusion. In: *Cell Fusion,* A.E. Sowers, Ed., pp. 145–166. Plenum Press, New York.

Burg, H.C.J., K.S. Ramulu, G.M.M. Bredemeijer, S. Roest, P. Dijkhuis, J.J. van Hoogen and A. Houwing. 1989. Patterns of phenotypic and tuber protein variation in plants derived from protoplasts of potato (*Solanum tuberosum* L. cv. Bintje). *Plant Sci.* 64:113–124.

Cheng, J., J.A. Saunders and S.L. Sinden. 1995. Colorado potato beetle resistant somatic hybrid potato plants produced via protoplast electrofusion. *In Vitro Cell. Dev. Biol.* 31:90–95.

Cocking, E.C. 1960. A method for the isolation of plant protoplasts and vacuoles. *Nature.* 187:927–929.

Diettrich, B., V. Schneider and M. Luckner. 1991. High variation in cardenolide content of plants regenerated from protoplasts of the embryogenic cell strain VII of *Digitalis lanata*. *J. Plant Physiol*. 139:199–204.

Fisk, H.J. and A.M. Dandekar. 1993. The introduction and expression of transgenes in plants. *Sci. Hort*. 55:5–36.

Leigh, R.A. and D. Branton. 1976. Isolation of vacuoles from root storage tissue of *Beta vulgaris* L. *Plant Physiol*. 58:656–662.

Ramulu, K.S., P. Dijkhuis and S. Roest. 1989. Patterns of phenotypic and chromosome variation in plants derived from protoplast cultures of monohaploid, dihaploid and diploid genotypes and in somatic hybrids of potato. *Plant Sci*. 60:101– 110.

Saunders, J.A. and G.W. Bates. 1987. Chemically induced fusion of plant protoplasts. In: A.E. Sowers, Ed., *Cell Fusion*, pp. 497–520. Plenum Press, New York.

Sawahel, W.A. and D.J. Cove. 1992. Gene transfer strategies in plants. *Biotech. Adv*. 10:393–412.

Shahin, E.A. and R. Spivey. 1986. A single dominant gene for *Fusarium* wilt resistance in protoplast-derived tomato plants. *Theor. Appl. Genet*. 73:164–169.

Shepard, J.F., D. Bidney and E. Shahin. 1980. Potato protoplasts in crop improvement. *Science*. 208:17–24.

Vyskot, B., J. Fajkus, P. Kuglík, B. Koukalová and V. Kuhrová. 1991. Genome modifications in protoplast-derived tobacco plants: phenotypic evaluation and RFLP analysis. *Biol. Plant*. 33:455–460.

*chapter twenty-four*

# Isolation, culture and fusion of tobacco and potato protoplasts

## Richard E. Veilleux and Michael E. Compton

Isolation of plant protoplasts followed by fusion and subsequent regeneration of somatic hybrids are remarkable techniques that allow the exchange of genetic information between sexually incompatible species (see chapter twenty-three). Success in application of these techniques was first achieved with species of tobacco (*Nicotiana tabacum* L.) for which Nagata and Takebe (1970) developed a protocol for cell wall regrowth and cell division, followed eventually by plant regeneration. Carlson et al. (1972) reported the first interspecific somatic hybrids. The list of species amenable to these techniques has grown to include many of the major crops. However, the source of explants for protoplast extraction, the choice of genotype and the specific conditions of isolation and medium changes must be customized for each species.

The objectives of somatic hybridization may be to develop hybrids with the complete genome of the two "parents" (true somatic or parasexual hybrids), partial genome transfer between species (asymmetrical hybrids) or cytoplasmic transfer to develop "cybrids" with the nucleus of one parent and cytoplasm of the other. Somatic hybrids between closely related species are generally polyploid and have the potential of fertility, depending on the relatedness of the two species. Potato has been the object of somatic hybridization with a range of related species from closely related tuber-bearing *Solanum* relatives to nontuberous *Solanum* and even to tomato (*Lycopersicon esculentum* Mill.), nightshade (*S. nigrum* L.) and tobacco. Both its plasticity in tissue culture and vegetative propagation have made it attractive to geneticists wishing to examine the potential of somatic hybridization in an important economic crop. Somatic hybrids between potato (*S. tuberosum* L.) and a wild, non-tuber-bearing relative, *S. brevidens* Phil., contained the additive number of chromosomes of the two parents (i.e., 72 with 48 from potato and 24 from *S. brevidens*). These somatic hybrids could be backcrossed to potato, resulting in fertile plants with considerable breeding potential as sources of disease resistance (Helgeson et al., 1986).

As the taxonomic divergence between species increases, the possibility of obtaining fertile somatic hybrids decreases and the likelihood of chromosome elimination, i.e., partial or complete spontaneous loss of chromosomes of one or the other parent, increases. Such chromosome elimination results in aneuploid regenerants or asymmetric hybrids. A recent example has been the regeneration of plants following protoplast fusion between tobacco and carrot (*Daucus carota* L.) reported by Kisaka and Kameya (1994). The somatic hybrids contained between 23 and 32 chromosomes instead of the additive number of 66 (48 from tobacco and 18 from carrot).

As opposed to somatic hybrids in which nuclear fusion has occurred, cybrids between even distantly related parents are more likely to be fertile (Rose et al., 1990). Nuclei of

0-8493-9409-0/96/$0.00+$.50
© 1996 by CRC Press, Inc.

protoplasts of one of the fusion partners are generally inactivated by irradiation or mutagenic treatment before fusion to prevent transmission of the nuclear genome deliberately. Protoplasts of the other fusion partner may be metabolically inhibited by iodoacetate treatment. In this way, unfused protoplasts of either parent will not regenerate and the preponderance of regenerants will be cybrids. Such a scheme has recently been successful to develop cybrids between potato and either of the two sexually incompatible wild species, *S. bulbocastanum* Dun. or *S. pinnatisectum* Dun. (Sidorov et al., 1994). complete lack of transmission of nuclear material of the irradiated protoplasts cannot be guaranteed such that asymmetric hybrids may result (Matabiri and Mantell, 1994).

The following exercises illustrate the basic procedures of protoplast isolation and fusion. The conditions under which donor plants are grown can greatly affect the success with protoplast yield and culture. This is especially true of potato. Therefore, the first exercise has been designed for tobacco protoplast isolation because it is much more tolerant of a range of conditions, making it easier to handle and more likely to succeed as a student lab. Protoplast fusion can be induced in many different ways (see chapter twenty-three). The exercise described in experiment 2 is a small-scale chemical fusion of protoplasts using polyethylene glycol and high $Ca^{2+}$ to facilitate membrane fusion between adjacent protoplasts.

## General considerations

### Tobacco

Most cultivars of tobacco will respond well to protoplast isolation and culture. However, Nagata and Takebe's (1970) initial report of successful cell division of isolated protoplasts had been hindered by inadvertent choice of a cultivar that was prone to seasonal fluctuations that prevented repeatability of experiments. We have often used 'Samsun', an older cultivar that has been used extensively in tissue culture research. Seeds can be obtained from Dr. Verne Sisson, Department of Crop Science, North Carolina State University, Box 7620, Raleigh, NC 27695.

### Potato

Response of potato to protoplast manipulation is under genetic control (Cheng and Veilleux, 1991) so care must be taken in selection of genotypes; some are completely recalcitrant in protoplast culture, whereas others respond easily. In a comparison of 36 cultivars or breeding lines for ease of manipulation in protoplast culture, Haberlach et al. (1985) listed *S. tuberosum* cvs. Russet Burbank, US-W 5328.4, US-W 9546.46, PI 423654, *S. etuberosum* Lindl. PIs 245924 and 245939, and *S. brevidens* Phil. 245763 as highly regenerable. (Of these, Russet Burbank is tetraploid, 2n=2x=48, and the rest are diploid, 2n=2x=24.) Many other genotypes yielded few or no protoplast calli (p-calli) or would not regenerate from p-calli. In vitro copies of regenerable cultivars or species may be obtained from the NRSP-6 Potato Introduction Project, Sturgeon Bay, WI 54235. Several genotypes should be used such that each student or student team will extract protoplasts from one genotype and then fuse their protoplasts with those extracted by another student or student team. It may be desirable to use one highly regenerable and one recalcitrant genotype for protoplast fusion. In this way, unfused protoplasts of one parent would not be expected to regenerate in the fusion experiments. Haberlach et al. (1985) list *S. tuberosum* PI 256977 (tetraploid), Kennebec haploid 731.3 (diploid), US-W 9587.24 (diploid) and 77-16 (diploid) as unable to form p-calli after protoplast isolation and culture.

## Growth of plants

### Tobacco

The plants should be approximately 50–60 days after planting at the time of protoplast isolation. The seedlings are slow to start so seeds can be planted in a single pot for the first month and then transplanted individually to 3.7 L nursery pots. One plant per student should be sufficient. Any of the soilless potting mixes is suitable. Tobacco is a heavy feeder so weekly fertilization with a 20:20:20 (N:P:K) liquid fertilizer is recommended (see also chapter twenty-eight). Light and temperature conditions in the greenhouse are not especially critical. Tobacco is susceptible to infestation with white fly under greenhouse conditions, so treatment with a systemic insecticide (e.g., Marathon, Olympic Horticultural Products, Mainland, PA — follow all label rates and precautions) shortly after transplanting is recommended. Otherwise, contamination is very likely in protoplast cultures.

### Potato

The best source of plant material for protoplast manipulation of potato is in vitro plantlets. Therefore, an exercise in making approximately 10–15 copies using single nodes of previously established in vitro plantlets of the selected genotypes on propagation medium (see also chapter nine) for each student should be planned approximately three weeks before the protoplast fusion lab. Obviously, adequate in vitro multiplication will have been necessary 3–4 weeks prior to this round of multiplication. In vitro potato can harbor bacterial contaminants that are not apparent during routine subculture on basal medium. However, such contaminants often become obvious during protoplast manipulations, resulting in failed experiments. A precaution against such "cryptic" contamination is to use the antibiotic cefotaxime in the propagation medium used just prior to the transfer when students will multiply their genotypes.

## Explant preparation

Fully expanded young leaves of tobacco yield the best protoplasts. Leaves should be harvested and brought to the lab on the day of the exercise. In contrast, in vitro potato cultures should be placed in the dark in a refrigerator (4 C) for 48 h prior to protoplast extraction. Examine cultures carefully and remove any obviously contaminated ones.

## Culture medium

For tobacco protoplast isolation, five media are needed (Table 24.1). Three of these are simple variations of protoplast isolation (PI) medium and can be made from the same preparation of PI (preplasmolysis = PI + 13% mannitol; flotation = PI + 20% sucrose; and rinse = PI + 10% mannitol). The other two media (enzyme solution and culture medium) have a different basal composition so must be prepared separately. Only the enzyme solution requires filter sterilization; the other four media can be autoclaved. It is advisable to dispense each medium into properly labeled $25 \times 150$ mm culture tubes with Magenta two-way caps in aliquots sufficient for each student or student team prior to autoclaving. If different students or student teams share a common supply of medium, contamination will nearly always result.

For potato protoplast isolation and fusion, nine different media (Tables 24.2 through 24.4) are required, including the protoplast propagation medium for the micropropagation step conducted 3 weeks before isolation. Four of these (propagation, preplasmolysis, enzyme and culture) are variations of Murashige and Skoog (1962) basal medium. Some media (preplasmolysis, enzyme, culture and $Ca^{2+}$) require filter sterilization. The other five

*Table 24.1*   Media Components for Protoplast Isolation from Tobacco Leaves

| | Amount (mg/L) | | |
|---|---|---|---|
| Component | Protoplast isolation (PI)[a] | Enzyme[b] | Protoplast culture (PC) |
| $Ca(H_2PO_4)\cdot H_2O$ | — | — | 100.0 |
| $CaCl_2\cdot H_2O$ | 1480.0 | — | 450.0 |
| $KH_2PO_4$ | 27.2 | — | — |
| $KNO_3$ | 101.0 | — | 2500.0 |
| $MgSO_4\cdot 7H_2O$ | 246.0 | — | 250.0 |
| $NaH_2PO_4\cdot 2H_2O$ | — | — | 170.0 |
| $(NH_3)_2SO_4$ | — | — | 134.0 |
| $CoCl_2\cdot 6H_2O$ | — | — | 0.025 |
| $CuSO_4\cdot 5H_2O$ | 0.025 | — | 0.025 |
| $H_3BO_3$ | — | — | 3.0 |
| KI | 0.16 | — | 0.75 |
| $MnSO_4\cdot 4H_2O$ | — | — | 13.2 |
| $NaMoO_4\cdot 2H_2O$ | — | — | 0.25 |
| $ZnSO_4\cdot 7H_2O$ | — | — | 2.0 |
| Sequestrene 330 | — | — | 28.0 |
| Sucrose | — | — | 10000 |
| Glucose | — | — | 18000 |
| Mannitol | — | 130000 | 100000 |
| Inositol | — | — | 100.0 |
| Nicotinic acid | — | — | 1.0 |
| Pyridoxine HCl | — | — | 1.0 |
| Thiamine HCl | — | — | 10.0 |
| 2,4-D | — | — | 0.1 |
| NAA | — | — | 1.0 |
| BA | — | — | 1.0 |
| Onozuka R10 cellulase | — | 5000 | — |
| Onozuka R10 macerozyme | — | 1000 | — |
| pH | 5.8 | 5.8 | 5.8 |
| Sterilization method | Autoclave | Filter sterilize | Autoclave |

[a] Basic components for the preplasmolysis, rinse and flotation media.

[b] Make fresh on the day of isolation.

*Table 24.2*   Components of the Potato Plantlet Propagation Medium to Be Added to Basal Salts of Murashige-Skoog (MS) Medium

| Component | mg/L |
|---|---|
| Sucrose | 20000 |
| Myo-inositol | 100 |
| $KH_2PO_4$ | 170 |
| Casein hydrolysate | 500 |
| Cefotaxime[a] | 250 |
| Agar | 7000 |
| pH 5.7 | |
| Autoclave | |

*Note:* See chapter two for composition.

[a] This antibiotic is optional but is especially recommended for propagation of stock plantlets one or more clonal generations before propagation for protoplast extractions to reduce bacterial contamination. It is not autoclavable so must be filter sterilized and added to the autoclaved medium as it cools.

*Table 24.3* Media Components for Isolation and Culture of Potato Protoplasts

| Component | Amount (mg/L) | | | | |
|---|---|---|---|---|---|
| | Preplasmolysis | Enzyme[a] | Rinse | Flotation | Culture |
| $NH_4NO_3$ | — | — | — | — | — |
| $KNO_3$ | 101.0 | 950.0 | — | — | 950.0 |
| $KH_2PO_4$ | 33.0 | 85.0 | — | — | 85.0 |
| $MgSO_4 \cdot 7H_2O$ | — | — | — | — | — |
| $MgSO_4$ | 6000.0 | 90.5 | — | — | 90.5 |
| $CaCl_2 \cdot H_2O$ | 1474.0 | 660.0 | 660.0 | 660.0 | 660.0 |
| $Na_2EDTA$ | — | 18.7 | — | — | 18.7 |
| $FeSO_4 \cdot 7H_2O$ | — | 13.9 | — | — | 13.9 |
| $MnSO_4 \cdot 4H_2O$ | — | 11.2 | — | — | 11.2 |
| $ZnSO_4 \cdot 7H_2O$ | — | 5.3 | — | — | 5.3 |
| $H_3BO_3$ | — | 3.1 | — | — | 3.1 |
| KI | — | 0.4 | — | — | 0.4 |
| $NaMoO_4 \cdot 2H_2O$ | — | 0.125 | — | — | 0.125 |
| $CoCl_2 \cdot 6H_2O$ | — | 0.0125 | — | — | 0.0125 |
| $CuSO_4 \cdot 5H_2O$ | — | 0.0125 | — | — | 0.0125 |
| KCl | — | — | 22000.0 | — | — |
| Glycine | 2.0 | 1.0 | — | — | 1.0 |
| Nicotinic acid | 0.5 | 0.25 | — | — | 0.25 |
| Pyridoxine HCl | 0.5 | 0.25 | — | — | 0.25 |
| Thiamine HCl | 0.1 | 0.05 | — | — | 10.0 |
| Inositol | — | — | — | — | 100.0 |
| Sucrose | — | — | — | 171000.0 | 10000.0 |
| Sorbitol | — | — | — | — | 50000.0 |
| Glucose | — | 18000.0 | — | — | 30000.0 |
| Mannitol | 90000.0 | 73000.0 | — | — | — |
| Casein hydrolysate | — | — | — | — | 500.0 |
| Glutamine | — | — | — | — | 100.0 |
| Serine | — | — | — | — | 10.0 |
| MES | 586.0 | — | — | — | — |
| Agar | — | — | — | — | — |
| NAA | — | — | — | — | 1.25 |
| 2,4-D | — | — | — | — | 0.25 |
| Zeatin | — | — | — | — | 1.0 |
| Onozuka R10 cellulase | — | 10000.0 | — | — | — |
| Onozuka R10 macerozyme | — | 1000.0 | — | — | — |
| pH 5.8 | 5.8 | 5.8 | 5.8 | 5.8 | 5.8 |
| Sterilization method | Filter sterilize | Filter sterilize | Autoclave | Autoclave | Filter sterilize |

[a] Make fresh on the day of isolation.

(propagation, flotation, rinse, CPW 13M and PEG 22.5) can be autoclaved. If the propagation medium contains cefotaxime, this component must be filter sterilized separately and added to the autoclaved propagation medium as it cools. As with the tobacco experiment, it is advisable to dispense these media into aliquots sufficient for each student or student team prior to autoclaving. Dispensing filter-sterilized media into autoclaved culture tubes is recommended.

*Table 24.4*   Media Components for Fusion of Potato Protoplasts

| Component | Amount (mg/L) | | |
|---|---|---|---|
| | CPW 13M | Fusogen PEG 22.5 | $Ca^{2+}$ washing |
| $KNO_3$ | 101.0 | — | — |
| $KH_2PO_4$ | 27.2 | 100.0 | — |
| $MgSO_4 \cdot 7H_2O$ | 246.0 | — | — |
| $CaCl_2 \cdot 2H_2O$ | 1480.0 | 1500.0 | 7400.0 |
| KI | 0.16[a] | — | — |
| $CuSO_4 \cdot 5H_2O$ | 0.025[b] | — | — |
| Glycine | — | — | 3800.0 |
| Sucrose | — | 18000 | 110000 |
| Mannitol | 130000 | — | — |
| PEG 8000 | — | 225000 | — |
| pH | 5.8 | 5.8[c] | 5.8 |
| Sterilization method | Autoclave | Autoclave | Filter sterilize |

[a] 8 mg/50 ml stock — use 1 ml.

[b] 2.5 mg/50 ml stock — use 0.5 ml.

[c] Use 1 N KOH buffer.

# Exercises

## Experiment 1. Protoplast isolation and culture of tobacco

### Materials

The following are needed for each student or student team:

* Five sterile 300-ml bleakers to surface disinfest leaves (one for ethanol, one for bleach and three for sterile water)
* 100 ml 20% commercial bleach with a drop of Tween 80
* Hemacytometer
* Centrifuge
* Fluorescein diacetate (FDA) stain, 5 mg FDA (product no. F 7378, Sigma Chemical Co., St. Louis, MO) in 1 ml acetone
* A 250-ml beaker containing a 63-μm filter (Such a filter can be constructed by ordering 63-μm wire mesh from the Newark Wire Mesh Co., Newark, NJ, cutting the wire mesh into squares just large enough to cover the cap end of a Nalgene 125-ml autoclavable animal watering bottle and fusing the wire mesh to the plastic on a hotplate covered with aluminum foil. After the mesh has embedded into the melted plastic, the excess mesh can be trimmed with a scissors and the bottom of the bottle can be cut off with a scalpel. The filter can be autoclaved in the beaker covered with foil and used repeatedly. Alternatively, nylon mesh of approximately the same size can be used, but most nylon cannot be autoclaved so must be sterilized by an ethanol soak.)
* Centrifuge tubes (15 ml)
* 10 ml enzyme solution
* 20 ml preplasmolysis medium
* 10 ml flotation medium
* 30 ml rinse medium
* 25 ml protoplast culture medium

* Calcofluor White stain, 0.1% Fluorescent Brightener 28 (product no. F 6259, Sigma Chemical Co.) in 0.7 M mannitol

## Procedure

*Day 1.* Surface disinfest one tobacco leaf by immersion in 70% ethanol for 30 sec and transfer to 20% bleach/Tween for 15 min. Rinse three times in sterile, distilled water. Place the tobacco leaf in a sterile petri dish and peel sections of the lower epidermis from the leaf and/or cut the leaf into 1-mm strips. Even though the peeling is tedious, it is more effective than cutting. With fine forceps, the epidermis can be stripped by lifting at a leaf vein and pulling across the lamina. This should be repeated until at least one half of the epidermis has been removed. Add 20 ml preplasmolysis medium and incubate the peeled leaf for 1–8 h. Remove preplasmolysis solution using a sterile pasteur pipette and replace with 10 ml enzyme solution. Wrap the petri dish with Parafilm, label and place in the dark overnight on a low-speed (40-rpm) shaker.

*Day 2.* Gently agitate the plate of protoplasts in enzyme solution to facilitate the release of protoplasts. Transfer protoplast suspension to the 63-μm filter in a 250-ml beaker using a pasteur pipette. Add 3 ml rinse medium to the plate with leaf debris, shake vigorously and combine with the protoplast/enzyme mixture in the 250-ml beaker. Use an additional 2 ml rinse medium to rinse protoplasts off the filter into the beaker by holding the filter with forceps as you rinse. Transfer the protoplast/enzyme mixture to a 15-ml sterile centrifuge tube and centrifuge at $50 \times g$ for 10 min. Remove the supernatant and resuspend the pellet in 10 ml flotation medium, add 1 ml rinse medium to the top of the protoplasts in flotation medium dropwise so as not to mix the two media, but rather to have distinct layers of medium (dripping the rinse medium down the side of the tube works better than splashing it on the surface of the flotation medium) and recentrifuge for 10 min at $50 \times g$. Viable protoplasts should float to the interface of the two media. Remove the band of green protoplasts with a pasteur pipette, transfer it to a fresh centrifuge tube, add 10 ml rinse medium and recentrifuge at the same speed. Protoplasts should sink to the bottom and form a pellet. Rinse the protoplasts again in 10 ml rinse medium. (This second rinse can be omitted if the protoplast pellet is small.) Remove the supernatant and resuspend the protoplasts in approximately 1 ml culture medium.

Place a drop of protoplast suspension on a hemacytometer and estimate protoplast density (see chapter twenty-five). To culture the protoplasts, dilute to 50,000 cells/ml with protoplast culture (PC) medium and transfer to a sterile petri dish. Stain a protoplast sample with FDA as described in chapter twenty-five. Viable protoplasts fluoresce green; dead cells fluoresce red because of the autofluorescence of chlorophyll. Record the frequency of viable protoplasts in a sample of 100–200 cells. Seal, label and culture in the dark at 28 C overnight.

*Day 3.* Move culture to low light (10–20 μmol·m$^{-2}$·sec$^{-1}$) at a 16-h photoperiod (e.g., under cool white fluorescent light covered by a layer of cheesecloth) for two days.

*Day 5.* Move culture to higher light intensity (50–75 μmol·m$^{-2}$·sec$^{-1}$) by removing the layer of cheesecloth.

*Day 7.* Examine cultures for cell wall regrowth by staining a sample with Calcofluor White stain and observing under a fluorescence microscope (see chapter twenty-five for details). Is there any evidence of cell wall regrowth or cell division evident by blue fluorescence? If so, record the plating efficiency of the protoplast cultures, i.e., the number of dividing cells divided by the total number of cultured cells, in a sample of 100–200 cultured protoplasts. Is there evidence of contamination?

### Anticipated results

By day 7, the protoplasts should fluoresce blue after staining with Calcofluor White. Protoplasts that have died prior to wall resynthesis do not fluoresce. Some protoplasts should have divided already such that two- or four-celled colonies are visible. If cell colonies can be found, the protoplasts can be diluted with fresh culture medium with lower mannitol to observe the development of p-calli. After approximately 8–10 weeks, p-calli can be transferred to shoot regeneration medium.

### Questions

* Why do protoplasts float to the interface of the flotation and rinse solutions while the debris sinks to the bottom of the sucrose solution during centrifugation?
* Why do viable protoplasts fluoresce green when stained with FDA stain?
* Report the viability of freshly isolated protoplasts and the plating efficiency after 7 days. Explain why there is a difference between frequency of viable protoplasts and plating efficiency.
* Would you expect the plants regenerated from individual protoplasts derived from a single leaf to be genetically identical? Explain.

## Experiment 2. Protoplast fusion of potato

### Materials

The following materials are needed for each student or student team:

* All the materials as described in experiment 1 except solutions
* Mineral oil (autoclaved)
* Microscope coverslips (autoclaved in a glass petri dish wrapped in foil)
* 10 ml preplasmolysis medium
* 10 ml enzyme medium containing either rhodamine isothiocyanate (RITC) or fluorescein isothiocyanate (FITC) (10 μl of a 3-mg/ml stock in ethanol)
* 30 ml rinse medium
* 10 ml flotation medium
* 10 ml CPW 13M
* 1 ml fusogen (PEG 22.5)
* 5 ml $Ca^{2+}$ wash medium
* 10 ml culture medium

### Procedure

*Day 1.* Procedures for the isolation and fusion of potato protoplasts are outlined in Figure 24.1. In the late afternoon, cut in vitro shoots and leaves of the potato plantlets into fine pieces using two scalpels in a $100 \times 15$ mm petri dish and add preplasmolysis medium for 1–8 h. Remove preplasmolysis medium with a pasteur pipette and add the enzyme medium with either FITC or RITC (one for one fusion partner and one for the other) if a fluorescence microscope is available. Incubate in the dark at room temperature on a shaker (60 rpm) for 4–16 h.

**Figure 24.1** Experimental procedures for fusing potato protoplasts. A. Propagate potato plantlets in vitro until 15 copies of each genotype are obtained. B. Cut shoots and leaves into fine pieces using two scalpels. C. Incubate cut pieces in plasmolysis medium for 1–8 h followed by 4–16 h in enzyme medium. $D_1$. Pour protoplast suspension through a sterile 63-μm filter into a 250-ml beaker. $D_2$. Transfer protoplast suspension into a sterile centrifuge tube and centrifuge. E. Resuspend pellet in 10 ml flotation medium overlayed with 1 ml of rinse medium and centrifuge (protoplasts collect at the interface of the two media during centrifugation). F. Rinse protoplasts and resuspend in CPW 13M at a density of $1 \times 10^6$ per milliliter. G. (1) In the center of a sterile petri dish add 1 small drop of sterile mineral oil. (2) Position a sterile coverslip over the droplet. (3) Add 3 drops of each of the two protoplast suspensions. H. Fuse protoplasts by gradually adding 6 drops of Fusogen PEG 22.5 medium. I. Wash protoplasts with $Ca^{2+}$ washing medium. Replace washing medium with culture medium. J. Incubate protoplasts in dim light at 25 C. K. After 2–3 weeks embed multicellular colonies in medium with agarose. L. Regenerate shoots and transfer to MS basal medium for rooting.

*Day 2 — Isolation.* The following morning, gently swirl the petri dishes to loosen protoplasts from debris; then transfer protoplasts in enzyme media to a sterile 63-μm filter in a 250-ml beaker; add 3 ml rinse solution to the petri dish, swirl vigorously and add this to the enzyme solution. Rinse the filter with an additional 2 ml rinse medium. Remove the filter and pour the protoplast suspension into a sterile centrifuge tube. Centrifuge at 50 × *g* for 5 min.

*Remove supernatant with a pasteur pipette.* Resuspend the pellet in 10 ml flotation medium. Add 1 ml (25 drops) rinse medium dropwise to the top of the sucrose solution by dripping it gently down the side of the centrifuge tube so that the two layers remain intact. Centrifuge at 50 × *g* for 10 min.

Collect protoplast band at the interface between rinse and sucrose media with a pasteur pipette. Transfer protoplasts to a sterile 15-ml centrifuge tube, add rinse medium to 10 ml and spin at 50 × *g* for 5 min. Remove rinse medium with a pasteur pipette, being careful not to disturb the pellet. Adjust the protoplast density to $1 \times 10^6$ per milliliter in CPW 13M medium. If insufficient protoplasts have been obtained to reach this density, adjust both fusion partners to the same density as close to this as possible.

*Day 2 — Fusion.* Students should be paired such that each student in a pair has isolated protoplasts from different parents with different stains. The students can then share their protoplasts with each other and retain the remainder for control plates. Place a small drop of sterile mineral oil in the center of a plastic petri dish and gently lower a sterile coverslip onto the oil. Pipette 3 drops of each of the two protoplast suspensions (one incubated with RITC and isolated by one student in the pair and the other incubated with FITC and isolated by the other student in the pair) onto the coverslip. Leave undisturbed for 10 min to allow the protoplasts to settle and stick to the coverslip. Add 6 drops fusogen PEG 22.5, drop by drop, surrounding the protoplasts, leaving the last drop for the center. Incubate at room temperature for 20–25 min.

Gently add 3 drops of $Ca^{2+}$ washing medium every 5 min for 20 min to one side of the protoplast culture while removing 3 drops of fusion medium from the opposite side using a second pasteur pipette. Agglutination begins during this washing procedure. It is important that the protoplasts are disturbed as little as possible during this process.

Finally, replace the $Ca^{2+}$ washing medium with culture medium. Place 3 drops of culture medium at one side of the protoplast mixture. Then, using a second pasteur pipet, remove some of the washing medium from the opposite side; repeat two more times. Then flood the coverslip with culture medium and place a few drops of culture medium on the plate surrounding the coverslip to maintain humidity. Seal the plate and incubate in dim light at 25 C.

Prepare control dishes of the remaining unfused protoplasts by centrifuging for 5 min, removing the supernatant and replacing it with sufficient culture medium to adjust the density to $2.5 \times 10^5$ per milliliter. Dispense to the appropriate size petri dish, depending upon the amount of culture obtained — 4–6 ml in a $100 \times 20$ mm dish; 2–4 ml in a $60 \times 15$ mm dish; 1–2 ml in a $35 \times 10$ mm dish. Seal the plate and incubate in dim light at 25 C, as in experiment 1. After 2 and 7 days, the cultures should be checked for cell wall regrowth and cell division. At 7–10 days after isolation, plating efficiency of the fusions and control plates of each parent should be recorded. If the protoplasts are dividing and callus colonies are growing, the cultures should be diluted with fresh culture medium after approximately 2 weeks. This can be done by preparing double strength filter-sterilized culture medium and autoclaving an equal volume of double strength (0.6%) agarose in distilled water. The two can then be mixed to provide a volume of medium equal to what is in the plates of protoplasts. Then, the protoplasts in liquid culture medium are added to the fresh medium to embed the protoplasts in agarose while replenishing the medium simultaneously. After 2 more weeks, the number of visible callus colonies should be recorded for the fusion plates and the controls. Callus morphology often differs among clones for obvious morphological traits such as color, texture and size. The morphology of callus in each plate should be noted. Fast-growing or obviously different callus in the fusion plates may be indicative of putative somatic hybrids. The number of p-calli per plate should be recorded for fusions and controls after 4–6 weeks of culture. If there are too many to count, a sample region of equal size representative of each plate can be scored.

### Anticipated results

After the $Ca^{2+}$ washing medium has been added, protoplast fusion can be observed under an inverted microscope. If the microscope is equipped with fluorescence, FITC- (580 nm dichroic mirror with 510–560-nm excitation filter and 590-nm barrier filter) stained protoplasts should fluoresce green, RITC- (same mirror and filters) stained protoplasts should fluoresce red and fusion products should fluoresce yellow. Without a selection scheme for growing the fusion products exclusively, it would be difficult to recover somatic hybrids from these cultures. Using iodoacetate or irradiation to prevent unfused protoplasts from regenerating complicates the experiment unnecessarily for a student lab. However, it is possible that the hybrid p-calli will differ morphologically from the control p-calli, such that putative hybrids can be identified.

### Questions

* How would you select for fusion products?
* Why is mannitol used in protoplast culture?
* What frequency of fused protoplasts was observed?
* How did the plating efficiency compare between the fusion plates and the control plates?
* Was there a difference in p-calli frequency, morphology or growth rate between the control plates and the fusion plates?

## Literature cited

Carlson, P.S., H.H. Smith and R.D. Dearing. 1972. Parasexual interspecific plant hybridisation. *Proc. Natl. Acad. Sci. U.S.A.* 69:2292–2294.

Cheng, J. and R.E. Veilleux. 1991. Genetic analysis of protoplast culturability in *Solanum phureja*. *Plant Sci.* 75:257–265.

Haberlach, G.T., B.A. Cohen, N.A. Reichert, M.A. Baer, L.E. Towill and J.P. Helgeson. 1985. Isolation, culture and regeneration of protoplasts from potato and several related *Solanum* species. *Plant Sci.* 39:67–74.

Helgeson, J.P., G.J. Hunt, G.T. Haberlach and S. Austin. 1986. Somatic hybrids between *Solanum brevidens* and *Solanum tuberosum*: Expression of a late blight resistance gene and potato leaf roll resistance. *Plant Cell Rep.* 3:212–214.

Kisaka, H. and T. Kameya. 1994. Production of somatic hybrids between *Daucus carota* L. and *Nicotiana tabacum. Theor. Appl. Genet.* 88:75–80.

Matibiri, E.A. and S.H. Mantell. 1994. Cybridization in *Nicotiana tabacum* L. using double inactivation of parental protoplasts and post-fusion selection based on nuclear-encoded and chloroplast-encoded marker genes. *Theor. Appl. Genet.* 88:1017–1022.

Murashige, T. and F. Skoog. 1962. A revised medium for rapid growth and bioassays with tobacco tissue cultures. *Physiol. Plant.* 15:473–497.

Nagata, T. and I. Takebe. 1970. Cell wall regeneration and cell division in isolated tobacco mesophyll protoplasts. *Planta.* 92:301–308.

Rose, R.J., M.R. Thomas and J.T. Fitter. 1990. The transfer of cytoplasmic and nuclear genomes by somatic hybridisation. *Aust. J. Plant Physiol.* 17:303–321.

Sidorov, V.A., D.P. Yevtushenko, A.M. Shakhovsky, and Y.Y. Gleba. 1994. Cybrid production based on mutagenic inactivation of protoplasts and rescuing of mutant plastids in fusion products: potato with a plastome from *S. bulbocastanum* and *S. pinnatisectum.* Theor. Appl. Genet. 88:525–529.

*chapter twenty-five*

# Isolation of protoplasts from leaves of chrysanthemum and orchardgrass

## Robert N. Trigiano

The following laboratory exercise uses leaves of either chrysanthemum (*Dendrathema grandiflora* Tzvelev) or orchardgrass (*Dactylis glomerata* L.) to illustrate some aspects of protoplast technology. We have found that, although protoplasts are relatively easy to isolate from leaf tissue of these two species, it is more difficult to regenerate plants from chrysanthemum (Sauvadet et al., 1990) and nearly impossible from orchardgrass. As such, this laboratory is devised only to demonstrate some of the general principles of protoplast methodologies, i.e., donor tissue preparation, preculture, tissue digestion, purification, and assessment of purity and viability. We suggest that the laboratory exercise be completed without regard to aseptic conditions since callus and plants will not be regenerated. Additionally, some of the complex protoplast culture media discussed in chapter twenty-three are not required. The mechanics of the experiment may be completed in 1 day if some students can help out "off and on" for the entire day or if the instructor prepares some of the initial steps of the procedure in advance of the class meeting. The laboratory may also be conducted as a demonstration.

## Exercise

### Experiment Isolation of protoplasts from leaves

#### Materials

* 30–60 tillers of orchardgrass or about 15 young, expanding, light green leaves of chrysanthemum 'Iridon'
* Pectolyase Y-23 (Seishin Pharm. Co. Ltd., 4-13, Koamicho, Nihonbashi, Tokyo, Japan)
* RS "Onozuka" Cellulase (Yakult Honsha Co., Ltd. 1.1.19 Higashi Shinabashi, Minato-ku, Tokyo, 105, Japan)
* The following four isolation media:
  1. One-half strength SH salts (see chapter two) amended with 0.38 M (ca. 7%) mannitol, 3 mM MES, 2 mM $CaCl_2$, pH = 5.7;
  2. Same medium as (1) except use 0.49 M (ca. 9%) mannitol;
  3. Same medium as (1) except use 0.60 M (ca. 11%) mannitol; and
  4. Same medium as (1) except use 0.71 M (ca. 13%) mannitol

**Figure 25.1** Diagrammatic representation of the protocol for making protoplasts from chrysanthemum or orchardgrass leaves. A. Leaves of chrysanthemum and orchardgrass. B. Weigh leaves. C. Preculture in cold isolation medium. D. Cut tissue into small linear strips. E. Rinse cut tissue several times with isolation medium. F. Expose tissue to digestion medium for 3–4 h. G. Remove large pieces of undigested tissue with fine-tipped forceps. H. Pipette digestion medium and small debris into sieve apparatus ($H_1$) and express protoplasts through 40–100-μm sieve ($H_2$). I. Collect protoplasts by centrifugation. J. Rinse protoplasts with isolation medium. K. Determine the following: concentration of protoplasts with a hemacytometer ($K_1$), viability ($K_2$) and purity ($K_3$) of protoplasts with fluorescence.

* Digestion media: same as the four isolation media listed above except include 1% (w/v) RS cellulase and 0.4% (w/v) Y-23 pectolyase
* 1% calcofluor in water
* 0.05% fluorescein diacetate dissolved in acetone
* Hemacytometers
* Compound microscope equipped for epifluorescence and filters
* Nylon mesh sieves with 40–100-μm cross-sectional openings in beakers
* Low-speed table top centrifuge
* Two 25- or 50-ml glass or plastic centrifuge tubes
* Any solidified agar medium in petri dishes

### Procedure

The general procedure is outlined diagrammatically in Figure 25.1. The experiment, as listed below, is designed to investigate only the influence of isolation medium, specifically the osmotic pressure, on successful procurement of viable protoplasts. As alternative treatments, the concentration of hydrolytic enzymes may be varied. This is a little more difficult since the appropriate isolation medium must first be determined. We suggest that each student or student team work with either one or two of the isolation media listed above. Students should then share and analyze data, i.e., number of protoplasts, etc. The experiment may be considered as a completely randomized design for statistical analyses. Alternatively, if desired, only one replication of each treatment may be prepared and the class divided into four groups or teams. In this scenario, the experiment can be regarded as a demonstration.

Collect ten tillers of orchardgrass for each replication of isolation media and dissect the inner most two leaves for culture as outlined in chapter seventeen. Only use the bottom 2 cm of both leaves. For chrysanthemum, three to five leaves of any cultivar (see chapter twelve) for each replication of a treatment will be sufficient. Leaves do not need to be surface disinfested for this exercise. Record the fresh weight of the leaf tissue and place in a $60 \times 15$ mm petri dish. For each treatment, add 20 ml of ice cold (4 C) isolation medium and place in refrigerator for 30 min.

Using fine forceps and a scalpel with a #10 blade, cut the leaves into very thin (<1-mm) strips. Orchardgrass leaves are easily "feathered" by cutting parallel to the veins; chrysanthemum leaves should be cut along the long axis of the leaf. Care should be taken not to tear the leaves — a sharp cut seems to work best for protoplast formation and thinner strips of leaves will generally yield more protoplasts. After all the leaves are cut, use a pasteur pipette to remove most of the isolation medium and replace with more of the appropriate cold isolation medium. Repeat this rinsing process until the fluid is free of most small debris and green color. For chrysanthemum, the preparation should also be free of any odor.

Prepare digestion media about 1 h before use. Dissolve 200 mg (1%) of RS cellulase and 80 mg (0.4%) y-23 Pectolase in 20 ml of each of the four isolation media — use 100-ml beakers and allow at least 15–30 min of vigorous stirring. These are the digestion media and may be added directly to the cut leaf tissue. Note: if this experiment is to be conducted aseptically, then the enzyme solution must be filter sterilized into cooled autoclaved medium as described in chapter two. Wrap the petri dishes with several layers of Parafilm and incubate at room temperature in the dark. The cultures should remain stationary — shaking the petri dishes at this time is detrimental to protoplast formation.

Using a compound microscope, observe the preparation for release of protoplasts. Focus on the scalpel blade marks on the bottom of the petri dish — often the smaller, more cytoplasmically dense protoplasts will sink to the bottom. Also focus on the cut edges of the leaves to observe plasmolyzed cells and the formation of protoplasts (Figure 25.2). Repeat this operation and record your observations every $1/2$ h for a total of 3 or 4 h.

At the end of digestion period, place the petri dishes on a rotary shaker set for the slowest speed possible for 5 min; the medium should hardly move within the petri dishes. Afterward, remove as much as possible of the undigested leaf tissue using a fine pair of forceps. The leaf strips will be very soft and have a tendency to adhere to the forceps blades, but can easily be removed by running the forceps blades through some solidified medium in a petri dish.

Gently swirl the contents of the petri dish and, using a wide-bore (mouth) glass pipette equipped with a bulb or pump, carefully remove the liquid from the petri dish. While holding the tip of the pipette on the inside of the sieve housing, very gently express the medium down the side. This procedure is important because the protoplasts are very fragile and may be damaged if not handled very carefully. Use a nylon mesh sieve with between 40- and 100-μm cross-sectional openings in a beaker. At times, the surface tension between the medium and the sieve screen may not permit the passage of medium into the collection beaker. This is very easily solved by gently "scratching" the surface of the nylon sieve with the tips of the forceps blades — the liquid should now flow into the beaker.

**Figure 25.2**   Protoplasts formed from a strip of a chrysanthemum leaf exposed to release medium with cell wall degrading enzymes for 2 h.

Remove the sieve housing and very gently pipette the medium and protoplasts from the collection beaker into either a 25- or 50-ml plastic centrifuge tube. The medium may appear somewhat "cloudy" — no problem, this is great! This is the first sign, other than direct observation, that many protoplasts have been formed. Centrifuge the suspension at $100 \times g$ for 5 min. The following two cautionary notes before centrifugation:

1.   Be sure that the "centrifuge or rotor is balanced". Insert a tube of equal weight in the rotor station exactly opposite of your tube containing protoplasts.
2.   Do not confuse $100 \times g$ with 100 rpm (revolutions per minute) — they are not the same. Check the specifications of the centrifuge (the manufacturer usually provides a conversion chart or a mathematical formula to calculate relative force, $g$, from rpm) and adjust rpm to achieve $100 \times g$.

Decant or pipette the medium from the tube without disturbing the green pellet of protoplasts on the bottom. Carefully pipette 20 ml of isolation medium down the inside of the centrifuge tube. Dislodge the pellet with a few gentle jets of medium from a pasteur pipette. Centrifuge as before and repeat this washing step twice. This step is important if culturing the protoplasts. All hydrolytic enzymes must be removed since regeneration of the cell wall will not occur in their presence.

### Determining the yield of protoplasts
After the final wash with isolation medium, resuspend the pellet in 5 ml of fresh isolation medium. Determine the number of protoplasts in the suspension by using a hemacytometer. Sum the number of protoplasts in five squares (Figure 25.3) on an individual counting surface. Only count those protoplasts that are completely within the boundaries of the counting surface; those touching a line should not be included. Repeat counts from at least three samples and find the mean. Calculate the number of protoplasts per milliliter using the following equation: $X \times 2,000 = P$; where $X$ = the mean of three counts and $P$ = the number of protoplasts. For example, if $X = 50$, then there are 100,000 protoplasts per milliliter. This is an estimated number of protoplasts plus or minus about 10%. Since the

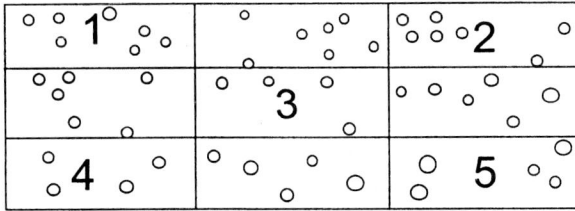

**Figure 25.3**  Representation of the counting areas on a hemacytometer, omitting the fine lines. Sum the protoplasts in the five numbered areas — do not count any protoplasts that touch any of the boundary lines of the squares — and multiply by 2,000 to obtain an estimate (± 10%) of the protoplasts per milliliter. In this example, there are 27 protoplasts pictured in the counting areas (1–5) and, therefore, an estimated 54,000 protoplasts/ml.

protoplasts were suspended in 5 ml, the estimate of the total number of protoplasts from the preparation equals: 5 ml × 100,000 protoplasts/ml = 500,000 protoplasts in the tube.

Now calculate the estimated yield of protoplasts per gram of tissue. For example, if 0.5 g of tissue were used, then 500,000 protoplasts per 0.5 g = 1,000,000 protoplasts per gram of tissue. The yield of protoplasts from the different treatments now can be compared since the data are all expressed on a per gram basis.

### Methods for evaluating the quality of protoplasts

When counting protoplasts you may have noticed that they were almost perfectly spherical; normally this would indicate that all of the cell wall material was digested by the enzymes. However, "patches" of cell wall may still remain. The best way to detect wall materials is with calcofluor, a chemical that binds with cellulose, a major plant cell wall polymer. The stain is made by mixing 1 ml of 1% aqueous calcofluor with 20 ml of the appropriate isolation medium. Pipette 0.25 ml of protoplast suspension into 1 ml of the stain solution. Swirl to mix, mount a few drops of the mixture on a glass slide and cover with a #1 or #1 $^1/_2$ coverslip. First, observe protoplasts with transmitted visible light using a 10 or 20× objective lens. Next, examine the protoplasts using an ultraviolet light (UV) source with the following filters and mirror: 400–440 nm excitation; 470-nm barrier; and 455-nm dichroic mirror (equivalent to Nikon filter assembly BV-2A). If the wall is not completely digested, the calcofluor bound to cellulose will fluoresce a brilliant blue that will not fade with time. Some protoplasts may still retain wall fragments, which are visible as irregularly shaped blue glowing patches. If true protoplasts (i.e., the cell minus the wall), the field will appear dark and the protoplast may not be discernible except for perhaps some red autofluorescence of chloroplasts. Observe at least 50 protoplasts and calculate the percentage that are free of cell wall. Incidentally, this is the same optical brightener that makes white clothing "bright" in natural light or fluoresce with black light.

The protoplast isolation procedure often damages or kills a percentage of cells; therefore, it is advisable to assess the condition of the protoplasts. Mount a small drop of the protoplast suspension on a glass slide, coverslip and observe using 20 and 40× objective lens. The cytoplasm of some or a few may appear to be streaming or moving in a circle — look for the small "particles" (organelles) in the cytoplasm. This phenomenon is termed *cyclosis* and only occurs in living cells. However, a cell that does not exhibit cyclosis is not necessarily dead. Also, look for "bulges or budding" of the plasmalemma (cell membrane) that may indicate damaged, dying or fused protoplasts.

There are a number of presumptive tests to determine viability of cells. The tests are presumptive because they do not actually measure which cells are "living" and those cells that are "dead", but usually assess some physiological or physical properties associated

**Figure 25.4**  Determining viability of protoplasts using fluorescein diacetate. a. Protoplasts viewed with transmitted visible light. b. Protoplasts "stained" with fluorescein diacetate and viewed with UV light. The fluorescence light is green.

with living cells. Most of these tests are based on enzyme activity or permeability of the plasmalemma to certain indicator chemicals, such as dyes or stains.

Fluorescein diacetate often has been used to estimate the percentage of viable proto-plasts in a preparation (Widholm, 1972). This technique uses fluorescence and is very efficient; many protoplasts can be examined in a short time. Fluorescein diacetate normally does not fluoresce, but once across the cell membrane, an esterase in the cytoplasm cleaves the acetate groups from the main portion of the molecule. The resulting moiety will fluoresce green when exposed to the proper wavelength UV light.

Mix 0.25 ml of 0.05% fluorescein diacetate (dissolved in acetone) with 20 ml of the appropriate isolation medium to make the "stain" solution. Pipette 0.25 ml of the proto-plast suspension into 1.0 ml of the stain solution. Swirl gently to mix and mount some of the protoplasts on a glass slide. First, look at the protoplasts using a 10 or 20× objective lens. Now, view the protoplasts using the UV source and the following filters and mirror: 420–490 nm excitation; 520-nm barrier; and a 510-nm dichroic mirror (equivalent to Nikon filter assembly B-3A). The cytoplasm of "living" protoplasts should fluoresce green (Figure 25.4); vacuoles in the protoplasts will appear dark. Protoplasts in which the cytoplasmic esterase is not active are presumably dead and will appear to be dark. Occasionally, the autofluorescence of chlorophyll or other compounds will impart a reddish hue to the protoplasts. Observe at least 50 protoplasts and calculate the percentage of living cells. Multiply the estimate of number of protoplasts by this percentage to obtain an estimate of the number of living protoplasts. From the example above, there are $5 \times 10^5$ protoplasts per milliliter multiplied by 0.95 viable equals $4.75 \times 10^5$ living protoplasts per milliliter. A good isolation will yield between 90 and 95% viable protoplasts.

As an alternative to epifluorescence microscopy to determine viability, Evans blue stain may be used with visible light (Gaff and Okong 'O-Ogola, 1971). Mix 2 ml of 1% aqueous stain with 10 ml of the appropriate isolation medium. Pipette 0.25 ml of the protoplast suspension into 1.0 ml of the stain mixture and allow to stand for about 15 min. The dye will be excluded from the protoplasts with intact and functional plasmalemma (presumed living cells), whereas those protoplasts whose membrane are dysfunctional will turn blue and are considered nonviable.

*Anticipated results*

Preculturing and cutting the donor tissue in isolation medium should minimize damage to cells. Rinsing the donor tissue before digestion will remove cellular debris and chemical compounds released from broken cells during the cutting operation. Some of these chemicals are thought to be detrimental to the remaining cells.

A very few protoplasts should be released within the first hour and will be difficult to find. However, focusing on a cut edge of the leaf tissue (Figure 25.2) should show cells with various degrees of cell wall degradation. Also, plasmolysis of cells should be easily observed. As the experiment progresses, more and more protoplasts will be released and can easily be seen on the bottom of the petri dish or floating in the medium. The brief shaking at the conclusion of the experiment helps to maximize the release of protoplasts. Because so many factors influence the yield of protoplasts, there is not a way to accurately foretell which release medium will produce the best results. Typically, this protocol can be optimized for both orchardgrass and chrysanthemum leaves to yield between $3 \times 10^5$ and $5 \times 10^6$ protoplasts per gram of tissue. Although some protoplasts should be produced throughout the range of osmotic pressures developed in the four media, protoplasts in some of the media may be very fragile and may not have survived the centrifugation steps. It is not unusual to observe in the lowest osmotic pressure medium that intact, viable protoplasts will not be produced. Therefore, when approaching the question of how to isolate protoplasts from a "new species", it is often useful to initially try a wide range of media and enzyme concentrations.

*Questions*

* Why would preculturing the donor tissue in isolation medium at 4 C facilitate production of protoplasts?
* Why are protoplast isolation and release media made with relatively high concentrations (%) of mannitol? Could other osmotically active sugar alcohols be used?
* Why are protoplasts isolated from relatively young, meristematic tissues? *Hint:* Consider cell wall properties also.
* Why is it important to remove all of the enzyme solution before trying to culture protoplasts?
* Why is it important to determine both the yield of protoplasts from the donor tissue and the concentration of protoplasts in the culture medium?
* Are the cell wall degradation enzymes ("cellulase" and "pectinase") used in this experiment pure? If not, what other hydrolytic and oxidative enzymes may be present as contaminants, and are these beneficial (aids in liberating protoplasts) or detrimental (may cause harm or death to the cells or protoplasts)?
* What is fluorescence?

# Literature cited

Gaff, D.F. and O. Okong 'O-Ogola. 1971. The use of nonpermeating pigments for testing the survival of cells. *J. Exp. Bot.* 22:756–758.

Sauvadet, M.-A., P. Brochard and J. B. Gibod. 1990. A protoplast-to-plant system in chrysanthemum: differential response among several commercial clones. *Plant Cell Rpt.* 8:692–695.

Widholm, J. 1972. The use of fluorescein diacetate and phenosafranine for determining viability of cultured plant cells. *Stain Technol.* 47:189–194.

*chapter twenty-six*

# Isolation and culture of protoplasts from embryogenic yellow poplar suspension cultures

*Scott A. Merkle*

Protoplast cultures have a number of potential applications, including direct gene transfer and generation of somatic hybrids or cybrids via protoplast fusion. Although the application of protoplasts for gene transfer has largely been obviated by the development of techniques that make use of intact tissues or cells (e.g., microprojectile bombardment), there are still some cases where somatic hybridization or cybridization may be highly desirable. To be employed for any of these applications, once protoplasts are isolated and manipulated, they must be able to rebuild their cell walls and undergo division. Furthermore, they must maintain their morphogenetic competence in order for plants to ultimately be regenerated. There have been a number of reports of protoplast isolation from hardwood forest trees, but very few of these have resulted in plant production from the protoplasts. A critical factor determining the ability of protoplasts to regenerate whole plants is the starting material from which the protoplasts are isolated. Embryogenic cultures that proliferate as proembryogenic masses (PEMs) (Halperin, 1966) are often chosen as protoplast source material. PEMs, which may be thought of as very early stage proembryos caught in a repetitive cycle of proliferation, are an excellent source of protoplasts for a number of the following reasons: (1) cells of PEMs grown in suspension are already habituated to culture and thus are less vulnerable to the shock of in vitro culture manipulations than tissues not previously exposed to culture conditions; (2) since cells of PEMs are highly proliferative, they rapidly rebuild cell walls and begin dividing following protoplast isolation; and (3) many of the cells that compose PEMs are programmed to make somatic embryos, and these cells maintain their morphogenetic competence through protoplast isolation and culture. Yellow poplar (*Liriodendron tulipifera* L.) embryogenic cultures have the rare property among hardwood forest tree embryogenic cultures of proliferating as PEMs. A brief introduction to yellow poplar, as well as exercises for initiating embryogenic cultures and producing somatic embryos and plantlets are presented in chapter twenty-one of this book. Using yellow poplar PEMs as starting material, we demonstrated that protoplasts isolated from these PEMs could be induced to rebuild their cell walls, undergo cell division and eventually produce new PEMs from which somatic embryos and plantlets could be regenerated (Merkle and Sommer, 1987). The objective of the following laboratory exercise, which is designed for advanced students, is to give students experience with isolating and culturing protoplasts from embryogenic suspension cultures. Refer to Figure 26.1 for a diagrammatic representation of this exercise. The exercise will require about 4 weeks to complete.

0-8493-9409-0/96/$0.00+$.50

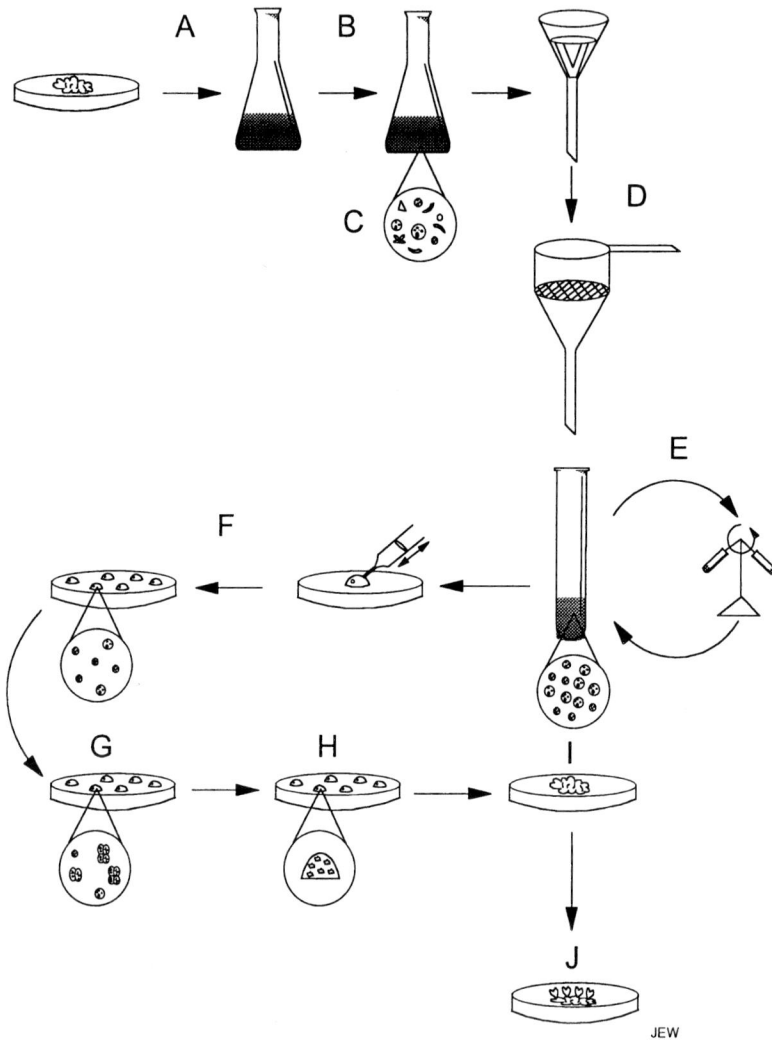

**Figure 26.1** Diagrammatic representation of yellow poplar protoplast isolation and culture. A. Embryogenic suspension culture initiated by inoculating PEMs from solid culture into flask of liquid induction medium. B. Protoplast isolation begins with inoculation of suspension-cultured PEMs into flask of digestion medium containing cell wall digesting enzymes. C. After at least 12 h of incubation, flask contains protoplasts, undigested cells and debris. D. To clean up protoplasts, digestion mix is filtered through Miracloth followed by a 25-μm stainless steel sieve. E. Filtrate is repeatedly centrifuged and resuspended in wash medium, and, finally, purified protoplasts are resuspended in regeneration medium. F. A drop of protoplasts is pipetted into the center of a petri dish, warm agarose-supplemented regeneration medium is added, and the mixture is taken up and redistributed as droplets on the petri dish, which are allowed to solidify into beads before the petri dish is flooded with liquid regeneration medium. G. Within 4 days, some of the protoplasts in the agarose beads rebuild their cell walls and begin to divide to form colonies. H. Cell colonies continue to divide to form PEMs. I. Beads containing PEMs are plated on induction medium allowing PEMs to grow out onto the induction medium. J. Protoplast-derived PEMs transferred to basal medium produce somatic embryos.

# Exercise

## Experimental protocol

### Materials

The following materials will be needed for each student or team of students:

* Embryogenic yellow poplar suspensions, 3 weeks following last transfer (see instructions below for initiating suspension cultures)
* Spoonulas
* 50-ml Erlenmeyer flasks
* 10-ml syringe with luer lock
* Millex GV 0.22-μm filter units (Millipore Corp., Bedford, MA)
* Macerase (Calbiochem, San Diego, CA) or other pectinase
* Cellulysin (Calbiochem, San Diego, CA) or other cellulase
* Bovine serum albumin
* Glass funnels
* Screw-top test tubes
* Bench-top centrifuge
* "Cellector" tissue sieve pan with 25-μm opening size screen (Bellco Glass, Inc., Vineland, NJ)
* Miracloth (Sigma, St. Louis, MO)
* Water bath or heating block
* 35 × 10 mm plastic petri dishes (Corning-Costar, Cambridge, MA)
* GA-7 (Magenta Corp., Chicago, IL) vessels
* SeaPlaque low melting point agarose (FMC Corp., Midland, ME)
* Inverted microscope or compound microscope

### Procedure

*Initiation of suspension cultures.* Embryogenic yellow poplar suspension cultures can be easily initiated from PEMs growing on semisolid induction medium (see chapter twenty-one). Prepare liquid induction medium (see Table 21.1), leaving out the Phytagar and dispense 40-ml aliquots into 125-ml Erlenmeyer flasks. Inoculate the flasks with about 0.5 g of PEMs and incubate in the light (16 h/day) on a gyratory shaker at 90 rpm. Cultures should be maintained by transferring about a spoonula of PEMs to fresh liquid induction medium every 3 weeks.

*Media preparation.* Prepare 100 ml of wash medium, which consists of 100 ml of induction medium given in Table 21.1 excluding sucrose and casein hydrolysate, but amended with 10.9 g mannitol, 50 mg $CaCl_2 \cdot H_2O$ and 10 mg 2[*N*-morpholino] ethanesulfonic acid (MES) as a buffering agent. Place 10 ml of wash medium into a 25-ml beaker or other very small container, add 100 mg Cellulysin, 50 mg Macerase and 10 mg bovine serum albumin and dissolve using a stir bar to make digestion medium. Filter sterilize the digestion medium into a sterile 50-ml flask using a syringe fitted with a Millex GV filter unit. Prepare regeneration medium by starting with 100-ml of induction medium without sucrose and plant growth regulators (PGRs) and supplemented with 50 mg $CaCl_2 \cdot H_2O$, 9 g glucose, 25 mg each of sucrose, xylose, and fructose, 0.1 mg 2,4-D and 0.025 mg BA. Pipette 10 ml of regeneration medium into a 50-ml flask. To the 10 ml of regeneration medium, add 0.25 g of SeaPlaque agarose. Autoclave the medium with agarose and, soon

**Figure 26.2**  Freshly isolated yellow poplar protoplasts from PEMs.

after removal from the autoclave, place the flask in a 38 C water bath to keep the medium liquified. Filter sterilize the remaining 90 ml of regeneration medium into a sterile flask.

*Protoplast isolation.*  Use a spoonula to aseptically transfer about 0.5 g of yellow poplar PEMs into the flask of digestion medium. Place the flask in a dark incubator/shaker at about 30 C and shake at about 50 rpm overnight (at least 12 h).

Following the overnight incubation of PEMs in digestion medium, check for the release of protoplasts in the flask using an inverted microscope. If an inverted microscope is not available, aseptically remove a small sample from the flask using a sterile pasteur pipette, put a drop of suspension on a slide and use a compound microscope to check for protoplasts. Protoplasts are easy to distinguish from intact cells because of their spherical shapes (Figure 26.2). If there are few or no protoplasts, you can either allow the digestion to continue or start over with new material. If there are a substantial number of protoplasts, they are now ready to be purified and then cultured.

Line a glass funnel with two layers of sterile Miracloth and slowly drip the digested PEMs and protoplasts through the Miracloth using a sterile pasteur pipette, allowing them to collect in a sterile test tube. Then, using another pipette, slowly drip the filtered suspension through a 25-μm pore size stainless steel sieve, collecting the filtrate in a sterile screw-cap test tube. You may need to wet the screen first with some sterile wash medium for the suspension to begin dripping through the screen. Close the screw-cap test tube and centrifuge the filtrate at $100 \times g$ for 5 min. Discard the supernatant and resuspend the pellet in a few milliliters of filter-sterilized wash medium. Repeat the centrifugation and resuspension steps twice more except, after the third centrifugation, resuspend the pellet in a few milliliters of filter-sterilized regeneration medium instead of wash medium. Once the pellet, which should now be composed of almost pure protoplasts, is resuspended, take a small sample of the suspended protoplasts and count them in a hemacytometer (refer to chapter twenty-five). Yields should range from 5–10 million protoplasts/g of PEMs.

*Protoplast culture.*  Protoplasts can be cultured in agarose beads using the following method. Dilute the protoplasts to about 20,000–50,000 per milliliter with regeneration medium and pipette 0.2 ml drops of suspended protoplasts into the middle of plastic petri

**Figure 26.3**  Rapidly dividing yellow poplar cell colonies derived from protoplasts following 1 week of culture in agarose beads.

dishes (1 drop per dish). Pipette an equal amount of the 38 C agarose-supplemented regeneration medium into the drop of protoplasts and immediately take up the mixture with the same pipette and redistribute it as small (about 0.05-ml) droplets around the perimeter of the petri dish. After the droplets have gelled into beads, flood the petri dishes with sufficient liquid regeneration medium so that the beads are about half submerged. Seal the petri dishes with Parafilm and stack inside of a GA-7 vessel to slow desiccation. Incubate in the dark at 30 C.

*Anticipated results*

Individual protoplasts should begin to lose their spherical shape within 3–4 days, indicating that cell wall regeneration has occurred. (See chapter twenty-five for alternate methods of observing wall regeneration and viability.) The first cell divisions should be observed within 5–6 days (Figure 26.3), colonies with eight or more cells should be visible under the microscope within 2 weeks and protoplast-derived microcalli will be large enough to be seen without a microscope in 3–4 weeks. At this point, the regeneration medium should be diluted with an equal volume of liquid yellow poplar induction medium, and 3 weeks later the agarose beads with calli can be transferred to either liquid or solid induction medium. Within a few weeks following this transfer, PEMs will expand out of the agarose beads and continue to proliferate. The PEMs should be indistinguishable from the original material from which the protoplasts were isolated. Protoplast-derived PEMs can be treated just like other PEMs and should be capable of producing somatic embryos and plantlets using the protocol outlined in experiment 2 of chapter 21.

*Questions*

* Why is BSA included in the digestion medium?
* Why is mannitol included in the digestion medium?
* Why is mannitol unnecessary in the regeneration medium?
* Leaf mesophyll protoplast isolation protocols use a sucrose gradient to separate the newly isolated protoplasts from undigested material and debris by floating the protoplasts away from the other material. Why do you think this may not work with protoplasts isolated from embryogenic cells?

## *Literature cited*

Halperin, W. 1966. Alternative morphogenic events in cell suspensions. *Am. J. Bot.* 53:443–453.

Merkle, S.A. and H.E. Sommer. 1987. Regeneration of *Liriodendron tulipifera* (family Magnoliaceae) from protoplast culture. *Am. J. Bot.* 74:1317–1321.

*chapter twenty-seven*

# Haploid cultures

**Sandra M. Reed**

Haploid plants are of great value to plant breeders and geneticists. Because of the presence of only one set of chromosomes, recessive mutations are easily identified. In addition, doubling the chromosome number of a haploid, to produce a dihaploid, results in a completely homozygous plant. Theoretically, the genotypes present among a large group of dihaploids derived from an $F_1$ hybrid represent, in a fixed form, the genotypes expected from an $F_2$ population. Use of dihaploids in breeding programs can thus greatly reduce the time required for development of improved cultivars. To be most useful, the researcher needs to be able to produce a large number of haploids from many different genotypes.

Haploids have been available for genetic studies for many years. Prior to the 1960s, they were mostly obtained spontaneously following interspecific hybridization or through the use of irradiated pollen, but usually only infrequently and in very small numbers. Haploid methodology took a giant step forward 30 years ago when Guha and Maheshwari (1964) found that haploid plants could be obtained on a regular basis and in relatively large numbers by placing immature anthers (male) of *Datura innoxia* Mill. into culture. This work was rapidly expanded using tobacco (*Nicotiana tabacum* L.), which became the "model species" for anther culture experiments. To date, haploids have been produced via anther culture in over 170 species, and several good reviews provide lists (Maheshwari et al., 1982; Bajaj, 1983; Heberle-Bors, 1985). While the work has been much more limited, haploids have also been obtained from in vitro culture of the female gametophyte.

Although much of the terminology used in this chapter has been discussed in previous chapters, the in vitro induction of haploids involves a few specialized terms that need to be defined before further consideration of the topic. A haploid is a plant with the gametic or "n" number of chromosomes. Dihaploids, or doubled haploids, are chromosome doubled haploids or "2n" plants. When anthers are cultured intact, the procedure is called anther culture; in contrast, microspore culture involves isolating the microspores from the anthers before culture and is sometimes referred to in the literature as pollen culture. The development of haploid plants via anther or pollen culture occurs through a process commonly known as androgenesis.

As with the other in vitro techniques discussed in this book, many variables can and do affect the success of anther culture. This chapter will begin with a discussion of these variables and be followed by a consideration of androgenesis and the different pathways by which haploid plants arise. Lastly, some of the basics of anther culture methodology will be summarized.

# Factors affecting anther culture

## Genotype

The choice of starting material for an anther culture project is of the utmost importance. In particular, genotype plays a major role in determining the success or failure of an experiment. Haploid plant production through anther culture has been very limited or nonexistent in many plant species and, furthermore, within a species, differences exist for ability to produce haploid plants. For example, some corn (*Zea mays* L.) cultivars are completely unresponsive in anther culture, whereas a few haploids can be obtained from others (Wan and Widholm, 1993). Even within an amenable species, such as tobacco, some genotypes produce haploids at a much higher rate than do others. Because of this genotypic effect, it is important to include as much genetic diversity as possible when developing protocols for producing haploid plants via anther culture.

## Condition of donor plants

The age and physiological condition of donor plants often affect the outcome of anther culture experiments. In most species, the best response comes from the first set of flowers produced by a plant. As a general rule, anthers should be cultured from buds collected as early as possible during the course of flowering.

Various environmental factors that the donor plants are exposed to may also affect haploid plant production. Light intensity, photoperiod and temperature have been investigated, and at least for some species, found to influence the number of plants produced from anther cultures. Specific optimum growing conditions differ from species to species and are reviewed by Powell (1990). In general, the best results are obtained from healthy, vigorously growing plants.

## Stage of microspore development

The most critical factor affecting haploid production from anther culture is the stage of microspore development; for many species, anthers are only responsive during the uninucleate stage of pollen development. In contrast, optimum response is obtained in tobacco from anthers cultured just before, during and just after the first pollen mitosis (late uninucleate to early binucleate microspores).

In developing a protocol for anther culture, one anther from each bud is usually set aside and later cytologically observed to determine the stage of microspore development. Generally, anthers within a bud are sufficiently synchronized to allow this one anther to represent the remaining cultured anthers. Measurements of physical characteristics of the flower, s ch as calyx and corolla length and anther color, shape and size, are also recorded. Results o the experiments are analyzed to determine which microspore stage was the most responsive. The physical descriptions of the buds and anthers are then examined to determine if this microspore stage correlates to any easily identified inflorescence, flower or anther characteristic(s). For example, in tobacco, buds in which the calyx and corolla are almost identical in length usually contain anthers having microspores at or near the first pollen mitosis. A researcher wishing to produce a maximum number of haploid plants of tobacco would collect only buds fitting this physical description.

## Pretreatment and preincubation

For some species, a pretreatment following collection of buds, but before surface disinfestation and excision of anthers, has been found to be beneficial. Yields of tobacco haploids are often increased by storing excised buds at 7–8 C for 12 days prior to anther excision and culture (Sunderland and Roberts, 1979). For other species, temperatures from 4–10 C and durations from 3 days to 3 weeks have been utilized. For any one species, there may be more than one optimum temperature–length of treatment combination. In general, lower temperatures require shorter durations, whereas a longer pretreatment time is dictated for temperatures at the upper end of the cold pretreatment range mentioned. High temperature pre-incubations have been effective in some species. For example, haploid plant production was increased in rape (*Brassica campestris* L.) by culturing the anthers at 35 C for 1–3 days prior to culture at 25 C (Keller and Armstrong, 1979).

## Media

Androgenesis can be induced on a simple medium such as that developed by Nitsch and Nitsch (1969) for tobacco and a few other species. For most other species, the most commonly used media for anther culture are MS (Murashige and Skoog, 1962) and N6 (Chu, 1978) or variations on these media. In some cases, complex organic compounds, such as potato extract, coconut milk and casein hydrolysate, have been added to the media. For many species, 58–88 mM (2–3%) sucrose is added to the media, whereas other species, particularly the cereals, have responded better to higher (up to 435 mM or about 15%) concentrations of sucrose. The higher levels of sucrose fulfill an osmotic, rather than a nutritional, requirement. Other sugars, such as ribose, maltose and glucose, have been found to be superior to sucrose for some species.

For a few species, such as tobacco, it is not necessary to add plant growth regulators (PGRs) to the anther culture media. Most species, however, require a low concentration of some form of auxin in the media. Cytokinin is sometimes used in combination with auxin, especially in species in which a callus phase is intermediate in the production of haploid plants.

Anther culture media is often solidified using agar. Because agar may contain compounds inhibitory to the androgenic process in some species, the use of alternative gelling agents has been investigated. Agarose has been reported to be a superior gelling agent for anther culture of many of the cereal species. Alternatively, the use of liquid medium has been advocated by some researchers as a way to avoid the potentially inhibitory substances in gelling agents. In this technique, the anthers are placed on the surface of the medium, forming a so-called "float culture".

## Androgenesis

### Development of haploids

Haploid plants develop from anther culture either directly or indirectly through a callus phase. Direct androgenesis mimics zygotic embryogenesis with the exception that a true suspensor is not observed in androgenic embryos (see chapter sixteen). At the globular stage of development, most of the embryos are released from the pollen cell wall (exine). They continue to develop, and after 4–8 weeks, the cotyledons unfold and plantlets emerge from the anthers. Direct androgenesis is primarily found among members of the tobacco (Solanaceae) and mustard (Cruciferae) families.

During indirect androgenesis, the early cell division pattern is similar to that found in the zygotic embryogenic and direct androgenic pathways. After the globular stage, irregular and asynchronous divisions occur and callus is formed. This callus must then undergo organogenesis for haploid plants to be recovered. The cereals are among the species that undergo indirect androgenesis.

The early cell divisions that occur in cultured anthers have been studied (for review, see Reynolds, 1990). For species cultured during the uninucleate stage, the microspore either undergoes a normal mitosis and forms a vegetative and a generative nucleus or divides to form two "similar-looking" nuclei. In those cases where a vegetative and a generative nuclei are formed in culture, or where binucleate microspores are placed into culture, it is usually the vegetative nucleus that participates in androgenesis. The only species in which the generative nucleus has been found to be actively involved in androgenesis is black henbane (*Hyoscyamus niger* L.). When similar-looking nuclei are formed, one or both nuclei may undergo further divisions. In some cases, the two nuclei will fuse, producing homozygous diploid plants or callus. Since diploid callus may also arise from somatic tissue associated with the anther, diploids produced from anther culture cannot be assumed to be homozygous. To verify that plants produced from anther culture are haploid, chromosome counts should be made from root tips or other meristematic somatic tissues. Because haploids derived from diploid species are expected to be sterile or have greatly reduced fertility, pollen staining, which is much quicker and requires less skill than chromosome counting, can also be used to identify potential diploids. Pollen staining, however, will not identify plants that have reduced fertility because they have a few extra or missing chromosomes, i.e., aneuploids; aneuploidy occasionally occurs as a result of tissue culture.

## Problems associated with anther culture

Problems encountered in plants during or as a result of anther culture range from low yields to genetic instability. Many of the major horticultural and agronomic crops do not yield sufficient haploids to allow them to be useful in breeding programs. With some species, particularly the cereals, there have been problems with a high frequency of albino plants recovered from haploid cultures. Both low yields and albino plant production have responded somewhat to modifications of media, stage of microspore development and other factors affecting anther culture. Further improvements may be expected with continued experimentation.

Tobacco has been one of the few crops in which yields of haploids have been high enough to test for genetic stability following anther culture. Large numbers of dihaploids have been produced from inbred cultivars of tobacco and compared in the field. Since these dihaploids arose from nearly homozygous plants, they were expected to be nearly identical to each other and to the cultivar from which they were derived. Instead, results of several experiments (reviewed by Reed, 1996), have shown that tremendous variability exists between these tobacco dihaploids. At least some of this genetic variability appears to be due to an increase in amount of nuclear DNA without a concomitant increase in chromosome number (DNA amplification). Unfortunately, since most of the variability is negative in nature, the usefulness of anther culture for tobacco-breeding work is questionable.

## General anther culture procedures

### Collection of floral buds

Floral buds may be collected from plants grown in the field, greenhouse or growth chamber. Entire inflorescences or individual buds are harvested and kept moist until ready for culturing. If buds are to be pretreated (i.e., low temperature), they should be wrapped in a moistened paper tissue and placed into a small zipper-type plastic bag.

## Disinfestation, excision and culture

Disinfestation typically uses a 5% sodium or calcium hypochlorite solution for 5–10 min. The beds are then rinsed thoroughly in sterile distilled water. Anthers are aseptically excised in a laminar flow hood, taking care not to cause injury. If the anther is still attached to the filament, the filament is carefully removed.

If a solid medium is used, the anthers are gently pressed onto the surface of the medium (just enough to adhere to the medium), but should not be deeply embedded. When using a liquid medium, the anthers are floated on the surface. Care must be taken when moving liquid cultures so as not to cause the anthers to sink below the surface.

For most species, disposable petri dishes are utilized for anther cultures. For a species with large anthers, like tobacco, the anthers from four to five buds (20–25 anthers) may be cultured together on one $100 \times 15$ mm diam. petri dish. For species with smaller anthers, or for certain experimental designs, smaller petri dishes or other containers may be more useful. Petri dishes are usually sealed and placed into an incubator; the specific temperature and light requirements of the incubator may depend on the species being cultured.

## Determining stage of microspore development

For most species, stage of microspore development can be determined by "squashing" an entire anther in acetocarmine or propionocarmine and then observing the preparation under a low power objective of a light microscope. The early uninucleate microspore is lightly staining with a centrally located nucleus. As the uninucleate microspore develops, its size increases and a large central vacuole is formed. As the microspore nears the first pollen mitosis, the nucleus is pressed up near the periphery of the microspore. Staining will still be fairly light. Pollen mitosis is of short duration, but it may sometimes be observed; it is recognized by the presence of condensed chromosomes. The product of the first pollen mitosis is a binucleate microspore containing a large vegetative and a small generative nucleus. The vegetative nucleus is often difficult to recognize because it is so diffuse and lightly staining. However, this stage may be definitively identified by the presence of the small densely staining generative nucleus. As the binucleate microspore ages, the intensity of the staining increases and starch granules begin to accumulate. Eventually, both nuclei may be hidden by the dark-staining starch granules.

## Handling of haploid plantlets

For species undergoing direct androgenesis, small plantlets can usually be seen emerging from the anthers 4–8 weeks after culture. When these get large enough to handle, they should be teased apart using fine-pointed forceps and can then either be placed on a rooting medium (usually low salt, with small amount of auxin) or transplanted directly into a small pot filled with soilless potting mixture. The callus produced in species that undergo indirect androgenesis must be removed from the anther and placed onto a regeneration medium containing the appropriate ratio of cytokinin to auxin.

To produce dihaploid plants, it is necessary to double the chromosome number of the haploids, and for many species a colchicine treatment is used. Published procedures for producing polyploids from diploids can be modified for use with anther culture–derived haploids. For example, it may be possible to use a colchicine treatment designed for small seedlings with haploid plants directly out of anther culture. Alternatively, established procedures using larger plants may be used. For tobacco, midveins of fully expanded leaves are excised and cultured on a regeneration-promoting medium (Kasperbauer and Collins, 1972). On the average, depending primarily on genotype, 50% of the plants regenerated from this media will be dihaploids.

In summary, haploids of many plant species can be produced using anther culture. Yields of haploids differ greatly depending on species and are also affected by cultural conditions, such as media formulation, stage of microspore development at time of culture and use of a low- or high-temperature pretreatment. Haploids may either arise directly from the anthers or may be produced indirectly through a callus intermediate.

The primary reason for producing haploids is for use in plant breeding, where they can hasten the development of superior cultivars. While anther culture is not effective in all species, for many it is the only or most efficient method of producing haploids. Continued research into cultural conditions and media requirements should eventually lead to improved anther culture protocols for many important crop species.

## Literature cited

Bajaj, Y.P.S. 1983. In vitro production of haploids. In: *Handbook of Plant Cell Culture,* Vol. 1, *Techniques for Propagation and Breeding.* D.A. Evans, W.R. Sharp, P.V. Ammirato and Y. Yamada, Eds., pp. 228–287. Macmillan, New York.

Chu, C. 1978. The N6 medium and its applications to anther culture of cereal crops. In: *Proceedings of Symposium on Plant Tissue Culture.* pp. 51–56. Science Press, Peking.

Guha, S. and S.C. Maheshwari. 1964. In vitro production of embryos from anthers of *Datura. Nature.* 204:497.

Heberle-Bors, E. 1985. In vitro haploid formation from pollen: a critical review. *Theor. Appl. Genet.* 71:361–374.

Kasperbauer, M.J. and G.B. Collins. 1972. Reconstitution of diploids from leaf tissue of anther-derived haploids in tobacco. *Crop Sci.* 12:98–101.

Keller, W.A. and K.C. Armstrong. 1979. Stimulation of embryogenesis and haploid production in *Brassica campestris* anther cultures by elevated temperature treatments. *Theor. Appl. Genet.* 55:65–67.

Maheshwari, S.C, A. Rashid and A.K. Tyagi. 1982. Haploids from pollen grains — retrospect and prospect. *Am. J. Bot.* 69:865–879.

Murashige, T. and F. Skoog. 1962. A revised medium for rapid growth and bioassays with tobacco tissue cultures. *Physiol. Plant.* 15:473–497.

Nitsch, J.P. and C. Nitsch. 1969. Haploid plants from pollen grains. *Science.* 163:85–87.

Powell, W. 1990. Environmental and genetical aspects of pollen embryogenesis. In: *Biotechnology in Agriculture and Forestry,* Vol. 12, *Haploids in Crop Improvement.* Y.P.S. Bajaj, Ed., pp. 45–65. Springer-Verlag, Berlin.

Reed, S.M. 1996. DNA amplification in doubled haploids of *Nicotiana.* In: *In Vitro Haploid Production in Higher Plants,* Vol. 2, S.M. Jain, S.K. Sopory and R.E. Veilleux, Eds. pp. 111–125. Kluwer Academic Publisher, Dordrecht.

Reynolds, T.L. 1990. Ultrastructure of pollen embryogenesis. In: *Biotechnology in Agriculture and Forestry,* Vol. 12, *Haploids in Crop Improvement.* Y.P.S. Bajaj, Ed., pp. 66–82. Springer-Verlag, Berlin.

Sunderland, N. and M. Roberts. 1979. Cold-treatment of excised flower buds in float culture of tobacco anthers. *Ann. Bot.* 43:405–414.

Wan, Y. and J.M. Widholm. 1993. Anther culture of maize. In: *Plant Breeding Review,* Vol. 11. J. Janick, Ed., pp. 199–224. John Wiley & Sons, New York.

## chapter twenty-eight

# Production of haploid tobacco plants using anther culture

*Sandra M. Reed*

Haploids are plants that have the gametic, or n, number of chromosomes. They are very useful in genetic and breeding studies because recessive genotypes are easily identified. Moreover, completely homozygous plants can be obtained quickly by doubling the chromosome number of the haploids to produce dihaploids ($2 \times$ n number of chromosomes); these dihaploids are very useful in plant breeding programs (see chapter twenty-seven).

For many years, few haploids were available to geneticists and breeders. This situation began to change in the 1960s with the discovery that haploids of *Datura* species and tobacco (*Nicotiana tabacum* L.) could be produced by culturing anthers. Today, haploids of over 170 species have been produced via anther culture. In no species has anther culture been more successful than tobacco, where large numbers of haploids can be obtained from many different genotypes. Because it is so easy to obtain tobacco haploids, it is the ideal species to demonstrate and learn anther culture methodology.

The following laboratory exercises illustrate the technique of anther culture using tobacco. Students will become familiar with the following techniques and concepts: (1) basic anther culture procedures, (2) determining the optimum stage of microspore development for anther culture and (3) the use of cold pretreatments for enhancing haploid production from anther culture. The general procedure used in both experiments is illustrated in Figure 28.1.

Directions are given for student teams. It should take 30–60 min of laminar flow hood use for each team to complete the culturing stage of each of the experiments. Preparation of media, staging of microspores and evaluation of cultures will require additional laboratory time, but minimum usage of the flow hood. Depending on class size, facilities and time available, two to four students per team are recommended.

## General considerations

### Growth of plants

A small quantity of tobacco seeds can be obtained from Dr. Verne Sisson, Department of Crop Science, North Carolina State University, Box 7620, Raleigh, NC 27695, and any cultivar may be used for the experiments. After completing the anther culture experiments, allow a few plants to continue to flower. Approximately 1 month after the flowers open, open-pollinated seed capsules will turn brown and dry. Seed can be collected and stored for several years under cool, dry conditions. A single capsule will contain 1–2000 seed that can be used as a source of plants for future experiments.

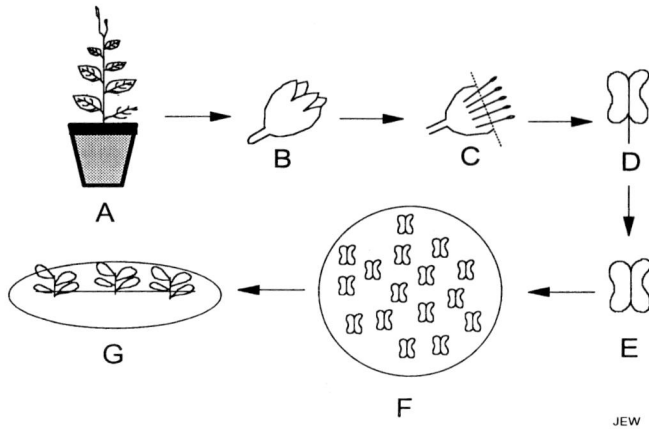

**Figure 28.1** Diagrammatic representation of the anther culture protocol for tobacco. A. Flowering diploid plant. B. Floral bud excised from plant. Calyx and corolla are of approximately equal lengths. C. Floral bud with calyx and corolla removed. Buds normally contain five anthers, which are shown still attached to filaments. D. Isolated stamen with anther and filament. E. Anther removed from filament. F. Petri dish with cultured anthers. G. Plantlets emerging from anthers.

Fill a 10-cm-diam. clay pot with vermiculite and water from the bottom by placing the pot in a pan of water. Lightly sow seed on the surface of vermiculite and keep the pot in a saucer of water until the seeds germinate in about 1 week. Continue to water seedlings from the bottom of the pot, but do not allow the vermiculite to be continually soaked nor to dry out completely. Seedlings may be grown in the greenhouse or in a growth chamber with 22–26 C and 16/8-h light/dark cycle.

After about 6 weeks, the seedlings will be large enough to transplant to individual 10- or 15-cm-diam. clay pots, and any of the soilless potting mixtures may be used. A slow-release all-purpose fertilizer may be incorporated into the medium at this time, or pots can be watered twice weekly with an all-purpose liquid fertilizer. Plants should be grown in a greenhouse. Supplemental light is not required during the winter, but if provided, may accelerate growth of the plants and hasten flowering. Five plants per team of students should be sufficient for both experiments included in this chapter. Plants should flower 2–4 months after transplant to 15-cm-diam. pots, depending on growing conditions, cultivar and size of pot (plants can be grown in a smaller pot to encourage more rapid flowering).

## Explant preparation

Collect buds soon after the plants begin to flower and transport to the laboratory in plastic bags containing moistened paper tissues. Surface disinfest the buds in groups of 20–25. Begin by placing the buds in 70% ethanol for 1 min, then transferring to 20 % commercial bleach for 5 min, followed by two rinses in sterile distilled water. Gently agitate the solution during the disinfesting procedure to facilitate complete coverage of the buds with the bleach solution.

Place a single bud in a sterile 100 × 15 mm plastic petri dish. Holding the bud gently with one set of forceps, remove the calyx (green) and corolla (colored) using another set of sharp fine-pointed forceps. Using the fine-pointed forceps, push the anthers away from the filaments (they should separate easily). Taking care not to squeeze the anthers, lift them gently with the forceps and place on the surface of the agar-solidified media (Figure 28.1). Wrap each petri dish with Parafilm-M and place in a 25 C incubator or growth chamber under constant light or 16/8-h light regime.

Cultures can be scored for number of haploids produced approximately 8 weeks after culture initiation. Using fine-tipped forceps, gently separate plants in order to get accurate counts. Unless these plants are to be used for other experiments, cultures can be scored outside of aseptic conditions.

The media that will be used in these two experiments is that of Nitsch and Nitsch (1969, see chapter two for composition). Each liter of media will be supplemented with 20 g (58 mM) sucrose and 8 g of agar. Plant growth regulators (PGRs) will not be added. Pour the medium into a number 100 × 15 mm sterile plastic petri dishes.

## Exercises

### Experiment 1. Effect of stage of microspore development on yield of haploid plants

As described earlier in chapter twenty-seven, the stage of microspore development at the time of culture is a critical factor affecting androgenesis. Students will be culturing anthers from buds of different developmental stages, determining stage of microspore development and if physical characteristics of the tobacco buds and anthers are correlated to stage of microspore development and the ability to produce haploid plants.

#### Materials

The following items are needed for each team of students:

* 100 ml of 20% commercial bleach
* Nine 100 × 15 mm petri dishes containing Nitsch and Nitsch (1969) medium (see chapter two for composition)
* Lighted incubator providing approximately 25 C
* Acetocarmine stain (45 ml glacial acetic acid, 55 ml of distilled water, 0.5 g carmine); place in beaker covered with aluminum foil and gently boil solution for 5 min under a fume hood. Filter the solution (may require several days and filter papers); the solution should be dark red.

#### Procedure

Each team should collect 36 tobacco flower buds. The length of the buds required will differ depending on cultivar, growing conditions, etc; therefore, the following general guidelines should be used for selection. Collect 12 buds having a calyx and corolla of approximately the same length (medium length buds). Select 12 buds in which the corolla is extended 5–7 mm beyond the calyx (long length buds). The remaining 12 buds should be the largest of the buds in which the corolla is not visible at all beneath the calyx (short length buds). Using these buds, the mid-uninucleate through mid-binucleate microspore stages should be represented regardless of the cultivar of tobacco used for the experiment.

Using an indelible marker, divide the bottom of each of the nine petri dishes containing culture medium into four quadrants and label from 1–36, which will correspond to the culture numbers. Each quadrant will be used for four of the anthers from a single bud. The remaining or fifth anther from each bud will be placed on a microscope slide that should also be marked with the quadrant number. For each culture number, a record should be kept for bud length (short, medium or long), stage of microspore development and number of plantlets produced.

Surface disinfest buds following general directions provided above while keeping the groups of three bud lengths separate. Excise anthers and place into culture as described previously. Squash the fifth anther of each group on a separate microscope slide in a small

**Figure 28.2**   Microspores of tobacco. a. Late uninucleate microspores. Vacuole can be seen on right side of cell. b. Anaphase of first pollen mitosis. c. Early binucleate microspore. Large diffuse vegetative nucleus and smaller, more darkly staining generative nucleus can be distinguished. d. Mid-binucleate microspore. Staining is darker, obscuring vegetative nucleus, but the generative nucleus is still visible.

drop of acetocarmine stain, remove debris and cover with coverslip. Heat the slide gently over steam or an alcohol burner for a few seconds and observe the stages of pollen development under the microscope. Using the photographs provided in Figure 28.2a–d and the descriptions of microspore development presented earlier in chapter twenty-seven, determine and record stage of microspore development for each anther. If there is not enough time to examine the anthers the day the cultures are initiated, place them in small vials with a fixative composed of three parts 95% ethanol and one part glacial acetic acid (make this solution under a fume hood). After 2 days, replace fixative with 70% ethanol. Leave the vials at room temperature for up to a week, then place in refrigerator. Anthers can be removed from fixative or alcohol at any time during this procedure, squashed and examined following the previous directions.

### Anticipated results

Four to six weeks after initiating the anther cultures, small plantlets will be seen emerging from some of the anthers (Figure 28.3). Even under the best of conditions, not all tobacco anthers will produce haploids. Numbers of plantlets emerging from each anther will vary from none to a few to many — there is no way to predict the results of any given experiment.

Anthers from short length buds will produce few, if any, haploid plantlets. Uninucleate microspores will have been observed from the anthers of this size bud. Anthers from the long length buds will also produce few, if any, plantlets. This size bud will correspond to a mid-to-late binucleate microspore stage. All or almost all of the plantlets produced in

**Figure 28.3** Haploid tobacco plantlets emerging from an anther.

this experiment will come from the anthers in which the calyx and corolla were equal in length. This size bud corresponds to the late-uninucleate to early-binucleate microspore stage, which produces the most haploid plants in tobacco.

*Questions*

* What is the purpose of trying to correlate a physical characteristic of buds and/or anthers to stage of microspore development?
* How would plantlets have been handled if you had been trying to produce dihaploid plants?
* How can you be sure that the plants you obtained are haploids?

## Experiment 2. Effect of cold pretreatment on yield of haploids from anther cultures of tobacco

Subjecting buds to several days of low temperatures after collection, but before culture, increases yields of haploid plants in tobacco and several other species. This experiment will demonstrate the effect that cold pretreatment of flower buds (anthers) has on yields of tobacco haploid plants.

*Materials*

Each student team will require the materials listed in experiment 1 and the following:

* Ten 100-mm petri dishes containing Nitsch and Nitsch (1969) media
* Refrigerator or incubator providing temperature of 7–8 C

*Procedure*

Each team needs to collect 40 tobacco buds, all of which should have a calyx and corolla of equal, or almost equal, length. Twenty of these buds should be placed in a zipper-type plastic bag with a water-moistened tissue, sealed and stored in a 7–8 C incubator or refrigerator for 10–12 days. The anthers from the remaining 20 buds should be placed into culture, following directions given under general considerations. After 10–12 days, culture anthers from cold pretreated buds in a similar fashion. For both sets of buds, the anthers from four buds can be placed into the same dish of medium (20 anthers per dish of medium).

Eight weeks after initiating cultures, count the number of plantlets from both the untreated and cold pretreated buds. Calculate the mean number and standard deviation (see chapter six) of plants/anther for both groups of buds.

### Anticipated results

The mean number of plants/anther should be significantly greater for the buds that received a cold pretreatment; however, differences between treatments will not be as great as observed in experiment 1. It may be advisable to pool results from all the teams in the class in order to see the effect of the cold pretreatment and statistical analysis.

### Questions

* Why were buds collected when the calyx and corolla were approximately equal in length?
* What temperatures and lengths of treatment would you test if you were trying to work out an anther culture procedure for a new species?

## Literature cited

Nitsch, J.P. and C. Nitsch. 1969. Haploid plants from pollen grains. *Science.* 163:85–87.

*chapter twenty-nine*

# Transformation

*Melina López-Meyer, Ignacio E. Maldonado-Mendoza and Craig L. Nessler*

Transformation can be broadly defined as the process of introducing DNA into cells. The importance of this technique comes from the fact that DNA, the molecule that carries the blueprints for life, can be cut, spliced and otherwise altered in vitro. Genetic manipulation techniques allow scientists to isolate a single DNA sequence from one organism and join it to that of a completely different organism to form what is now known as recombinant DNA.

The first recombinant DNA molecules were generated at Stanford University in 1972 by utilizing the cleavage properties of restriction enzymes and the ability of DNA ligase to join DNA strands together. The isolation of these two kinds of enzymes in the late 1960s was a major milestone in the development of genetic engineering. In 1973 DNA fragments were joined to the plasmid pSC101. These recombinant molecules behaved as replicons, i.e., they could replicate when introduced (transformed) into *Escherichia coli* cells.

Typically, when a particular DNA sequence is to be transformed into a living cell, it is first inserted in a vector. Vectors are used to assist in the transfer, replication and sometimes expression of a specific DNA sequence in a target cell.

There are certain essential features that vectors should possess. Ideally, they should be fairly small DNA molecules to facilitate their isolation and handling. They must also have an origin of replication, so that their DNA can be copied and thus maintained in the cell population as the host organism grows and divides. It is also important for a vector to have some type of selectable marker, usually resistance to an antibiotic, that will enable the vector to be selected from a large population of cells that have not taken up the foreign DNA. Finally, the vector must also have at least one unique restriction endonuclease recognition site to serve as the site of DNA insertion during the production of recombinants.

Bacteria and some other organisms contain relatively small, circular DNA molecules that are separate from the bacterial chromosome and replicate independently. These molecules are called plasmids. Although plasmids are generally dispensable, i.e., not essential for cell growth and division, they often provide a selective advantage to the host organism such as antibiotic resistance. Because of these features, plasmids are extensively used as vectors, particularly in the "construction" of complex recombinant molecules.

In addition to plasmids, bacteriophages have also been used as vectors. Bacteriophages are viruses of bacteria that are useful as vectors because they can act as carrier molecules for relatively large pieces of DNA.

Not all bacteria can be easily transformed, and it was not until the early 1970s that transformation was demonstrated in *E. coli*. Today, the protocol for transforming *E. coli* is very simple. It is carried out by mixing plasmid DNA with spheroplasts (prepared from cells that were pretreated with calcium chloride), incubating the mixture on ice and then giving a brief heat shock to facilitate the uptake of DNA into the cells.

Although the initial steps of gene manipulation and transformation methodologies were developed for prokaryotic systems (which still represent a major part of this technology), their potential application to eukaryotic systems such as plants was soon realized. The problems of introducing recombinant DNA into plants, however, are much different than those encountered with bacterial systems, and the methods available for introducing DNA into bacterial cells are not easily transferred to other cell types. It was not until 1983 that the first report of transformed plant cells was published (Fraley et al., 1983).

Several different approaches since developed for the transformation of vascular plants are discussed in this chapter. However, the most widely used system, which is based on the *Agrobacterium tumefaciens* tumor-inducing (Ti) plasmid, is described in the greatest detail. Laboratory exercises that follow this chapter illustrate both *Agrobacterium*- and particle bombardment–mediated transformation systems.

## Agrobacterium transformation system

*Agrobacterium tumefaciens* is a soil-inhabiting bacterium that is responsible for crown gall disease of many plants. The bacterium infects through a wound in the stem of the plant, and a tumor develops at the junction of root and stem (crown). The agent responsible for crown gall formation is not the bacterium itself, but a plasmid known as the Ti plasmid. Ti plasmids are large, ranging in size from 140–235 kb (1 kb = 1000 base pairs). During infection, a small portion of Ti plasmid DNA (15–30 kb), called T-DNA, is transferred to the plant cell nucleus, where it becomes covalently inserted into the nuclear DNA. In this manner, the T-DNA becomes stably maintained in the genome of transformed cells.

T-DNA carries the genes responsible for tumor formation and for the synthesis of unusual amino acid derivatives known as opines. The biological relevance of opines is that they can be used as the sole carbon and/or nitrogen source for the inducing *Agrobacterium* strain. The genes responsible for opines catabolism reside in the Ti plasmid. Strains of *Agrobacterium* can be classified according to the type of opines that are synthesized by the crown gall tumors that they induce. The most common types of opines are octopine and nopaline.

The genes responsible for transfer of T-DNA are also contained on the Ti plasmid and are called virulence genes (*vir* genes). *Agrobacterium* infection requires wounded plant tissue because *vir* genes are induced by phenolic compounds released by the injured plant cells. Induction of the *vir* genes results in the transfer of T-DNA (Figure 29.1).

The only regions of T-DNA that are absolutely required for its transfer and integration into the plant genome are the border regions. These are short repeat sequences of 25 base pairs. Any DNA sequence inserted between the border repeats will be transferred to and integrated into the plant genome. Therefore, Ti-based plasmids are excellent vectors for introducing foreign genes into plants.

In order to use Ti-based plasmids as vectors, the genes responsible for tumor formation must be removed. Ti-based plasmids lacking tumorigenic functions are known as disarmed vectors. Cells infected with such plasmids do not produce tumors and are consequently much easier to regenerate into normal, fertile plants by standard tissue culture techniques (see chapter thirty-one).

Even though Ti plasmids can be disarmed, they are too large to be conveniently used as vectors. Thus, smaller vectors (based on the features of the Ti plasmid) have been constructed that are suitable for manipulation in vitro.

**Figure 29.1** *Agrobacterium*-based plant cell transformation. A. *Agrobacterium tumefaciens* is a Gram negative soil bacterium. B. Wounded plant tissues release phenolic compounds, which induce the *vir* genes. *Vir* gene products mobilize and transfer T-DNA into the plant cell. C. T-DNA is incorporated into chromosomal DNA of the plant cell. Genes in the T-DNA encode proteins, which make opines that feed other bacteria near the wound as well as compounds that induce tumors formation.

## Cointegration and binary vector system

Two strategies have been developed for using Ti-based plasmids: a methodology called cointegration or the binary vector system. The cointegration technique is based on in vivo recombination of two plasmids. One plasmid carries desirable DNA sequence; the other plasmid contains *vir* genes and the border repeats of the T-DNA. After recombination, a very large recombinant Ti plasmid is obtained, which now can be used to transform plants.

The most frequently used approach for *Agrobacterium*-based plant transformation is the binary vector system, which uses two separate plasmids: one to supply the disarmed T-DNA and a second Ti plasmid, which includes the virulence functions (encoded by *vir* genes). The plasmid carrying the disarmed T-DNA is called mini-Ti plasmid. This plasmid bears the gene construct that will be inserted into the plant genome, along with a eukaryotic selectable marker between T-DNA border sequences, so that both genes will be inserted as a unit. When a mini-Ti plasmid is then placed into an *Agrobacterium* strain containing a plasmid with virulence functions, the *vir* gene products are able to drive the transfer of T-DNA into plant cells, even though T-DNA is located on a separate DNA molecule. This approach has been used extensively because mini-Ti plasmids are very easy to manipulate using standard recombinant DNA techniques.

The main drawback to the *Agrobacterium*-based system is that it is generally only applicable to dicotyledonous plants. Unfortunately, many of the most economically important crop plants are monocotyledonous cereals including wheat (*Triticum aestivum* L.), maize (*Zea mays* L.), rice (*Oryza sativa* L.), barley (*Hordeum vulgare* L.), and oats (*Avena sativa* L.). Nevertheless, limited success has been achieved using the Ti plasmid with these plants, and further work may extend the range of plant species that are amenable to Ti-mediated gene transfer.

## Plant viruses as vectors

Plant viruses have several special features as potential vectors for plant genetic engineering. For example, purified viral RNA/DNA is directly infectious to plants. Thus, simply rubbing a solution of viral nucleic acid on the leaf of a susceptible plant is often enough to obtain infection. In addition, some viruses can spread to every cell in the plant, which eliminates the regeneration step required in the *Agrobacterium* system. Also, relatively large amounts of virus can be produced from infected plants. This can result in the accumulation of very high levels of viral and foreign proteins being expressed from recombinant viruses. Furthermore, some viruses are able to infect crop plants, such as cereals, in which case the *Agrobacterium* system cannot be used.

Regardless of these potential advantages, the virus-based transformation systems also have some major disadvantages. In particular, viral nucleic acids do not become integrated into the plant genome like the *Agrobacterium* T-DNA. Thus, recombinant DNA will not be inherited by the next generation through the seed. In addition, viral infections usually debilitate plants to varying degrees, although mild virus strains might be used in this kind of approach. Another major drawback of using plant viruses as vectors is the limited amount of recombinant DNA that they can carry. Despite these disadvantages, plant viruses still represent a potentially useful way to introduce genetic information into plants.

## Direct transformation systems

In addition to natural transformation systems, such as *Agrobacterium* and plant viruses, alternative methods to deliver DNA directly into plant cells have been developed in recent years. Plant cells have a rigid wall, which is a barrier to DNA uptake. All direct methods of DNA transfer must somehow penetrate this barrier, either by removing the wall or by penetrating through it. For example, cellulolytic and pectolytic enzymes can be used to remove the cell wall. This produces protoplasts, which can be made to take up DNA as described below and then cultured to regenerate their walls and eventually form intact plants.

Direct DNA uptake into plant cells is possible by adsorbing DNA molecules onto the surface of protoplasts using calcium phosphate or polyethylene glycol (PEG). The frequency of transformation using this technique is very low. However, because protoplasts can be produced in great numbers, sufficient transformants can be generated to make this approach practical. Protoplast transformation is a useful technique that is not limited by *Agrobacterium* host range and, thus, can be used on monocots.

Another direct method of introducing DNA into plant cells is a technique called electroporation. Electroporation is based on the fact that high electrical pulses create transient holes in the cell membrane through which large molecules, including DNA, can enter the cell. This technique has several drawbacks, which include the following: (1) the percentage of stable transformed cells is low, (2) cell viability drops dramatically after the electrical shock and (3) for some plants, regeneration is difficult after electroporation. Nevertheless, this technique has been useful for the transformation of some monocotyledonous species such as maize and rice.

Microinjection is an alternative to using protoplasts for the direct transformation of plant cells. In this technique DNA is directly injected into the plant cell nucleus. This is accomplished by holding the cell on a glass tube by mild suction and injecting the DNA through a very finely drawn glass micropipette. Microinjection requires a mechanical micromanipulator and a high-quality microscope. Although it has been shown to work well for plant cells, microinjection requires a great deal of patience, as well as technical ability and training, in order to carry out the injections in the proper way.

Particle bombardment is a relatively recent development that has proved to be extremely useful in transforming plant cells. The technique involves literally shooting DNA into cells using specialized "particle guns" or "gene guns" (see chapter thirty-two). To accomplish particle bombardment, DNA first is used to coat microscopic tungsten or gold particles known as microprojectiles, which are then accelerated toward the cells either by firing a gunpowder charge, releasing a burst of compressed gas or accelerating the particles through an electric field. Upon contact with plant target tissue, DNA-coated particles penetrate through the cell walls and, in some cases, stable transformation events occur. As with protoplast-based transformation, successful particle-mediated transformation depends both on generating a large number of target cells as well as their efficient selection from among an even greater number of untransformed cells.

## Reasons for plant transformation

The following are two primary reasons for transforming plants: (1) to insert specific DNA sequences that confer novel desired characteristics to recipient plants (i.e., pest and herbicide resistance, improvement of yield, etc.) and (2) to answer basic questions concerning the regulation of plant gene expression.

## Creating transformed plants with useful characteristics

Genetic manipulation of plants has actually been practiced by humans for thousands of years by the process of selective breeding. This approach has been extremely successful, and it will continue to play a major role in modern agriculture. However, classic plant-breeding programs rely on being able to carry out genetic crosses between individual plants. The recipient and donor plants must be sexually compatible, and, thus, it is generally not possible to combine genetic traits from widely divergent species. The advent of recombinant DNA techniques and transformation methods in plants has removed this constraint and has given agricultural scientists a powerful new way of incorporating defined genetic changes into plants.

An example of the application of these techniques is the production of plants resistant to certain insect predators. For instance, the bacterium *Bacillus thuringiensis* synthesizes a glycoprotein that is highly toxic to some Lepidopterans (moths and butterflies). Other varieties of *B. thuringiensis* produce toxins that act specifically on Dipterans, such as mosquitoes and blackflies, and on Coleopterans (beetles). The genes for these so-called BT proteins have been characterized, and their introduction into plants could have dramatic effects on predators, such as caterpillars of the cabbage white butterfly in brassicas. The gene for *B. thuringiensis* var. *berliner* toxin has been introduced in tobacco (*Nicotiana tabacum* L.) and confers resistance to tobacco hornworm infestations (Vaeck et al., 1987).

This approach depends not only on the introduction of proper protein coding sequences into the plant genome, but also the expression of that sequence to produce the toxic protein. Thus, the "transgene" has to be correctly transcribed and translated.

To express a protein coding sequence conferring a desirable characteristic in a transgenic organism, it has to be coupled to a promoter to direct its transcription. A promoter is a specific DNA sequence to which an RNA polymerase complex binds to initiate mRNA synthesis. The mRNA is subsequently translated in the cytoplasm to produce a protein that is actually the molecule that confers a specific characteristic or phenotype.

Promoters can be constitutive or inducible. When DNA sequences are introduced into plants to produce a high amount of a specific protein, constitutive promoters are usually preferred because they drive expression in all plant tissues and are independent of the development state of the transformed plant. Inducible promoters are able to induce expression only under certain conditions; consequently, they can be viewed as "molecular switches" that will promote transcription under a specific set of circumstances.

## Analysis of gene regulation

Many plant genes are highly regulated; this means that they are only transcribed in specific tissues and/or at certain stages of development. Still other genes are only expressed when a plant encounters a particular environmental stimulus. Differential expression of genes is controlled by specific regulatory sequences (promoters) that are generally outside the DNA protein coding sequence located upstream (i.e., 5′ to the transcription start site).

In order to study how the expression of a specific gene is controlled, its regulatory sequences can be fused to a reporter gene. The main feature of a reporter gene is that its expression can be easily monitored. Once plants are transformed with a plasmid containing a specific regulatory sequence and a reporter gene, the expression of the reporter gene is analyzed in different tissues to learn about the tissue-specific expression, or through different developmental stages of the plant to know how the gene is expressed during the development of the plant. The effects of different environmental stimuli on gene expression also can be analyzed by exposing the transformed plants to the stimuli and then monitoring the expression and activity of the reporter gene.

## Conclusion

Since the first report of plant cell transformation was published in 1983, many genes have been introduced into different plant species and additional transformation techniques have been developed. Considerable success has been achieved in transforming monocotyledon species through direct DNA transfer, and further work will undoubtedly extend the number of species capable of being transformed and regenerated. The special emphasis on transforming agronomically important crop plants will certainly continue because of the great impact that genetically engineered crops will likely have on agriculture.

In addition, detailed studies of plant gene regulation have been facilitated by the study of regulatory sequences in transgenic plants. This approach continues to be very useful for understanding how specific regulatory motifs control specific aspects of gene expression. Gene manipulation and transformation techniques will surely continue to give new insights into the processes by which plants grow and develop.

# *Literature cited*

Fraley, R.T., S.G. Rogers, R.B. Horsch, P.R. Sanders, J.S. Flick, S.P. Adams, M.L. Bittner, L.A. Brand, C.L. Fink, J.S. Fry, G.R. Galluppi, S.B. Goldberg, N.L. Hoffman and S.C. Woo. 1983. Expression of bacterial genes in plant cells. *Proc. Natl. Acad. Sci. U.S.A.* 80:4804-4807.

Nicholl, D.S.T. 1994. *An Introduction to Genetic Engineering.* Cambridge University Press, Cambridge.

Old, R.W. and S.B. Primrose. 1989. *Principles of Gene Manipulation.* Chapter 12, pp. 222–254. Blackwell Scientific Publications, Oxford.

Vaeck, M., A. Reynaerts, H. Hofte, S. Jansens, M. De Beuckeller, C. Dean, M. Zabeau, M. Van Montagu and J. Leemans. 1987. Transgenic plants protected from insects attack. *Nature.* 328:33–37.

# Transformation of tobacco and carrot using Agrobacterium tumefaciens and expression of the β-glucuronidase (GUS) reporter gene

*Ignacio E. Maldonado-Mendoza, Melina López-Meyer and Craig L. Nessler*

*Agrobacterium tumefaciens*, a Gram negative soil bacterium, has become the most commonly used organism to achieve plant transformation by mimicking the naturally occurring process in which DNA is transferred from the bacterium to plant cells. This transformation protocol uses a binary vector system and, thus, requires two plasmids, as described in chapter twenty-nine.

The first plasmid is the tumor-inducing (Ti) plasmid normally present in *Agrobacterium* cells. This plasmid encodes the virulence (*vir*) genes responsible for the excision, transport and insertion of DNA sequences contained within the borders of the transfer-DNA (T-DNA) into the plant genome. These functions are carried out by the gene products (proteins) encoded by the *vir* genes that can act in "trans" to transfer a T-DNA. Acting in trans simply means that the T-DNA region does not have to be on the same plasmid as the *vir* genes. Therefore, the piece of DNA that we wish to transfer into a plant can be easily manipulated in *Escherichia coli* and then placed back into the *Agrobacterium*.

The second plasmid used in this exercise is pBIN19 developed by Michael Bevan (1984). The pBin19 plasmid has the direct repeats found in the borders of the T-DNA, which define the region of DNA that will be transferred into the plant genome. But, unlike the wild-type plasmid, pBIN19 has had its Ti genes removed so that transformed cells will not undergo uncontrolled cell division. Our gene of interest (promoter-GUS fusion) has been inserted between the T-DNA borders of pBIN19 along with a gene conferring kanamycin resistance, which will be used as a selectable marker. This means that only plant cells that were successfully transformed with the T-DNA will be able to survive in the presence of the selective agent (kanamycin), while untransformed cells will not. Thus, the pBin19 plasmid acts as a carrier for the genetic information that we want to insert into the plant genome. In the *Agrobacterium*-based system, both plasmids work together to permit the transfer of the desired piece of genetic material and its integration in a stable, heritable form into the plant genome.

In this chapter we will make use of a reporter gene system to study gene expression in transgenic plants. The term reporter gene is applied to a gene whose activity is not found in the species in which it will be used. Thus, it can "report" gene activity without back-

0-8493-9409-0/96/$0.00+$.50

ground activity from the organism itself. Reporter genes generally are isolated from a different species than the one in which it will be used and can be employed to monitor regulatory elements of genes (i.e., promoters). During the past decade the bacterial β-glucuronidase gene (*uidA*), originally isolated from *Escherichia coli*, has become the major reporter for the analysis of plant gene expression (Jefferson et al., 1987). The activity of this gene, affectionately known as GUS, cleaves the β-glucuronide bond in a variety of useful substrates, some of which produce colored or fluorescent products. To use the GUS gene fusion system, typically the GUS-coding region (the region that encodes a functional enzyme) is coupled to a gene regulatory element (i.e., promoter region) (see Figure 30.2 for more detailed information on the vectors used in these exercises). Once this fusion is introduced into plant cells, GUS activity mimics the expression of the original gene and reports its activity. With few exceptions, plants normally lack appreciable GUS activity so that detection of the GUS enzyme either in whole tissue or tissue homogenates can be used as a sensitive background-free measure of gene expression.

The wide acceptance of GUS as a reporter gene is mainly due to the fact that it can be quickly assayed without using radioactive substrates. GUS assays are extremely sensitive, and it is possible to obtain both quantitative (i.e., level of expression) and qualitative (i.e., localization of expression in tissues and organs) data with this reporter gene (Martin et al., 1992). GUS is used for a wide range of studies in transgenic plants. We and others have found this system to be very useful for plant promoter analyses (Burnett et al., 1993; Stockhaus et al., 1989; Lin et al., 1991) and in the development and optimization of transformation/regeneration protocols for different plant species (Maldonado-Mendoza and Nessler, 1993).

Design of successful transformation protocols has been made possible by advances in molecular biology, combined with the development of in vitro culture methods for regeneration of different important crop species. The experiments described in this chapter were designed not only to show how transformation protocols work, but also to illustrate how we can use these powerful tools to study gene regulation and alter the expression of any gene by using specific regulatory elements (promoters).

The following laboratory exercises will help students become familiar with the following concepts: (1) *Agrobacterium*-based transformation, (2) the use of reporter and marker genes as a way to confirm plant transformation, (3) the expression of a gene under the regulation of an inducible and a constitutive promoter and (4) different patterns of plant gene expression.

## General considerations

### Growth of tobacco plants

*Nicotiana tabacum* L. 'Xanthi' has been chosen for these experiments, although other cultivars may be suitable. All plant material used for the *Agrobacterium* transformation experiment should be grown under sterile conditions. Tobacco seeds are very small, but easy to handle. They can be surface disinfected in an 1.7-ml microcentrifuge tube. One or two plants will be enough for each student. Surface disinfest the seeds by soaking them in 5% Triton X-100 for 3 min, then rinse repeatedly with sterile water until foam is no longer seen. Soak the "wetted" seeds in 70% ethanol for 1 min and transfer to a dilute commercial bleach solution (1:10 with sterile water) for 3 min. Rinse the seeds three to five times with sterile water to remove all traces of chlorine. Resuspend the seeds in 0.5 ml of water and pipette them into a Magenta GA-7 culture vessel, each containing 75 ml of half-strength MS medium (see chapter two) supplemented with 20 g (59 mM) sucrose and 2.5 g/L Phytagel. Place the Magenta vessels in a growth chamber illuminated with 50–100 $\mu mol \cdot m^{-2} \cdot sec^{-1}$

fluorescent light at 25–28 C. Because 'Xanthi' is day neutral, the day length in the chamber is not important — continuous illumination works well. When the plants have reached approximately 5 cm in height (about 1 month), transfer individual plants into Magenta vessels to avoid competition for space and nutrients. Plants will be ready when their leaves reach the lid of the Magenta boxes (1 month after the transfer).

## Germination of carrot seeds

Carrot (*Daucus carota* L.) seeds Denver's 126 (Northrup King Seed Co.) or Imperator S8 (Southland Garden Seed) can be used and purchased in any local supermarket. Seeds can be surface sterilized using a 10% solution of commercial bleach (0.5% NaOCl) with 1 drop of Tween 20 per each 500 ml. After 10 min, the seeds should be washed four times in sterile deionized water and then placed on MSr medium (half-strength MS medium macro salts, complete strength of microsalts, 2 g/L [5.8 mM] of sucrose, 8 g/L (22.2 mM) of glucose, MS vitamins with an additional 0.4 mg/L thiamine and 8 g/L Difco Bactoagar). Place seeds in an incubator with a day cycle of 16 h light (25–50 $\mu$mol·m$^{-2}$·sec$^{-1}$) and 8 h darkness at 25–28 C.

## Preparation of A. tumefaciens

The following two different plasmids will be introduced in *A. tumefaciens* (strain LBA4404): (1) pBI121 and (2) pCAH-1678 (Burnett et al., 1993). Both plasmids may be purchased from Clontech Laboratories, Palo Alto, CA. Introduction of the two plasmids into *A. tumefaciens* strain LBA4404 can be done by electroporation or triparental mating, as described below. For this procedure prepare a culture of the bacteria in a 500-ml Erlenmeyer flask containing 250 ml of Luria Broth (LB — see below for composition). Grow the culture at room temperature for 2–3 days with continuous shaking at 150–200 rpm. Harvest the cells by low speed centrifugation (3000 rpm for 10 min) and wash five times by resuspending the pellet in sterile water and recentrifuging. Resuspend the washed cells in about 3 ml of a 10% glycerol solution and aliquot 100 $\mu$l into 0.5-ml microcentrifuge tubes. Freeze the tubes in liquid nitrogen immediately, and store in –80 C freezer. These cells are now ready to be used for electroporation, but can be used for this purpose for up to 1 year if stored at –80 C.

    *Electroporation.*   To electroporate the *Agrobacterium* cells, slowly thaw a tube of the LBA4404 electrocompetent cells and immediately place on ice. Add 0.5 $\mu$l of the plasmid solution (diluted to 1 $\mu$g/$\mu$l) to the chilled cells and transfer to a prechilled electroporation chamber (2 or 5 mm width). Set the electroporation apparatus to the following conditions: 2.5 kV, 200 $\Omega$ of resistance, 25 $\mu$F capacitance and set the capacitance extender to 960 $\mu$F. Discharge the apparatus and transfer the bacteria into a disposable culture tube (13 × 100 mm) containing 1 ml of LB medium using a sterile pasteur pipette. Incubate the culture tube for 1 h in an orbital shaker at 150 rpm at room temperature. After this "recovery" period spread 250 ml of the bacterial suspension on LB plates containing 50 mg/L of kanamycin. Place the plates in an incubator at 28 C or in a cabinet at room temperature and let them grow until colonies easily are seen (2–3 days). *Agrobacterium* cannot grow at temperatures much above 30 C and should not be placed in a 37 C incubator used to grow *E. coli.*

    Because the pBI121 and pCAH-1678 plasmids contain a kanamycin resistance gene, only those bacterial cells carrying the plasmid will survive and grow into colonies. It is important not to skip the recovery step in liquid LB medium because the bacteria need time to synthesize neomycin phosphotransferase, the protein conferring kanamycin resistance, before they come in contact with the antibiotic. Otherwise, the bacteria may not grow on LB + kanamycin plates. Pick a single colony and streak it out on a fresh LB + kanamycin plate. Incubate the plate for 2–3 days at 28 C. Take a single colony from this plate and place it into a flask of liquid LB medium with 50 mg/L kanamycin. A 125-ml erlenmeyer flask with 50 ml of medium is enough to transform two to three plants. Start the liquid culture 2–3 days before the bacteria are needed for plant transformation.

LB is composed of 5 g sodium chloride (85.5 mM), 5 g yeast extract, 10 g bactotryptone, 10 ml of a 1 M magnesium sulfate solution (10 mM) per liter; the pH is adjusted to 7.5 with 1 M NaOH. If the LB is used in petri dishes, 16 g/L agar is added prior to autoclaving. Antibiotics are filter sterilized and added after autoclaving when the media has cooled enough to be handled with bare hands (40–50 C).

*Triparental mating.* In the case that an electroporation machine is not available, an alternative way to obtain the *Agrobacterium* constructions is by using a method called triparental or tripartite cross. It involves the use of an *A. tumefaciens* strain, which contains the Ti plasmid harboring the virulence genes responsible for plant transformation. The recombinant plasmid pBIN19 is present in one *E. coli* strain and a conjugation-proficient plasmid in another. When the three strains are mixed, the conjugation-proficient "helper" plasmid transfers to the strain carrying the recombinant plasmid, which is then mobilized and transferred to the *Agrobacterium* (*A. tumefaciens* strain LBA4404, pBIN19 in MC1022 *E. coli* strain, and the conjugative plasmid pRK2013 in HB101 *E. coli* cells can be purchased from Clontech.)

Start a liquid culture of LBA4404 *Agrobacterium* cells in M9 sucrose with 250 µg/ml streptomycin and grow at 28–30 C. The culture should be 2 days old, and the cells grow in a fine suspension in this medium. One day before the mating experiment, both *E. coli* cultures should be started. Grow HB101 cells (harboring helper plasmid pRK2013) in LB medium, and start the pBIN19/MC1022 culture in LB medium with 50 µg/ml kanamycin. Grow at 37 C.

To perform the mating, combine 1 drop of each three cultures on an LB or M9 plate without antibiotics. Grow overnight at 28–30 C. Scrape off the resultant culture and place into 2 ml of LB medium with 50 µg/ml kanamycin and 250 µg/ml streptomycin. The amount should be the size of an inoculation loop. Grow overnight at 28–30 C.

Remove 50 µl from the overnight culture and dilute into 1 ml of LB liquid medium and continue dilutions to $10^{-6}$ and plate on M9 sucrose medium with 50–100 µg/ml kanamycin and 250 µg/ml streptomycin. Grow overnight at 28–30 C. The colonies should contain your recombinant plasmid in *Agrobacterium* by now!

M9 medium is very good at selection against *E. coli*. To make 1 L of 10× M9 salts, dissolve in water 60 g of $Na_2HPO_4$, 30 g of $KH_2PO_4$, 5 g of NaCl, and 10 g $NH_4Cl$. Adjust the pH to 7.4 with 1 N HCl or NaOH, bring to 1 L and autoclave. Also prepare separate stock solutions dissolving in water as follows: 20% sucrose, 1 M $MgSO_4$, 1 M $CaCl_2$, 1 mg/ml thiamine HCl. Autoclave all solutions.

Finally, to make 1 L for pouring into petri dishes, add 15 g of Bactoagar to 890 ml of MilliQ water, autoclave 20 min, cool down to 50–60 C and then add each component taking care to swirl after each addition. First, add 100 ml 10× M9 salts, then add the following components in this order: 2 ml $MgSO_4$, 10 ml 20% sucrose, 1 ml thiamine HCl, 0.1 ml $CaCl_2$, swirl fast at this step, some precipitate will form. At the end, add the antibiotics. If the medium looks dark brown, the cells will not grow. Medium should be light brown with a slight blue/clear color.

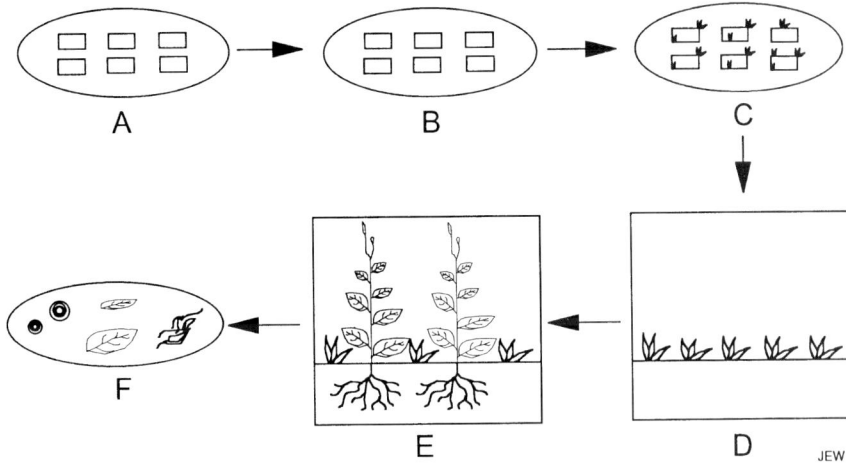

**Figure 30.1**   Outline of the tobacco transformation protocol. A. Leaf sections incubated with *A. tumefaciens* strain LBA4404. B. 2-day incubation in $^1/_2$ MS medium. C. Callus proliferation and shoot formation after 4–6 weeks in TR medium with kanamycin and cefotaxime. D. Shoots (0.5–1 cm) transferred to a Magenta box containing $^1/_2$ MS medium with kanamycin and cefotaxime. E. After 5–6 weeks kanamycin-resistant shoots are rooted and ready to analyze. F. Harvest kanamycin resistant plants and test for transformation.

## Exercises

### Experiment 1. Transformation and regeneration of tobacco plants expressing the GUS reporter gene under the regulation of constitutive and inducible promoters

Tobacco is prepared from the dried leaves of *N. tabacum* L., an annual herb of the potato family (Solanaceae), native to tropical America. Tobacco contains nicotine, which is a health-threatening, addictive drug. However, it is used as the active ingredient in certain insecticides and to kill intestinal worms in farm animals.

Tobacco provides one of the most easy-to-use systems for plant transformation because of its high transformation and regeneration frequencies and relatively short life cycle. Since the first reports of *Agrobacterium*-based transformation, tobacco has been a model system for such studies. In spite of the fact that many important crops like corn and soybeans can now be transformed and regenerated, tobacco is still widely used in a variety of studies dealing with the expression of plant genes. This is particularly true with genes isolated from species with very long life cycles, such as trees, and for genes isolated from species that cannot yet be regenerated.

This experiment introduces concepts underlying the methodology of plant transformation. The basic concepts introduced in this experiment are also applicable to the second experiment in this chapter. Refer to Figure 30.1 for a diagram outlining this exercise. Experiment 1 will require approximately 10–12 weeks for completion.

*Note:* The following experiments must be conducted in strict accordance with APHIS guidelines. All facilities, including laboratory, growth rooms and greenhouses, must be inspected and approved by an APHIS representative as well as the appropriate office of your institution. The instructor and students must follow all procedures. *A. tumefaciens* is a plant pathogen. Even though you will be using a "disarmed" strain (LBA4404) that cannot form plant tumors, it should be handled as if it could "escape" into the environment. All containers should be sterilized after use. Disposable tubes, petri plates etc. should be placed in biohazard bags and autoclaved before discarding.

*Materials*

The following items are needed for each student or team of two to four students for the first part of this experiment.

* Two 125-ml erlenmeyer flasks containing 50 ml of sterile half-strength MS medium
* Two 50-ml sterile Falcon™ tubes or sterile-capped centrifuge tubes for concentration of bacteria
* One to two sterilely grown tobacco plants in magenta boxes per each *A. tumefaciens* strain (two to four in total)
* Six 100 × 15 mm petri dishes containing one-half strength MS ($1/2$ MS) solid medium
* Six 100 × 15 mm petri dishes containing TR medium (MS basal medium containing 0.1 mg/L (0.5 μM) NAA, 0.2 mg/L (0.88 μM) BA, 30 g/L (87.6 mM) sucrose, 2.5 g/L Phytagel, pH 5.8, amended with 100 mg/L kanamycin and 200 mg/L cefotaxime
* Ten GA-7 Magenta vessels or other similar size vessels containing 75 ml of $1/2$ MS medium supplemented with 100 mg/L kanamycin and 200 mg/L cefotaxime

The final part of the experiment (test of transformation) requires the following items per each group of students:

* 40 ml of GUS-assay buffer containing 0.1 M phosphate buffer pH 7.0, 5 mM potassium ferricyanide, 5 mM potassium ferrocyanide, and 10 mM EDTA
* 40 mg of X-gluc (5-bromo-4-chloro-3-indolyl β-d-glucuronic acid)
* 400 μl of *N,N*-dimethylformamide

*Procedure*

Pour each of the bacterial suspensions, pBI121 and pCAH-1678, strains into a separate 50-ml Falcon tube and centrifuge at 3000 rpm for 10 min. Discard the supernatant and resuspend the pellet in about 10–15 ml of $1/2$ MS medium. Using sterile technique, remove each leaf from a plant and place them in the sterile petri dishes. Add the bacterial suspension and cut the leaves into 1.5–2-cm² sections while submerged in the bacterial suspension. Incubate sections with bacteria for at least 10 min. Blot the leaf sections on sterile paper towels and place them on $1/2$ MS medium in petri dishes. It is important to eliminate the excess liquid by blotting to avoid growth of the bacteria in the medium and all over the tissue. Seal the petri dishes with Parafilm and incubate at room temperature in the dark for 2–3 days. During this period the T-DNA is transferred from the bacteria to the plant cells. In addition, the transformed plant cells begin to express the gene, which will make them resistant to kanamycin selection.

Once the transformation stage is completed, transfer the tissue pieces to plates containing tobacco regeneration medium (TR) with 100 mg/L kanamycin and 200 mg/L cefotaxime. Although kanamycin will efficiently select against untransformed plant cells, you may recall that the *Agrobacterium* cells also contain a plasmid (pBIN19) with a kanamycin resistance gene. The cefotaxime is therefore added to inhibit growth of the bacteria. Fortunately, this antibiotic does not affect the growth of plant cells and can be used to prevent bacteria from overgrowing the leaf sections.

Place the petri dishes in a culture room or incubator with 75–100 μmol·m⁻²·sec⁻¹ light at 20–25 C. After 4–6 weeks, a small amount of callus tissue will develop and shoots will start to emerge along the cut surfaces of the explant. When the shoots reach 0.5–1 cm in length, transfer individual shoots into Magenta boxes (9–10 shoots per box) containing

$^1/_2$ MS medium with 100 mg/L kanamycin and 200 mg/L cefotaxime. Try to remove any callus tissue at the base of the shoots during this step. Grow the shoots out under the same conditions for 3–4 weeks. By then, transformed shoots will start forming roots. If a shoot turns white and does not form roots, it is probably an "escape", i.e., untransformed. After 5 or 6 weeks, kanamycin-resistant plants should be evident as healthy, green plants.

At this point the plants can be examined for GUS expression. Also, if it is desirable to get seeds from these plants, they can be transferred to pots containing soil mix of perlite and bactosoil 1:3. Transfer the plants by gently removing the media from the roots and place them in previously watered pots. Cover them with an inverted Magenta vessel for a few days to prevent wilting. Water regularly and, when flower buds appear, cover them with paper bags to avoid cross pollination. Tobacco will self-pollinate under these conditions and the seeds can be harvested when the capsules have dried and turned from green to brown.

To test for transformation and expression of the introduced genes in the tobacco plants, we will use the X-gluc staining method, which produces only qualitative, but nevertheless spectacular results. X-gluc is a colorless substrate that is cleaved by the GUS enzyme. The product of this enzymatic reaction forms a blue indigo dye that precipitates at the site of the enzymatic cleavage. The blue color thus permits the histochemical localization of GUS activity and tells us which cells, tissues and organs expressed the GUS gene.

Dissolve 40 mg of X-gluc in 400 µl of *N,N*-dimethyl formamide. Once X-gluc has gone into solution, add it to 40 ml of GUS-assay buffer. Place 10 ml of this solution into each petri dish (four in total). Select at least one plant from each type of transformant: pBI121.1 and pCAH-1678. Using a razor blade or scalpel, dissect the following material from each plant: (1) small (1–2 cm) pieces of each plant organ including young and old roots, young and fully expanded leaves, (2) thin cross-sectional slices at different points along the stem and (3) longitudinal sections of buds. Immerse the tissue pieces into the assay solution. In addition, take a leaf section from each of the remaining plants. Number each leaf section with a black marker and place them in a petri dish containing the X-gluc assay solution. Incubate for 1 h at 37 C. Check for color development under a high power dissection microscope. If blue staining is still not visible, seal the plate with Parafilm and let the reaction continue overnight at 37 C. Sometimes the blue stain cannot be visualized because chlorophyll masks the indigo dye. To remove the chlorophyll, immerse the tissue in a 1:50 dilution of commercial bleach for 20 min or until tissues are no longer green. Observe the distribution of blue staining under the microscope. Thin sections made with a razor blade can be mounted on a microscope slide in water with a coverslip and viewed under a compound microscope. Make notes on your observations of each type of transgenic plant (pBI121.1 or pCAH-1678), as well as any differences found between plants transformed with the same plasmid.

### Anticipated results

After the initial 2–3 days of incubation with *Agrobacterium*, you may find bacteria overgrowing the tissue. If this happens, start over again. The next time be sure that you blot the tissue very well before placing it into the $^1/_2$ MS medium. It is also possible that your cefotaxime was added when the medium was too hot, or that it was added at too low a concentration. You may want to prepare a new batch of plates to eliminate this possibility. Once the tissue is covered by bacteria, it is almost impossible to remove, even after washing and rinsing the tissues in antibiotic solution.

Four weeks after transfer to regeneration medium (TR) the leaf tissue may assume a "puckered" shape, and callus may have formed around the cut edges of the leaf sections. Small buds may also be apparent by the last 2 weeks of this period.

After buds are transferred to plant growth regulator (PGR)-free medium, roots will probably start emerging from the base of the shoots after 2–3 weeks. Nevertheless, you should wait until the primary root branches before continuing to the next stage. Sometimes some untransformed plants (escapes) are able to form a primary root, but this root will not develop extensive root hairs nor will it branch. Besides showing good root growth, kanamycin-resistant plants will grow taller, healthier and greener than nonresistant plants. At the end of this period (5–6 weeks), you should be able to distinguish which plants are actually transformed and which are not. Untransformed shoots will look sick, pale green to almost white in color and be very stunted. You should find up to 50% of the plants growing in the rooting medium to be transformed.

During the analysis of all transformed plants, you will notice that each one can exhibit a different level of GUS expression. Low or high expression levels will be indicated by pale or dark blue staining, respectively. This phenomenon is generally ascribed to the "position effect", which is thought to result from the random positioning of the inserted T-DNA in the plant genome. Some of these random insertion sites appear to favor the expression of inserted genes, whereas others seem to suppress their expression.

You should also notice differences in the histological staining of both types of plants. The pCAH-1678 contains the promoter region of a 3-hydroxy-3-methylglutaryl coenzyme A reductase (HMGR) gene from *Camptotheca acuminata* (Decne), a Chinese medicinal tree (see Figure 30.2). This gene is involved in the production of mevalonate, which is a precursor for a large number of the plant terpenoids. One site of terpenoid synthesis, and thus of GUS staining, should be the glandular hairs on the surface of the leaves and stem. Normally, tobacco accumulates defense compounds in these glands, which require large amounts of mevalonate for their synthesis. To check the pattern of GUS expression in these plants see Burnett et al. (1993). Plants transformed with the pBI121 vector have GUS expressed under the control of plant viral promoter, the 35S promoter from the cauliflower mosaic virus. This is considered to be a more constitutive promoter, so you should see widespread expression in virtually every part of the plant.

*Questions*

* What characteristics should a gene have to be used as a reporter? Can a highly expressed gene in a species be used as a reporter gene in the same species? Why or why not?
* How can you explain the differences seen in GUS staining between individual transformed plants and plants transformed with different promoter-GUS fusions?
* List three desirable characteristics that could be given to a transgenic plant using these transformation techniques and explain how they would work.
* How could a promoter conferring tissue-specificity (i.e., high expression in the leaf hairs) be used to genetically modify a crop plant?

## Experiment 2. Effect of wounding on expression of the GUS gene under the regulation of different promoters

Although this part of the exercise is optional, it illustrates not only an additional way to test for transformation, but also it represents an easy way to analyze quantitatively differences between specifically promoter-driven activity. It can be done in parallel to the test for transformation; however, it may be preferable to identify plants that express high levels of GUS activity and use them in this experiment.

A. PROMOTER-GUS FUSIONS

**Figure 30.2**    A. Physical map of the promoter-GUS fusions used in these exercises. B. Diagram of the binary vector system.

*Materials*

* Flask containing 50 ml $^1/_2$ MS medium
* One fully expanded transformed tobacco leaf per each plasmid used
* Cork borer (#3 or #4) or paper-hole punch
* 1.7-ml microcentrifuge tubes
* Liquid nitrogen or ultrafreezer at –80 C (if a –80 C freezer is not available, a freezer at –20 C can be used)
* Pellet-pestle grinder or any other type of tissue homogenizer
* Microcentrifuge
* 20 ml of lysis buffer containing 50 mM $NaPO_4$ (pH 7.0), 0.1% Triton X-100, 10 mM ß-mercaptoethanol (BME), 10 mM EDTA, 0.1% lauryl sarkosyl

* 1 ml substrate buffer containing 2 mM 4-methylumbelliferone glucuronide (MUG); prepare by dissolving 7.6 mg of MUG in 10 ml of lysis buffer
* 20–40 ml stop buffer solution (0.2 M $Na_2CO_3$)
* Fluorimeter

*Procedure*

Set up two petri dishes containing 10–15 ml of $1/2$ MS medium (one per each promoter-GUS fusion). The experiment will require prolonged incubation times (up to 72 h), which is enough time for bacterial or fungal contamination to occur. Since contamination would affect the results, it is advisable to set this experiment up in a laminar flow hood using sterile techniques. Starting from the apex, note the size of leaves as you near the base. Select the largest leaf nearest the apex that represents the youngest fully expanded leaf. Cut 5–7-mm disks from the leaf using a hole punch or cork borer. Avoid cutting disks from large veins. Try to obtain all of the disks from the same leaf because basal GUS expression varies depending on the developmental stage of the leaf. Float 18 disks in the $1/2$ MS medium in each petri dish. Incubate the leaf disks at room temperature under continuous light 25–50 $\mu mol \cdot m^{-2} \cdot sec^{-1}$. Remove three leaf disks, place them into a 1.7-ml microcentrifuge tube labeled as "zero" and freeze them in liquid nitrogen or in –80 C freezer. Do this at the beginning of the experiment, and this will be your zero time control. Collect, label and freeze three disks at 12, 24, 48 and 72 h after wounding. Once all disks have been collected you are ready to quantitate their GUS activity.

GUS activity can be quantitated by fluorimetry. The protocol described in this exercise involves the use of a fluorochrome substrate MUG. MUG produces the fluorescent compound 4-methylumbelliferone after GUS enzymatic cleavage. The amount of fluorescence detected in this assay is a measure of relative GUS activity and, thus, the level of GUS expressed by the promoter used in the construct.

Take the microcentrifuge tubes containing the frozen tissues, add 100 $\mu l$ of lysis buffer to each tube and then grind the tissue using a pellet pestle grinder or any available tissue homogenizer. The tissue should be thoroughly homogenized. Spin the tubes for 3–5 min in a microcentrifuge at full speed (10,000–14,000 rpm) to pellet the cuticle and other debris. Transfer the supernatant to another labeled tube without disturbing the pellet (particulate material will reflect light and produce artificially high readings). Discard the tube containing the pellet. Pipette 25 $\mu l$ from the supernatant tube to two additional tubes. Add 1 ml of stop buffer to one tube which should be labeled "time _ blank" where _ is 0, 12, 24, 48 or 72 h. This sample will be used to calibrate the fluorimeter before reading each experimental sample. Add 25 $\mu l$ of substrate buffer containing 2 mM MUG (10 ml lysis buffer, 7.6 mg MUG) to each tube. Mix well by vortexing, centrifuge briefly, and incubate at 37 C for 30 min. After incubation, add 1 ml of stop buffer to the experimental tubes and mix well. Assay samples in fluorimeter according to the instrument instructions. If the sample exceeds the linear range of the fluorimeter, it may be diluted with reaction stop buffer and read again. Draw a graph for each transformed plant assayed plotting the time of induction (Y axis) vs. units of fluorescence per h (X axis).

*Anticipated results*

This experiment involves the induction of GUS expression by wounding, which stimulates the HMGR promoter. Thus, there should be an increase in GUS activity with elapsed time (Burnett et al., 1993). In contrast, the constitutive 35S promoter should not show any wound induction, but should show the same levels of expression throughout the experiment.

*Questions*

* Why does GUS activity increase over time during the wounding experiment in the pCAH-1678 transgenic plants and not in the pBI121 transgenic plants?
* When do you think it would be desirable to use a constitutive promoter and when would you use an inducible promoter to engineer a plant?
* How can you explain the fact that the same construct produces transformed tobacco plants with different levels of gene expression?

## *Experiment 3. Somatic embryogenesis from carrot hypocotyls transformed using* A. tumefaciens.

Carrot is an edible biennial plant which produces a modified, fleshy storage root. This storage organ develops from the root and hypocotyl during the first year of growth, and the stored food is mobilized the following year when the carrot plant "bolts" to make flowers and set seeds. Carrots are hardy plants that can tolerate a wide range of climate and soils. They belong to the celery family (Apiaceae), which is abundantly represented in the Northern Hemisphere.

Somatic embryogenesis in carrot has been used as a model system for the study of plant development for over 30 years (Steward et al., 1964). The isolation of large quantities of staged carrot somatic embryos is relatively simple, which is not always true of seed-derived embryos.

This procedure was designed to generate transformed carrot plants by coupling the *Agrobacterium*-based transformation system to the process of somatic embryogenesis. The exercise also illustrates the application of transformation technology as a tool to study a specific biological process, in this case plant development. Development involves temporal and spatial changes in cellular phenotype. In addition to morphological changes, specific enzyme activities appear and disappear during morphogenesis. These biochemical changes are specifically regulated as sets of gene functions that are simultaneously activated and inactivated.

This experiment will require approximately 26 weeks for completion, although it can be shortened to 20 weeks (see diagram Figure 30.3). This exercise is adapted from a paper by Thomas et al. (1989).

*Materials*

* Four magenta boxes containing 75 ml of MSr medium
* Bacterial suspensions of pBI121 and pCAH-1678 *Agrobacterium* strains prepared as in experiment 1 (resuspend in PGR free B5 medium)
* Eight 100 × 15 mm petri dishes containing 2B5 medium (B5 medium containing 2 mg/L (9 μM) 2,4-D and 8 g/L of Difco Bactoagar)
* Eight 100 × 15 mm petri dishes containing 2B5 medium amended with 100 mg/L kanamycin and 400 mg/L carbenicillin
* Twelve 125-ml erlenmeyer flasks containing suspension culture medium (40 ml of 2B5 liquid medium supplemented with 100 mg/L kanamycin and 400 mg/L carbenicillin
* Six 50-ml sterile Falcon tubes or any similar size sterile-capped centrifuge tubes
* One 1-L erlenmeyer flask containing 500 ml of sterile PGR-free B5 medium
* Six 125-ml erlenmeyer flasks containing 40 ml of plain B5 medium
* Klett–Summerson spectrocolorimeter (optional)

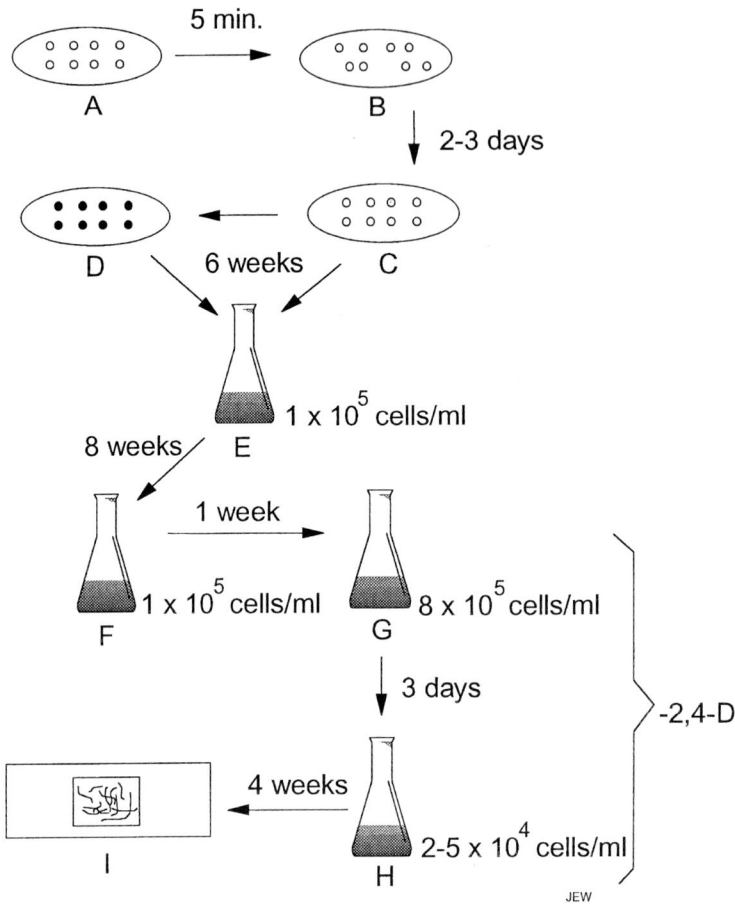

**Figure 30.3** Outline of the carrot transformation protocol. A. Hypocotyl sections incubated with *A. tumefaciens* strain LBA4404. B. Incubation for 2–3 days in 2B5 solid medium. C. Transfer of the sections into 2B5 medium containing kanamycin and carbenicillin and growth for 6 weeks. D. Second transfer to fresh plates and 6 more weeks of growth or E. Direct transfer of callus tissue into suspension culture medium. F. Culture in liquid medium (initial cell density $1 \times 10^5$ cells/ml) for 8 weeks. G. Transfer to fresh suspension culture medium and incubation for 7 days followed by removal of 2,4-D by washing 4 times in B5 medium, transfer and incubate for 3 days at high cell density ($8 \times 10^5$ cells/ml) in B5 medium. H. Dilution of the cells to low density ($2–5 \times 10^4$ cells/ml) and incubation for 4 weeks without shaking. I. Harvest embryos and examine for transformation.

*Procedure*

Slice 1-week-old carrot hypocotyls into 10–15 mm sections and place about 20 hypocotyl sections into each petri dish containing 2B5 medium (B5 medium of Gamborg et al. (1968) — see chapter two) with 2 mg/L (9 μM) 2,4-D and 8 g/L Difco Bactoagar). Keep the tissue in the dark at 20–22 C for 2–3 days before the experiment.

Place the 2-day precultured hypocotyls in *Agrobacterium* solution for 5 min (prepared as described in experiment 1). Blot the excess liquid from the tissue, and place the hypocotyls onto 2B5 solid medium at 20–22 C for 2–3 days to allow for T-DNA transfer and the development of kanamycin resistance. Once the *Agrobacterium* cells are evident upon the explants, they should be transferred to 2B5 medium plates containing kanamycin and carbenicillin. Use two petri dishes per bacterial strain. Place the cultures in the dark at 20–22 C.

After approximately 6 weeks, kanamycin-resistant calli will develop at the cut edges and inside the hypocotyls. Transfer the callus tissue, trying to avoid taking the original explant tissues, to fresh 2B5 medium plates with kanamycin and carbenicillin and grow them for 6 more weeks under the same conditions. You need only use two petri dishes per bacterial strain. To shorten the time for this exercise, the kanamycin-resistant calli from the first transfer can be placed directly into suspension culture medium, although the number of truly transformed calli may be reduced.

Following the second transfer to solid media, calli are inoculated into the liquid suspension culture medium (2B5 medium with kanamycin and carbenicillin). Start at least three flasks per each bacterial strain. Grow the cultures at 20–22 C in the dark with continuous shaking at 150 rpm. No more than 1 g fresh weight of callus per flask should be used as the initial inoculum. Initial cell density should be about $1 \times 10^5$ cells/ml. Subcultures should be made every 8 weeks by diluting 1:40 in fresh suspension culture medium, which results in a good approximation to the optimum initial cell density.

The number of cells at any time point of growth can be estimated by measuring the turbidity of the cell cultures with a Klett-Summerson spectrocolorimeter. Readings in this apparatus are expressed in arbitrary units. For example, Klett 100 (K100) corresponds to about $2 \times 10^6$ cells. The relationship is linear up to K150 using this apparatus. If this instrument is not available, use the estimate for cell density described above.

After 8 weeks of growth in suspension culture, transfer cells to fresh medium. Seven days after this transfer, collect the cells in a sterile 50-ml Falcon tube and centrifuge at low speed (3000 rpm for 5 min). Decant the supernatant and wash the pelleted cells by resuspending in fresh B5 medium. Repeat this operation three more times.

Take an aliquot of cells and incubate them in the X-gluc reagent as described in the previous exercise. Observe the stained cells under a high-power dissection microscope. Compare the results obtained between different transformed cultures and between the two kinds of promoter-GUS fusions.

To initiate embryogenesis, cell suspension cultures should be transferred to B5 medium without 2,4-D. Start two flasks per bacterial strain with an initial cell density of $8 \times 10^5$ cells/ml. Incubate the cultures for 3 days under the same conditions and then dilute them to low density — i.e., $2–5 \times 10^4$ cells/ml in 20 ml of liquid B5 medium. Grow the cultures in phytohormone-free B5 media for 4 weeks at 20–22 C without shaking. Remove a sample from each flask and stain with X-gluc. Focus your observations on the developmental differences found in different stages of embryo development. Also note any variations in GUS expression in lines transformed with pBI121 vs. pCAH-1678.

To recover whole carrot plants from these cultures, allow them to grow in B5 medium until they reach the torpedo stage (see chapter sixteen). They can then be transferred to solid MSr medium containing kanamycin and carbenicillin and placed in the light to germinate. Later, when green, healthy plantlets are seen, transfer them to soil.

### Anticipated results

Four weeks after incubating *Agrobacterium*, infection callus tissue will start forming around the inoculated hypocotyls. By 6 weeks the kanamycin-resistant calli will be up to 1 cm in diam. A second transfer of calli to fresh medium is recommended because it reduces the possibility of continued growth of untransformed calli. We have found that all of the calli growing after the second subculture were resistant to kanamycin and truly transformed. Nevertheless, as mentioned earlier, the second transfer may be skipped to shorten the exercise.

Four weeks after transferring the cell suspensions into embryogenic conditions (low cell density, absence of 2,4-D) you should get a mixed population of somatic embryos. Globular embryos will appear as spherical, multicellular structures 250–350 μm in diam. Heart- and torpedo-staged embryos can reach sizes up to 1 mm.

GUS expression should be detectable throughout the developmental stages of somatic embryogenesis in the cell lines transformed with pBI121 which contains the constitutive promoter-GUS fusion. On the other hand, transformed plants containing the pCAH-1678, the HMGR promoter, will show variability in GUS staining during different developmental stages.

*Questions*

* What are some possible reasons for any differences in GUS expression seen in cell lines transformed with each *Agrobacterium* line? Are these differences correlated with embryo development?
* How could the somatic embryogenesis system be used to study gene regulation?
* How does the wound inducibility of the pCAH-1678 promoter relate to developmental expression of the promoter in carrot? Compare these results to the experiment in tobacco.

## Literature cited

Bevan, M.W. 1984. Binary *Agrobacterium* vectors for plant cell transformation. *Nucleic Acids Res.* 12:8711–8721.

Burnett, R.J., I.E. Maldonado-Mendoza, T.D. McKnight and C.L. Nessler. 1993. Expression of a 3-hydroxy-3-methylglutaryl coenzyme A reductase gene from *Camptotheca acuminata* is differentially regulated by wounding and methyl jasmonate. *Plant Physiol.* 103:41–48.

Gamborg, O.L., R.A. Miller and K. Ojima. 1968. Nutrient requirements of suspension cultures of soybean root cells. *Exp. Cell. Res.* 50:151–158.

Jefferson, R.A., T.A. Kavanagh and M.W. Bevan. 1987. GUS fusions β-glucuronidase as a sensitive and versatile gene fusion marker in higher plants. *EMBO J.* 6:3901–3907.

Lin, X.J., M. Rocha-Sosa, S. Rosahl, L. Willmitzer and W. Frommer. 1991. A detailed study of regulation and evolution of the two classes of patatin genes. *Plant Mol. Biol.* 17:1139–1154.

Maldonado-Mendoza, I.E. and C.L. Nessler. 1993. Biolistic transient and stable expression of β-glucuronidase (GUS) in embryogenic cell suspension cultures of *Papaver somniferum. Plant Physiol.* (suppl.) 102:152.

Martin, T., R.V. Wöhner, S. Hummel, L. Willmitzer and W.B. Frommer. 1992. The GUS reporter system as a tool to study plant gene expression. In: *GUS Protocols. Using the GUS Gene as a Reporter of Gene Expression.* Gallagher, S.R., Ed. Academic Press, San Diego.

Murashige, T. and F. Skoog. 1962. A revised medium for the rapid growth and bioassay with tobacco tissue cultures. *Physiol. Plant.* 15:473–479.

Stewart, F.C., M.O. Mapes, A.E. Kent and R.D. Holsten. 1964. Growth and development of cultured plant cells. *Science.* 14:320–327.

Stockhaus, J., J. Schell and L. Willmitzer. 1989. Correlation of the expression of the photosynthetic gene ST-LS1 with the presence of chloroplasts. *EMBO J.* 8:2445–2451.

Thomas, J.C., M.J. Guiltinan, S. Bustos, T. Thomas and C.L. Nessler. 1989. Carrot (*Daucus carota*) hypocotyl transformation using *Agrobacterium tumefaciens. Plant Cell Rep.* 8:354–357.

*chapter thirty-one*

# Transformation of chrysanthemum leaf explants using Agrobacterium tumefaciens

*Robert N. Trigiano*

In the previous decade, many methodologies were developed to introduce or move a wide variety of genes from one species into another. Many of these techniques have been discussed in other chapters and will not be considered further in this chapter. However, in the following experiments, we will use *Agrobacterium tumefaciens*, a soil-inhabiting, pathogenic (disease-causing) bacterium occurring naturally in many areas of the world, as the agent/vector or shuttle that transfers gene(s) between the two species. This bacterium can be aptly described as nature's own and the original genetic engineer since, during pathogenesis (disease development), several genes located on a plasmid (a small circular piece of DNA) within the bacterium are delivered and incorporated into the genome of host plant cells. These genes cause an unusually high number of host cell divisions (hyperplasia) to occur and, as a result, a tumor or gall is formed. Hence, the common name of the disease is crown gall. We can take advantage of this natural gene delivery system. Via molecular biological techniques, genes that incite disease can be removed from the T-DNA region of the plasmid and replaced with gene(s) and promoters of our choice.

In recent years there has been increased interest in the transformation of ornamental plants, and chrysanthemum (*Dendranthema grandiflora* Tzvelev.) is no exception (see, e.g., de Jong et al., 1993; Lowe et al., 1993; Urban et al., 1994). Most of the interest in genetically transforming chrysanthemum thus far has been in the areas of creating new flower colors and conferring resistance to tomato spotted wilt virus. Before chrysanthemum could be genetically engineered, a reliable regeneration system had to be developed. The plant regeneration system described in chapter twelve is ideally suited in this regard for the following reasons: chrysanthemums are very susceptible to infection by certain strains of *A. tumefaciens*; shoots are formed directly (without an intervening callus phase); shoots develop from the cut edges of the explant, which are easily infiltrated with a bacterial vector; and the plant tissue has high regenerative capacity. For a more complete discussion of plant transformation with *A. tumefaciens*, refer to chapters twenty-nine and thirty.

A note of caution before beginning the experiments in this chapter. When working with any plant pathogen, in this case *A. tumefaciens*, precautions and care must be taken to limit contamination of the work area and "bystander plants" both in the laboratory and in the environment. Inoculated plants, contaminated instruments and cultures must never leave the laboratory. All cultures of bacteria (except stock cultures), plant tissues and plants exposed to the bacterium must be destroyed, by autoclaving, at the conclusion of the

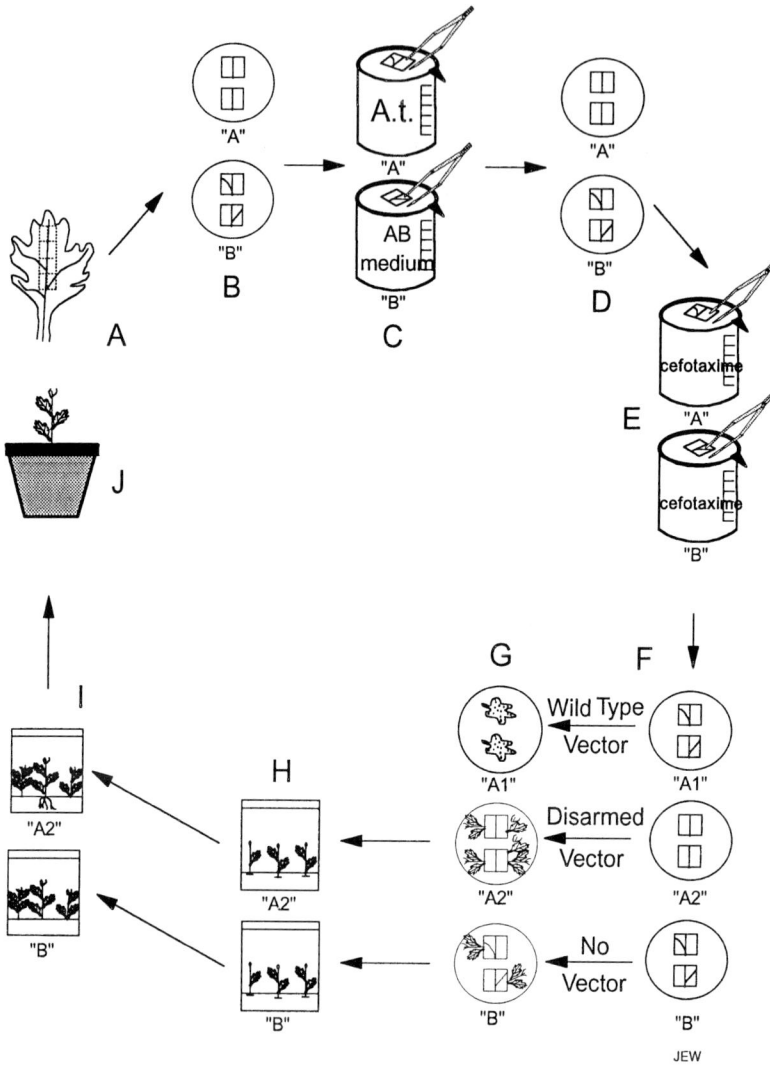

**Figure 31.1** Diagrammatic representation of transformation experiments using chrysanthemum leaves and *A. tumefaciens*. A. Leaves used for midrib explants. B. Preincubation of leaf sections for 2 days before inoculation with treatment. C. Treatments: "A" explants will be inoculated with bacterium (A.t.); "B" explants will be a control dish treated only with culture medium (AB). D. Co-culture with bacteria ("A") for 2 days. E. Transfer of explants from both treatments to medium containing cefotaxime to kill bacteria. F. Explants transferred to medium lacking PGRs, but amended with cefotaxime and kanamycin ("A2" and "B"). "A1" = inoculated with "wild-type" bacteria and "A2" = disarmed bacteria with NPTII gene; "B" without bacteria. G. "A1" transformed explants producing callus; "A2" transformed shoots; "B" shoots from control treatment. H. Rooting of putative transformed and "escape" shoots from "A2" and "B" on kanamycin-containing medium. I. Some of the shoots produced in "A2" are able to form roots (transformed) whereas others cannot (escapes) in the presence of kanamycin; none of the shoots of "B" produce shoots (all escapes). J. Growth of plant harboring the NPTII gene.

experiments. All instruments and glassware also should be decontaminated by autoclaving.

The experiments outlined in this chapter (Figure 31.1) are designed for smaller, more-advanced classes (fewer than 10 students) or students participating in special topics or

***Table 31.1***   Composition of AB Medium for Growth of
*Agrobacterium tumefaciens*

| 20× Salts (A) (500 ml) | 20× Buffer (B) (500 ml) | Vitamin (100 ml) |
|---|---|---|
| 10 g $NH_4Cl$ | 30 g $K_2HPO_4$ | 20 mg biotin |
| 12.5 g $MgSO_4 \cdot 7H_2O$ | 11.5 g $NaH_2PO_4$ | (filter sterilize) |
| 1.5 g KCl | | |
| 0.1 g $CaCl_2$ | | |
| 0.025 g $FeSO_4$ | | |

*Note:* Start with 400 ml of water and add 25 ml of each the 20× salts (A) and buffer (B). Dissolve 2.5 g of D-glucose, bring volume to 500 ml and add 7.5 g of agar. Autoclave, allow to cool, but not harden, and aseptically pipette 0.1 ml of biotin stock into the medium. Swirl to mix biotin and disperse agar and pour into vessels.

Derived from White, F.F. and E.W. Nester, *J. Bacteriol.* 141, 1134, 1980.

advanced individual study problems. Generally, the experiments require more-developed skills, careful attention to detail and closer supervision by the instructor. These experiments are adapted primarily from those published by Trigiano and May (1994).

## Exercises

### Experiment 1. Transformation of cells using "wild-type" A. tumefaciens

This experiment is designed as a paired variate (see chapter twenty) and uses the cultivars 'Goldmine' and 'Iridon'. We suggest that one-half of the students or teams of students work with one cultivar, while the other half works with the remaining cultivar.

#### Materials
The following items are required for each student or team of students:

* Liquid and solidified AB medium (Table 31.1)
* *A. tumefaciens* strain 281 (Hood et al., 1986) or any other virulent strain may be secured from any number of sources, including a Plant Pathology or Plant Sciences Department, but first ascertain whether or not a USDA permit for pathogens is needed
* 125-ml Erlenmeyer flasks
* 0.22-μm filters and syringes
* Autoclavable biological hazard bags
* Several pieces of sterile 9-cm filter paper
* Four sterile 50-ml beakers
* Six 60 × 15 mm petri dishes containing MS medium amended with 0.22 mg/L (1.0 μM) BA and 2 mg/l (11.5 μM) IAA (shoot-initiation medium)
* Six 60 × 15 mm petri dishes containing shoot-initiation medium and 100 mg/L cefotaxime
* Six 60 × 15 mm petri dishes containing MS medium without plant growth regulators (PGRs), but amended with 100 μg/ml cefotaxime
* Liquid MS medium with 100 μg/ml cefotaxime

*Note:* Gloves and a particle mask should be worn when working with antibiotics and PGRs. These antibiotic solutions should be sterilized using a syringe and a 0.22 μm filter and added after the basal medium has been autoclaved and cooled, but not hardened.

*Procedure*

Surface disinfest three leaves, excise four midrib sections from each leaf and culture two sections per petri dish containing MS medium supplemented with 1.0 µM BA and 11.5 µM IAA as described under experiment 1 in chapter twelve. Label petri dishes 1A and 1B; 2A and 2B; 3A and 3B; etc. — the sections in petri dishes with the same number are from the same leaf and sections within A and B dishes constitute the "pair" in the paired variate-designed experiment. By using this design, a specific result may be attributed to a treatment and not to some other variable, such as physiological state of the explant. Maintain the identity of each pair throughout the experiment and collect data for each pair. Incubate the petri dishes for 3 days in conditions described under experiment 1 in chapter twelve.

Twenty-four hours before leaf sections are to be inoculated with bacteria, start liquid cultures of *A. tumefaciens* strain A281–supervirulent (Hood et al., 1986) or other. Pipette 1.0 ml of sterile AB medium onto medium that has been inoculated with bacteria. Choose a petri dish that has a fairly well-grown "lawn" of bacterial colonies. Suspend the bacteria in the medium by gently rubbing the agar with a sterile glass rod or inoculation loop. Aseptically transfer approximately 0.3 ml of the bacterial suspension to 50 ml of AB medium in each of two 125-ml Erlenmeyer flasks. Incubate on a rotary shaker at 75–100 rpm at 25 C. Bacteria may be grown at room temperature, but may require 48 h for good growth. The suspension should appear cloudy, almost opaque, after 24 h. In any case, do not use a bacterial suspension that has been grown for more than 48 h. *Note:* many procedures will require that the suspension of bacteria be adjusted to a certain colony-forming units (cfu) concentration or optical density at a specified reference wavelength, usually between 500 and 600 nm (see Urban et al., 1994). Although in most experiments cfu should be standardized, this step is not necessary for the successful completion of this laboratory exercise in our experience.

Aseptically transfer about 10 ml of the bacterial suspension to sterile 50-ml beakers. Immerse all leaf sections from the petri dishes labeled "A" in the suspension of bacteria one at a time for 10 sec. Remove excess fluid and bacteria by blotting the sections on sterile, dry filter paper within 100 × 15 mm petri dishes. Return all leaf sections dipped in bacteria to their original petri dishes. Submerge the sections from "B" petri dishes in sterile AB medium in sterile 50-ml beakers without bacteria for 10 sec, blot dry on clean, sterile filter paper and return sections to their original petri dishes. The suspension of bacteria and filter paper should be autoclaved before discarding. Incubate all petri dishes as before for a maximum of 2 days.

Filter sterilize 25 ml of MS liquid medium containing 100 µg/ml cefotaxime (Sigma Chemical Co., St. Louis, MO) into each of two sterile 50-ml flasks. Vigorously agitate the leaf sections that have been co-cultivated with bacteria in the solution and blot dry with sterile filter paper. Use different beakers for washing bacteria-treated and bacteria-free sections. Incubate all leaf sections on MS medium amended with PRGs and 100 µg/ml cefotaxime for 9 days.

Transfer sections treated with bacteria to MS medium without PGRs, but containing 100 µg/ml cefotaxime; do the same for sections not treated with bacteria. Incubate all petri dishes for an additional 21 days, or a total of 35 days from the initiation of cultures.

*Anticipated results*

Leaf sections of 'Goldmine' not treated with bacteria ("B" petri dishes) should not have formed shoots, but may have a few adventitious roots and, perhaps, some scanty callus along the cut edges. Leaf sections of 'Iridon' not treated with bacteria ("B" dishes) should

**Figure 31.2**   Transformation of leaf explants of 'Iridon' using *A. tumefaciens* wild-type A281. Explant treated only with AB medium (without bacteria) produced shoots (a), whereas the explant treated with bacteria formed copious amounts of callus (b).

have produced plentiful shoots, a few roots and little or no callus at the cut edges (Figure 31.2a). Successful transformation of cells within the leaf explants of the corresponding pairs of both 'Goldmine' and 'Iridon' ("A" petri dishes) is indicated by the production of copious amounts of yellow-green callus that is typically not produced during shoot formation (Figure 31.2b). Callus growth may be patchy on some explants. Shoots and roots should not be present on explants of either cultivar in this treatment.

Callus and/or the explants treated with bacteria may be transferred to MS medium without PGRs supplemented with 100 µg/ml cefotaxime every 3–4 weeks. Callus should continue to grow (autotrophic for PGRs) and can be maintained for indefinite periods. At the conclusion of the experiment, autoclave all cultures including the petri dishes containing used medium.

> *Questions*
>
> * Why is the treatment of dipping leaf sections into sterile AB medium without bacteria included in the experiment?
> * Why is it necessary to maintain cefotaxime in the medium after co-cultivation with the bacteria?
> * Why is the callus produced on the leaf sections autotrophic with regard to PGRs?

## Experiment 2. Regeneration of transgenic chrysanthemum plants

The protocol used in this experiment is similar to the protocol used in experiment 1 except that *A. tumefaciens* "wild-type" A281 is replaced with EHA105 (Urban et al., 1994), a disarmed A281 lacking genes encoding PGRs but harboring a gene (and CMV-35s promoter) for neomycin phosphotransferase (NPTII). This gene when incorporated in the plant genome will confer resistance to kanamycin, an antibiotic. The experiment is designed as a paired variate (see experiment 1); two leaf sections co-cultivated with bacteria and two leaf sections not treated with bacteria from the same leaf will serve as the pairs. Only the cultivar 'Iridon' is used in this exercise.

**Figure 31.3**   Transformation of leaf explants of 'Iridon' using a disarmed A281 and harboring the NPTII gene. Green shoots (t) may have been transformed (putative) to contain the NPTII gene. White shoots (n) are not transformed.

*Materials*

In addition to the items listed under experiment 1, each student or team of students will require the following:

* Six 60 × 15 mm petri dishes containing MS medium without PGRs and supplemented with 100 µg/ml cefotaxime and 50 µg/ml kanamycin (Sigma)
* Disarmed strain of A281 harboring the NPTII gene (EHA105) (see suggested procurement procedures listed in experiment 1)

*Note:* Gloves and a particle mask should be worn when working with antibiotics. These solutions should be filter sterilized and added to medium after autoclaving and cooling.

*Procedure*

Follow the same procedures as outlined in experiment 1 except inoculate leaf sections with the disarmed A281 strain of *A. tumefaciens*. Instead of using PGR-free MS medium with only cefotaxime, use medium that has both cefotaxime and 50 µg/ml of kanamycin. Incubate sections as before for an additional 21 days. The number of white and green shoots formed in both treatments should be counted for each pair.

*Anticipated results*

Shoots should have formed after 35 days from both the bacteria-treated leaf sections and those treated with AB medium only. Shoots originating from sections treated with AB medium only should appear bleached or white. Experiments in our laboratory have shown that shoots formed from untransformed cells of 'Iridon' are sensitive to as little as 5 µg/ml kanamycin. If any green shoots are present in this treatment, they are probably escapes and should be retested on medium without PGRs, but containing kanamycin and cefotaxime. If sensitive, most of the newly produced leaves will be bleached white and the shoots will not produce roots.

Most shoots formed from explants treated with bacteria should be bleached, but putatively transformed shoots will remain dark green (Figure 31.3). Some of these will be escapes, and all green shoots should be excised and recultured to medium containing cefotaxime and kanamycin for an additional 3 weeks. If these shoots are escapes, then the new growth will be white and roots will not be formed. If the NPTII gene has been successfully incorporated, then the shoots will remain dark green and the shoots will produce roots. From our experience, about 2–5% of the shoots formed will be transgenic. Remember to autoclave all plants and materials from this experiment when finished.

## Questions

* Why are there "escape" shoots in the kanamycin treated explants?
* Why is it necessary to provide physiological/molecular evidence in addition to data from kanamycin selection to conclusively demonstrate transformation?

## *Alternative experiment 2. Regeneration of transgenic chrysanthemum plants*

A more complete version of the above experiment is to use the following four treatments of leaf sections that are incorporated into a randomized complete block design: (1) co-cultivation with bacteria, cultured on MS plus antibiotics; (2) without bacteria (immersed in AB medium), incubated on medium lacking antibiotics; (3) without bacteria, cultured on MS amended with cefotaxime; and (4) without bacteria, incubated on MS supplemented with kanamycin. Place one leaf section per $60 \times 15$ mm petri plate.

White and green shoots are obtained after 35 days with treatment 1. Green shoots should be excised and cultured on medium containing both antibiotics. Only green shoots are produced in treatments 2 and 3, and generally more shoots are produced in treatment 3 (cefotaxime) than treatment 2 (without the antibiotic). Green and white shoots are produced in treatment 4, but when the green shoots are transferred to fresh medium containing kanamycin, all new growth will be white and rooting will not occur.

## *Literature cited*

de Jong, J., W. Rademaker and M.F. Van Wordragen. 1993. Restoring adventitious shoot formation on chrysanthemum leaf explants following co-cultivation with *Agrobacterium tumefaciens*. *Plant Cell Tissue Organ Cult.* 32:263–270.

Hood, E.E., G.L. Helmer, R.T. Fraley and M.D. Chilton. 1986. The hypervirulence of *Agrobacterium tumefaciens* A281 is encoded in a region of pTiBo542 outside of t-DNA. *J. Bacteriol.* 168:1291–1301.

Lowe, J.M., M.R. Davey, J.B. Power and K.S. Blundy. 1993. A study of some factors affecting *Agrobacterium* transformation and plant regeneration of *Dendrathema grandiflora* Tzvelev (syn. *Chrysanthemum morifolium* Ramat.). *Plant Cell Tissue Organ Cult.* 33:171–180.

Trigiano, R.N. and R.A. May. 1994. Laboratory exercises illustrating organogenesis and transformation using chrysanthemum cultivars. *HortTechnology.* 4: 325–327.

Urban, L.A., J.M. Sherman, J.W. Moyer and M.E. Daub. 1994. High frequency shoot regeneration and *Agrobacterium*-mediated transformation of chrysanthemum (*Dendranthema grandiflora*). *Plant Sci.* 98:69–79.

White, F.F. and E.W. Nester. 1980. Hairy root: plasmid encodes virulence traits in *Agrobacterium rhizogenes*. *J. Bacteriol.* 141:1134–1141.

*chapter thirty-two*

# Construction and use of a simple gene gun for particle bombardment

*Dennis J. Gray, Ernest Hiebert, Cha-Min Lin,*
*Michael E. Compton and Victor P. Gaba*

As described in previous chapters (twenty-three and twenty-nine), there are several techniques that have been utilized successfully to introduce DNA into plant cells. Of these methods, particle bombardment, wherein microscopic metal particles coated with genetically engineered DNA are explosively accelerated into plant cells, has become the second most widely used vehicle for plant genetic transformation, after *Agrobacterium*-mediated transformation (Gray and Finer, 1993). Several distinct "particle guns" have been described, including the Biolistic PDS 1000/He (Kikkert, 1993), which is the only commercially available device. The most attractive of the noncommercial devices is the particle inflow gun (PIG) (Finer et al., 1992), which is based on a flowing helium device described by Takeuchi et al. (1992), since it can be fabricated from steel plate with readily available parts and offers performance on par with the Biolistic PDS 1000/He (Brown et al., 1994).

This exercise is a modified version of an article published by Gray et al. in *Plant Cell Tissue and Organ Culture* (1994) and is arranged into two parts. In the first part, complete directions for constructing a simplified version of the particle inflow gun are provided. The device is termed a Plastic PIG, because, although it is based on the PIG, the steel specimen chamber of the PIG is replaced with a plastic vacuum jar and other parts, which dramatically simplifies construction (Gray et al., 1994). In the second part, directions for operating the Plastic PIG are provided, including preparation of the DNA-particle mixture and methods to obtain transient expression of the β-glucuronidase (GUS) gene, a convenient reporter gene, in plant tissues. Fabrication of this particle gun is an ideal laboratory exercise or demonstration for the entire class because it can be assembled in less that 40 min using only simple hand tools from parts that are readily available from technical equipment supply companies and hardware stores. The only drawback to its construction is that the components will cost $400–500. However, since it offers performance on par with other devices and is durable, once constructed, it can be used in subsequent class exercises or research programs.

## General considerations

### Construction of the Plastic PIG

For convenience, specific brands and suppliers for the following required components are specified (Table 32.1); however, this does not constitute an endorsement of brands or suppliers, since others may be as suitable. Although 115 VAC versions of electrical parts

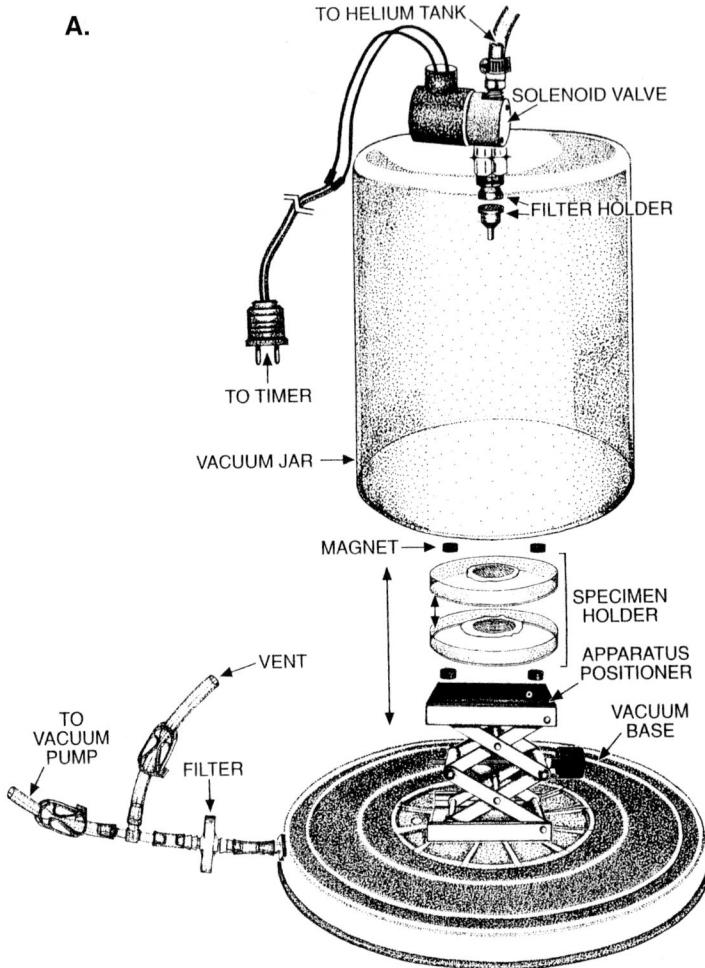

**Figure 32.1** Construction and setup of the Plastic PIG. (From Gray, D.J. et al., *Plant Cell Tissue Organ Cult.*, 37, 179–184, 1994. With permission.) A. Exploded view of Plastic PIG and specimen holders.

are specified here, 220 VAC valves and timers are available through the same suppliers. Similarly, the closest metric-equivalent fittings may be substituted for the English measurements given below. Electrical connection to supply current should be through a ground fault-protected circuit. The setup must be examined and approved by a qualified electrician before it is plugged into supply current source. Tools required to accomplish fabrication are listed in Table 32.2. Figures 32.1 and 32.2 detail construction and use of the Plastic PIG. Assembly steps are as follows:

1. Drill pilot hole in center top of vacuum jar. Enlarge hole with 7/16" masonry bit and thread the hole with 1/4"–18 NPT tap.
2. Drill pilot hole in center of brass cap, then enlarge to 1/4". Use 5/16"–24 NF tap to cut threads into hole.
3. Screw jam nut tightly onto close nipple, add Teflon tape to threads on short end of nipple and screw tightly into the #2 port of the solenoid valve. Screw the brass bushing, then the barbed fitting into the #1 port of the valve, using Teflon tape to seal.

**Figure 32.1** (continued from previous page). B. Detail of solenoid valve assembly.

4. Add a small amount of silicone glue to the exposed threads of the close nipple and screw the valve assembly snugly into the top of the vacuum jar. Add more silicone glue to the exposed threads protruding into the jar before screwing the brass cap tightly onto the nipple (see Figure 32.1B).
5. Use the 5/16"–24 NF die to cut carefully threads over the luerlock end of the plastic filter holder. Add Teflon tape to threaded end and screw the filter holder into the brass cap. Allow silicone glue on the assembled valve–vacuum jar unit to cure overnight before subjecting to vacuum.
6. Solder the two-prong electrical appliance cord onto exposed wires of valve and wrap securely with electrical tape. Plug cord into interval meter.
7. Clamp one end of plastic tubing to barbed fitting and the other end to the helium regulator.
8. Attach one end of vacuum-vent tubing to vacuum plate as shown in Figures 32.1A and 32.2 and the other end to the vacuum pump.
9. Place the apparatus positioner in the center of vacuum plate.

## Construction of specimen holders

Specimen holders consist of the top and bottom of autoclavable plastic petri dishes, in which holes have been drilled to accept screens (see Figure 32.1A and Table 32.1 for parts list).

1. Use hole saw on drill to cut 4-cm-diam. hole in center of petri dish tops and bottoms.
2. Autoclave screen material to preshrink. Cut into 5-cm-diam. disks. Use high-temperature silicone to attach screen to the *inside surface* of the bottom and the *outside*

*Table 32.1*   Parts List for the Plastic PIG and Specimen Holders

| Part | Catalog No. | Supplier |
|---|---|---|
| Vacuum chamber | 5305-0910 | Nalge Co., Rochester, NY |
| Solenoid valve | S24C-4V, NC 115/60V, 1/4" | Atkomatic Valve Co., Inc., Indianapolis, IN |
| Interval timer | G-08683-90 | Cole-Parmer Instrument Co., Niles, IL |
| Apparatus positioner | G-08056-10 | " |
| Vacuum pump | G-07055-04 | " |
| Gas pressure regulator | G-98200-70 | " |
| Barbed fitting | G-06362-40 | " |
| Tee connector | G-06455-15 | " |
| Tubing clamp | G-06833-00 | " |
| 13-mm plastic filter holder | 4312 | Gelman Sciences, Inc., Ann Arbor, MI |
| 0.2-μm air filter | 4464 | " |
| Plastic tubing | 01T252PE-RDT | Ark-Plas Products, Inc., Flippin, AR |
| Helium tank, lab grade | | Local supplier |
| Miscellaneous hardware | | Local hardware supply |
|   Metal hose clamps | | " |
|   Two-prong electrical appliance cord | | " |
|   Brass close nipple with flange, 1/4"–18 SAE | | " |
|   Brass cap, 1/4"–18 SAE | | " |
|   Brass bushing, 1/4" × 1/8" NPT | | " |
|   Steel jam nut, 9/16" SAE | | " |
|   Teflon and electrical tape, rosin-core solder, silicone sealing glue | | " |
| Specimen holders | | |
|   PC petri dishes | 5502-0010 | Nalge Co., Rochester, NY |
|   Spectra/mesh PP screen | 08-670-185 | Fisher Scientific, Pittsburgh, PA |
|   Magnets 3 mm diam. (6) | | Radio Shack, Inc. |
| Permatex high temp RTV silicone gasket maker | | Local auto parts store |

*Table 32.2*   Tools Required to Assemble Plastic
PIG and Specimen Holders

Electric hand-held drill
7/64" and 1/4" drill bits for metal
7/16" drill bit for masonry
5/16"–24 NF and 1/4"–18 NPT thread taps
5/16"–24 NF thread die
Adjustable hole saw
Hobby vise
Adjustable wrench
Pliers
Soldering iron

**Figure 32.2**   Setup of Plastic PIG and timer. (From Gray, D.J. et al., *Plant Cell Tissue Organ Cult.*, 37, 179–184, 1994. With permission.)

> *surface* of the top. The petri dish surface under the screen should be roughened with sandpaper prior to attachment.
>
> 3.  Cover glued surface with plastic wrap and add a weight to hold screen tightly in place. Allow to cure overnight.

For bombardment, specimens are held between screens by nesting the top piece of the dish over the inverted bottom, both of which are then clamped together with two to three pairs of magnets (Figure 32.1A).

## Preparation of plasmid-particle mixture

The plasmid used is pBI221 (Clontech Laboratories, Palo Alto, CA), which consists of a 3-kb HindII-EcoRI fragment of pBI121 containing the CaMV 35S promoter, GUS gene and NOS-terminator (Jefferson et al., 1987) cloned into pUC19. The plasmid can be maintained in DH5alpha bacterial host and purified according to the QIAGEN plasmid purification kit. Note that any plasmid that expresses the GUS gene in dicotyledonous plants may be substituted. Since instruction on plasmid purification and scaleup is beyond the scope of this book, we recommend that instructors who are unfamiliar with these procedures obtain purified plasmids, which are becoming increasingly available, from colleagues. Please see chapters twenty-nine and thirty for further discussion.

The following procedure was modified from Finer et al. (1992). Autoclave M17 tungsten particles (Bio-Rad Laboratories, Inc., Hercules, CA) and place 50 mg in 0.5 ml 95% ethanol, incubate 20 min then vortex well. Centrifuge at 10,000 rpm for 5 min and wash with 0.5 ml sterile water, repeat five times. Vortex, then pipette 25 µl of tungsten stock into eppendorf tube. Add 5 µl DNA (1 µg/µl) and mix well. Add 25 µl of 2.5 M $CaCl_2$, then quickly add 10 µl of 100 µm spermidine (free base) and vortex. Incubate on wet ice (important to maintain cold temperature) for 5 min. Without centrifugation, remove and discard about 50 µl of solution, leaving the particles. Use immediately.

## Operation cycle of the Plastic PIG

Clamp tissue in a specimen holder using magnets, centered on the positioner adjusted to the desired height. It is imperative that tissue not be "wet", i.e., not covered with a film of liquid, in order for the procedure to be successful. Place a 2-μl drop of DNA/particle mixture on the middle of a plastic filter holder screen and screw tightly into the vacuum jar. Place vacuum jar on base and evacuate the chamber to 90 kPa (27 in. Hg) vacuum. Set timer to 0.1 sec and fire valve, then vent the chamber and replace specimens on culture medium. The procedure, from DNA-particle preparation to bombardment and culture, should be accomplished as quickly as possible.

## Exercises

### Experiment 1. Comparison of helium pressures on transient expression of the GUS gene

Particle bombardment is affected by many parameters, which include biological (i.e., plasmid construct, type and physiological state of target tissue) and physical (i.e., particle size, vacuum level and helium pressure) factors. In this experiment, the effects of two tissue types and three helium pressures will be compared by assessing differences in transient GUS reaction. Since GUS reaction can be assessed as soon as 48 h after bombardment, this experiment can completed within 1–2 weeks.

#### Materials
The following materials are required for each student or team:

* Explant tissue, consisting of: (1) approximately 100 basal cotyledonary explants of melon, cultured for 4 days on embryo induction medium as described in chapter eighteen and (2) the innermost leaves from fresh cabbage, purchased from a supermarket (approximately one-half cabbage for each student or team)
* DNA/particle mixture, prepared continuously, immediately before every four bombardments
* Eight sterile specimen holders
* Three sterile moist chambers; i.e., empty $100 \times 15$ mm petri dishes containing moist filter paper (to incubate cabbage tissue after bombardment)
* 10 ml of X-gluc reagent (see chapter twenty-nine concerning *Agrobacterium*-mediated transformation)

#### Procedure
The experimental design compares three helium pressures (410 kPa [60 psi], 620 kPa [90 psi] or 830 kPa [120 psi]) and two explant tissues (melon cotyledons and cabbage leaves). Each treatment (pressure–explant combination) is replicated four times. Set the particle holder-to-specimen gap at 15 cm by adjusting the apparatus positioner. Arrange eight cotyledon explants as close together as possible, but not overlapping, in the center of a specimen holder and carefully center the specimen on the apparatus positioner relative to the bombardment pattern. To properly "aim" the gun, the actual particle pattern can be determined by first bombarding filter paper and viewing the resulting pattern. As rapidly as possible, perform a bombardment, as described above, and place the explants back onto the culture medium. Repeat four times. For cabbage, dissect young leaves into pieces approximately 2 cm in diam. and bombard. Place the four replicate leaves from one treatment into sterile moist chambers.

Incubate explants for 48 h and then place into X-gluc reagent in dark at 35 C for 12–24 h as described in chapter thirty. Count the number of distinct blue spots per replicate (specimens can be preserved indefinitely in cold 70% ethanol). Statistically analyze the data as a $2 \times 3$ factorial and perform a mean separation test as described in chapter six.

**Figure 32.3** Transient expression of GUS in four basal cotyledon quarters bombarded at 830 kPa with a 15-cm particle-to-specimen gap. Discrete GUS foci, shown in this photograph as black spots of various sizes, were actually dark blue in coloration. (From Gray, D.J. et al., *Plant Cell Tissue Organ Cult.*, 37, 179–184, 1994. With permission.)

*Anticipated results*

The GUS reaction seen in this experiment is due to the direct gene activity in bombarded cells. For such a response to occur, at least one intact DNA molecule must enter a cell and the cell must survive the impact. However, this response does not denote stable transformation (i.e., integration of the GUS gene into the chromosomes or genome of the cell). Dramatic differences in the number of blue spots will be seen because of different pressure treatments, with the highest pressure typically producing the greatest response in melon cotyledons, (Figure 32.3); however, different tissues would be expected to respond differently. Differences in GUS reaction due to tissue type are more difficult to predict because of wide variation in available cabbage tissue. However, when successful, bombarded cabbage leaf tissue demonstrates the "blast" pattern very well. Experiments comparing the other parameters mentioned above can be conducted in a similar manner.

While transient expression of GUS is relatively simple to achieve, obtaining stable expression is a more difficult task, which involves careful selection over time. Selection usually is accomplished by incorporating a selectable marker gene, such as NPTII (neomycin phosphotransferase II), which confers resistance to the antibiotics kanamycin and gentamycin (see chapter twenty-nine).

*Questions*

* What accounts for the differences in GUS expression due to helium pressure? Given these results, what might be expected by varying the particle holder-to-specimen gap?
* Similarly, why did the two explants exhibit different GUS reactions?

# Acknowledgment

This is Florida Agricultural Experiment Station Journal Series No. R-04835.

## Literature cited

Brown, D.C.W., L. Tian, D.J. Buckley, M. Lefebvre, A. McGrath and J. Webb. 1994. Development of a simple particle bombardment device for gene transfer into plant cells. *Plant Cell Tissue Organ Cult*. 37:47–53.

Finer, J.J., P. Vain, M.W. Jones and M. McMullen. 1992. Development of the particle inflow gun for DNA delivery to plant cells. *Plant Cell Rpt*. 11:323–328.

Gray, D.J. and J.J. Finer. 1993. Editorial introduction. Special section: development and operation of five particle guns for introduction of DNA into plant cells. *Plant Cell Tissue Organ Cult*. 33:219.

Gray, D.J., E. Hiebert, C.M. Lin, M.E. Compton and D.W. McColley. 1994. Simplified construction and performance of a device for particle bombardment. *Plant Cell Tissue Organ Cult*. 37:179–184.

Jefferson, R.A., T.A. Kavanagh and M.W. Bevan. 1987. GUS fusions: β-glucuronidase as a sensitive and versatile gene fusion marker in higher plants. *EMBO J*. 6:3901–3907.

Kikkert, J.R. 1993. The Biolistic PDS-1000/He device. *Plant Cell Tissue Organ Cult*. 33:221–226.

Takeuchi, Y., M. Dotson and N.T. Keen. 1992. Plant transformation: a simple particle bombardment device based on flowing helium. *Plant Molec. Biol*. 18:835–839.

*chapter thirty-three*

# Germplasm preservation

*Leigh E. Towill*

The concept of germplasm is inherently simple yet extremely complex in scope. Germplasm, as defined for a crop, represents all the different genotypes that could be used for improvement. Improvement is accomplished traditionally by making sexual crosses between individuals and then selecting plants with the desired characteristics, such as disease or stress tolerance or resistance. With the advent of more exotic techniques, for example, transformation and protoplast fusion, virtually any species may be modified by incorporating a gene or genes from unrelated species (see chapter twenty-nine). Thus, germplasm for a crop becomes more expansive. Nevertheless, we usually consider germplasm in its less cosmic size to be the gene pool from those individuals that are related and may intercross. Germplasm is valuable because it contains the diversity of genotypes that are needed to develop new and improved lines.

Germplasm diversity for any crop must be protected from loss to ensure its availablity for future plant improvement. This diversity was collected from centers of crop origin and from landraces developed over many years. In the case of crops, preservation of this diversity is not in the wild, but in what are termed repositories or genebanks (Table 33.1). Chapters in a book describing the U.S. National Plant Germplasm System (NPGS) (Janick, 1989) are an excellent source for those interested in the details of a comprehensive system.

A key theme for safe preservation of diversity is use of two sites. One site, the active collection or active genebank, has responsibilities for obtaining diversity (either by collection or by exchange with other genebanks worldwide), propagation, distribution to users, and characterization. The second site, the base collection or base genebank, is for long-term preservation and utilizes the best storage systems available. This site provides the crucial function of being a safety net should materials be lost at the active collection. It is important to have the distinction between these two sites firmly in mind in discussing the issue of preservation and in deciding what methods are useful for each collection.

This chapter describes germplasm preservation for crop species and emphasizes where and how in vitro culture and cryopreservation are used for efficient and effective management. Germplasm preservation for most crop plants does not involve in vitro culture but utilizes seeds, and a brief discussion is included to provide a complete picture of preservation. The laboratory exercises following this chapter emphasize cryopreservation of shoot tips.

## Preservation of seed-propagated species

Seeds are classified as being either desiccation tolerant (capable of retaining viability after being dried to virtually any moisture content — also termed *orthodox* seed) or desiccation sensitive (seeds losing viability after being dried below a critical limit, usually about

*Table 33.1*  Major Crops Maintained Clonally at Different Repositories within the U.S. National Plant Germplasm System

| Location | Species maintained clonally |
|---|---|
| Miami, FL/Mayaguez PR | Avocado, banana, plantain, coffee, cocao, mango, sugarcane |
| Corvallis, OR | Pear, filbert, mint, strawberry, raspberry, blackberry, currants, hops |
| Davis, CA | *Prunus* spp., *Vitis vinifera* and related species |
| Geneva, NY | Apple, *Vitis* spp. |
| Brownwood, TX | Pecans, walnuts |
| Hilo, HI | Guava, passion fruit, papaya, pineapple, macadamia, litchi, carambola |
| Griffin, GA | Sweet potato |
| Miami, FL/Mayaguez, PR | Avocado, banana, plantain, coffee, cocao, mango, sugarcane |
| Riverside, CA | *Citrus* spp. and related species, date palm |
| Orlando, FL | *Citrus* spp. and related species |

12–30% moisture — also termed *recalcitrant* seed). This classification is a great simplification and a number of intermediate types exist. Most temperate crop species have desiccation-tolerant seeds, and many studies have shown the importance of seed moisture and storage temperature for retaining viability (see review by Roos, 1989). Recent studies suggest that the optimum seed moisture content for obtaining the greatest longevity is related to the storage temperature. In a practical sense, seeds can be equilibrated with a 25% relative humidity at 5 C and then stored between –5 and –20 C.

Most desiccation-tolerant seed survive cryogenic storage (Stanwood, 1985). Temperatures less than about –130 C are desired for cryopreservation because of very low molecular kinetic energies, the absence of liquid water and extremely slow diffusion. Thus, many reactions leading to deterioration are minimal and longevities are postulated to be extremely long (centuries) and limited only by the buildup of genetic lesions resulting from background irradiation. In practice, cryogenic storage is in liquid nitrogen (LN, –196 C) or in the vapor above LN (ca. –150 to –180 C) because it is a relatively cheap and available cryogen.

Desiccation-sensitive seeds are produced by aquatic species, large-seeded species, species native to tropical areas, and some temperate zone species of trees. Coconut, cocoa, mango, nutmeg, and rubber plants are examples of economically important species with recalcitrant seed. To retain viability, recalcitrant seeds are stored at as low a temperature as possible under conditions that retain relatively high seed moisture levels and assure a supply of oxygen for respiration. These seeds have short life spans, ranging from a few weeks to several months. These are too short for adequate germplasm preservation, and these species usually are stored as vegetative plants.

In vitro technologies are becoming important in maintaining crops with desiccation-sensitive seed. In vitro plants may be maintained for active collections. Embryo culture may be used in preservation of desiccation-sensitive seeds. Isolated embryonic axes (cotyledons removed) from mature seeds can survive partial desiccation and cooling to low temperatures. The level of desiccation tolerated and whether or not this is sufficiently low to avoid damage from ice formation during cooling must be defined for each species. Upon retrieval from LN, axes are cultured and can develop into plants. Treatments involving application of cryoprotectants to excised embryos may also prove to be useful, but as yet no procedure has been developed for practical, reproducible cryopreservation of recalcitrant seeds. Axillary shoot tip or bud cryopreservation from in vivo or in vitro plants is also useful for base-collection storage of these species.

## Preservation of pollen

Pollen may also be used to store genes from crops (Towill, 1985). The major use of germplasm is for improvement of plants by sexual processes and, thus, requires making controlled crosses between desired individuals. Supplying pollen to the user facilitates this process; the user does not have to grow the plant and wait for flowering of the male parent. This is of obvious benefit for those species that often take many months to flower and, therefore, is of potential value for some species that are vegetatively propagated. Pollen preservation, thus, could be integrated into the active genebank, where it would be collected, stored and distributed upon request.

## Preservation of vegetatively propagated species

Many crops either have an extremely long juvenile phase (no flowers), produce seed with extremely short longevities, do not easily produce seed or do not produce viable seed (Towill, 1988). The vegetative plant for these species is stored either in the field or in greenhouses and is regularly propagated asexually to rejuvenate the line and prevent loss. These species are extremely diverse, comprising herbaceous and woody growth habits from tropical to temperate zones. These also differ greatly in stress tolerances to temperature, salt, moisture, toxic ions, anoxia, UV radiation, etc. as well as to disease.

### Active collections

It is impossible to grow all species at one location, and thus several repositories, termed clonal repositories, exist throughout the world. The vast number of different clones held precludes holding very many replicates, especially under field conditions. Maintenance is expensive and loss could easily occur because of disease or catastrophe.

The use of tissue culture in germplasm preservation usually is for species that are vegetatively propagated. In vitro culture of plants can be used to gain more efficient preservation (Withers, 1991). A number of advantages for using in vitro plants can be listed, including small space needed for storage, ease of shipping, year-round availability, potential for eliminating disease and for maintaining disease-free plants as well as rapid propagation and labor and cost savings (Towill, 1988). Many species can now be propagated in culture, but methods are often genotype specific and may require considerable modification. All in vitro lines held in one location are also susceptible to loss from catastrophe.

The advantages of maintaining in vitro versus in vivo plants often relate to what is desired by the user community. For example, if crosses are to be made between apple lines, apple budwood from ex vitro sources is preferred, since a grafted bud can lead to a flowering shoot within about 18 months. An in vitro apple plant, by comparison, might take more than 3 years to flower after transplanting. However, for potato, flowering times from planting a tuber or tranplanting an in vitro plant are similar. The advantage here for the in vitro system is ease in maintaining the disease-free status.

In vitro and in vivo systems are not mutually exclusive. To guard against loss, in vitro plants may serve at the same location as a backup to field or greenhouse lines. Again, recall that field replication is often minimal; often only two individuals are held for a clone.

Maintenance of in vitro cultures occurs in the active collection often under normal growth conditions. If material is requested by the user community, it can be quickly propagated and distributed. Backup materials at the active collection site may be maintained under conditions that minimize growth (Withers, 1991). This usually means holding the in vitro plant at lower temperatures — for example, potato at 10 C or mint at 4 C.

```
      in vitro or ex vitro plant
                  ↓                    temperature/light treatment
          conditioned plant
                  ↓                    excision
      isolated shoot tips/buds
                  ↓                    pretreatment
                                              {conc., time, temp.}
                  ↓                    application of cryoprotectants
                                              {conc., time, temp.}
                  ↓                    cooling {nucleation, rates}
                  ↓                    storage {time, temp.}
                  ↓                    warming {rates}
                  ↓                    removal of cryoprotectant
                                              {conc., time, temp.}
      isolated shoot tips/buds
                  ↓                    post-thaw treatment (culture)
                  ↓                    viability assessment
              growth
```

**Figure 33.1**  Generalized flow chart for cryopreservation of shoot tips.

Growth is retarded and cultures may be held for a few years, providing precautions are taken to avoid water loss. Elevated osmotic pressures and other growth inhibitors may also be used to reduce the frequency of transfer needed to maintain a viable culture. Empirical studies are required to apply this method to different species.

## Base collections: cryopreservation

It is prohibitively expensive to maintain duplicate collections for either ex vitro or in vitro plants at a second location. Hence, cryopreservation is desired for the base collection and is applied to axillary or apical shoot tips from either ex vitro or in vitro plants. Storage at low temperatures, such as –20 C or even –80 C, results in very short longevities with these propagules, for reasons which will not be discussed here. Cryogenic storage provides long-term storage at a reasonable cost with a low labor input. As with seed, storage is either in LN or in the vapor phase over LN.

It should be emphasized that, whereas cryopreservation for germplasm purposes uses shoot tips or buds, cryopreservation of protoplasts, cells, tissues and somatic embryos is also needed for many "tissue culture" operations to guard against loss, to assure line purity and performance and to avoid costly, frequent transfers.

Shoot tips have higher water content than seeds and cannot withstand extreme desiccation. In the hydrated state, exposure to low temperatures leads to damage due either to intracellular ice or to consequences of desiccation during extracellular ice formation. For these reasons, a more detailed protocol (Figure 33.1) is needed. Virtually all cryopreservation protocols follow these steps, but details vary with species. It is usually necessary to use cryoprotectants and to control cooling and warming rates. A theoretical treatment of rate phenomena is beyond the scope of this chapter. A number of excellent reviews address this and related topics (Mazur, 1984; Fahy et al., 1987).

## Pretreatment of the plant and isolated propagule

The physiological condition of the plant used for cryopreservation is important. Shoot tips isolated from plants that have been cold acclimated survive the cryogenic protocol in higher percentages than those not acclimated. This is useful only for those species that have the genetic capability to cold acclimate. Survival of excised shoot tips also can be enhanced by exposure to elevated sucrose levels. The exact effect of treatments on the plant or of the excised shoot tip is uncertain, but they probably alter intracellular contents of critical metabolites (particularly sugars), cell water content and physiological characteristics of the cell (e.g., membrane permeabilities).

## Cryoprotectant(s) application and cooling rate

After the pretreatment phase, shoot tips are exposed to cryoprotectants. The concentration of cryoprotectant(s) used and the time and temperature of exposure vary with the cooling strategy employed. Two cooling regimes are mainly used with shoot tips. The first is termed either slow or optimum rate cooling and employs cooling of the cryoprotected shoot tip at about 0.25–1.0 C/min to about –35 C prior to immersion in LN. For this method shoot tips are incubated in either 1–2 M concentrations of a suitable permeating cryoprotectant, such as dimethylsulfoxide (DMSO), or ethylene glycol. The second regime uses rapid cooling and is often termed a vitrification method. Here, shoot tips are exposed to much more concentrated solutions of cryoprotectants and then are immersed directly from about 0 C into –196 C.

Research with plant systems initially emphasized two-step cooling techniques to attain cryopreservation. Useful levels of survival were obtained for some species, but not for others. Reasons for this have not been elucidated. Summaries of this method can be found in recent reviews (Sakai, 1986; Towill, 1990). Attention now has focused on vitrification for cryopreservation.

The term *vitrification* refers to transformation of a liquid to a glass during cooling. Tissue culture literature has used the term vitrification in an entirely different sense to refer to plants that have an abnormal water-soaked or glassy appearance. However, *hyperhydricity* is now the preferred term for this abnormal appearance of plants in tissue culture.

There are increasing numbers of theoretical and practical studies examining vitrification in animal and plant systems (Fahy et al., 1987; Steponkus et al., 1992). Indeed, vitrification probably occurs naturally in extremely cold-hardy plants exposed to very low temperatures. Vitrification requires application of suitable concentrations of compounds, such that, with usually rapid cooling, the system forms a glass (i.e., the solution vitrifies). Information about physical aspects of the process is described in MacFarlane et al. (1992). Although it is difficult to prove that the cell itself vitrifies, circumstantial evidence suggests that this happens under defined conditions.

The advantages of vitrification are that it is a simple technique that does not require an expensive apparatus, it may avoid the damaging consequences of ice formation and it may be applicable to larger pieces of tissue than are usually used for two-step cooling methods. Survival after cryogenic exposure has been reported with several different vitrification procedures for protoplasts, cells, somatic embryos and shoot tips.

The general strategy for vitrification is to expose the shoot tips (or protoplasts, cells, somatic embryos, etc.) to low concentrations of the vitrification solution such that the permeable components have sufficient time to penetrate (the loading phase). Next, the shoot tips are exposed to much higher concentrations of the vitrification solution (the dehydration phase). Since water permeability is much greater than solute permeability, the shoot tip dehydrates and endogenous solutes and any permeating cryoprotectants are concentrated. Shoot tips are then rapidly cooled to facilitate glass formation.

It has become apparent that loading is not necessary for shoot tips from some species and, indeed, may be detrimental (Steponkus et al., 1992). The cell contents alone, when sufficiently dehydrated, apparently vitrify when cooled at a rapid rate. Here preculture for 1–3 days in a moderate osmotic solution (0.3–1.0 M) prior to dehydration is necessary to achieve survival after vitrification in such shoot tips.

A second major variation of vitrification uses an encapsulation/dehydration sequence (Fabre and Dereuddre, 1990). Shoot tips are encapsulated in an alginate gel, cultured in sucrose solutions, dehydrated in an air flow and then usually rapidly cooled. Encapsulated shoot tips vitrify under the conditions employed. Survival has been found in shoot tips from several unrelated species such as potato, grape, carnation and pear.

## Conclusion

Seeds and plants are used to preserve germplasm of most crops to assure that materials are available for future breeding to produce crops with specified characteristics. Whereas desiccation-tolerant seeds are simply stored, desiccation-sensitive seeds present challenges. Preservation of clones also is challenging because of the high costs of plant maintenance and difficulties in using cryopreservation. Increasingly, plant culture allows for efficient and effective preservation for these latter two groups. Many species now may be propagated and maintained in culture. Embryonic axes and shoot tips may be regrown in culture after cryogenic treatment. Some species are more problematic, but with advances in developing cryopreservation methods, such as vitrification, practical use is now feasible.

## Literature cited

Fabre, J. and J. Dereuddre. 1990. Encapsulation–dehydration: a new approach to cryopreservation of *Solanum* shoot tips. *Cryo-letters.* 11:413–426.

Fahy, G.M., D.I. Levy and S.E. Ali. 1987. Some emerging principles underlying the physical properties, biological actions, and utility of vitrification solutions. *Cryobiology.* 24:196–213.

Janick, J., Ed. 1989. *The National Plant Germplasm System of the United States. Plant Breeding Reviews,* Vol. 7, 230 pp. Timber Press. Portland, OR.

MacFarlane, D.R., M. Forsyth and C.A. Barton. 1992. Vitrification and devitrification in cryopreservation. In: *Advances in Low-Temperature Biology,* Vol. 1, P.L. Steponkus, Ed., pp. 221–278. JAI Press Ltd., Greenwich, CT.

Mazur, P. 1984. Freezing of living cells: mechanisms and implications. *Am. J. Physiol.* 247:c125–c142.

Roos, E.E. 1989. Long-term seed storage. *Plant Breed. Rev.* 7:129–158.

Sakai, A. 1986. Cryopreservation of germplasm of woody plants. In: *Biotechnology in Agriculture and Forestry,* Vol. 1, Trees I, Y.P.S. Bajaj, Ed., pp. 113–129. Springer-Verlag, Berlin.

Stanwood, P.C. 1985. Cryopreservation of seed germplasm for genetic conservation. In: *Cryopreservation of Plant Cells and Organs,* K.K. Kartha, Ed., pp. 199–226. CRC Press, Boca Raton, FL.

Steponkus, P.L., R. Langis and S. Fujikawa. 1992. Cryopreservation of plant tissues by vitrification. In: *Advances in Low-Temperature Biology,* Vol 1. P.L. Steponkus, Ed., pp. 1–62. JAI Press Ltd., Greenwich, CT.

Towill, L.E. 1985. Low temperature and freeze-/vacuum-drying preservation of pollen. In: *Cryopreservation of Plant Cells and Organs,* K.K. Kartha, Ed., pp. 171–198. CRC Press, Boca Raton, FL.

Towill, L.E. 1988. Genetic considerations for germplasm preservation of clonal materials. *HortScience.* 23:91–97.

Towill, L.E. 1990. Cryopreservation. In: *In Vitro Methods for Conservation of Plant Genetic Resources,* J.H. Dodds, Ed., pp. 41–70. Chapman and Hall, London.

Withers, L.A. 1991. In vitro conservation. *Biol. J. Linnean Soc.* 43:31–42.

*chapter thirty-four*

# Vitrification as a method to cryopreserve shoot tips

*Leigh E. Towill*

The following experiments demonstrate the use of vitrification as a method for cryopreservation. Both a solution-based method and an encapsulation/desiccation method are described. Shoot tips are emphasized because they are the propagules used for preservation of clones. Preservation of cell suspensions, callus, protoplasts or somatic embryos would follow a similar protocol. As with any method, details often differ with species, cell type and tissue complexity. These vitrification methods have been used successfully for several species of mint and various cultivars of potato, as well as apple, *Prunus* sp., endod, garlic, papaya, strawberry, carnation and sweet potato. The encapsulation/dehydration method has been extensively studied by Dereuddre et al. (Fabre and Dereuddre, 1990) and has been applied to several species.

## General considerations

### Growth of plants and preparation of media

Stock plants of either carnation, chrysanthemum (see chapter twelve), mint or potato (see chapter nine) are useful for these experiments since they grow rapidly in vitro, but almost any species may be utilized if a suitable shoot tip growth medium has been defined. In vitro–multiplied plants are preferred as sources of shoot tips, since the incidence of contamination will be reduced; however, greenhouse or growth-chamber plants can be used. If the latter are used, be sure that the surface disinfestation procedure is effective. A cold period (e.g., 2 weeks at about 2 C under short days) for the stock plant is often beneficial for those species that have the ability to cold acclimate (carnation and chrysanthemum). Details for micropropagation of plants by nodal sections and meristem tip or shoot tip culture are found in chapter seven.

Both liquid and solidified media are required. The growth medium used for excised shoot tips of potato and mint contains Murashige and Skoog (MS) minerals (see chapter two for composition) and vitamins, 100 mg/L (0.55 mM) myo-inositol, 2% (59 mM) sucrose and, if solidified, 0.7% agar (pH 5.7). The plant growth regulators (PGRs) for mint are 0.5 mg/L (2.2 µM) benzyladenine (BA) and 0.1 mg/L (0.4 µM) indolebutyric acid (IBA), and for carnation and potato are 0.01 mg/L (0.04 µM) BA, 0.001 mg/L (0.005 µM) naphthalene acetic acid (NAA) and 5 mg/L (14.4 µM) gibberelic acid (GA$_3$). These media facilitate shoot development and minimize callus formation, but may require modification for specific culture lines. If other species are used, prepare media as described in the respective literature for shoot tip culture. Media are autoclaved and stored in small aliquots in a refrigerator.

## Exercises

### Experiment 1. Cryopreservation of shoot tips using a solution-based vitrification method

This experiment (Figure 34.1) illustrates vitrification using two different vitrification solutions. Excising and manipulating shoot tips is time-consuming. It is advisable to excise a few shoot tips and follow the experiment to determine how long it takes to do each task. All solutions and glassware are sterile, and all manipulations should be done in a laminar flow hood. Cooling and warming the straws or ampoules need not be done in the flow hood.

#### Materials
The items below are needed for each student or team of students for the following two experiments:

* Solutions as described below
* Large, sturdy forceps with insulated tips
* Wide-mouth liquid nitrogen (LN) dewar
* 0.25- or 0.5-ml semen straws; in some cases plastic cryotubes are also satisfactory
* Straw closures or heat-sealing device for semen straws
* $12 \times 75$ mm glass test tubes
* Safety glasses or mask
* Timer

#### Vitrification methods
All the solutions below can be stored at 4 C for up to 4 weeks with no apparent loss in effectiveness.

*Method A.* PVS2 solution contains (wt/wt) 30% glycerol, 15% dimethylsulfoxide (DMSO) and 15% ethylene glycol (EG) and was devised by Sakai et al. (1991). The aqueous weight percentage (i.e., 40%) added contains MS salts (see chapter two for composition), vitamins and 0.4 M sucrose without PGRs. This solution is termed here 100% PVS2. Since this solution must be sterile, prepare the glycerol and EG portion at somewhat more concentrated levels in the growth medium and autoclave. After cooling, add DMSO (DMSO is sterile in a new bottle) and bring to the final volume with growth medium (0.4 M sucrose). The 100% PVS2 solution is diluted with growth medium (0.4 M sucrose without PGRs) to form a 60% (v/v) PVS2 and a 20% (v/v) PVS2 solution.

*Method B.* This method uses the formulation 50% EG, 15% sorbitol and 6% bovine serum albumin (BSA) (w/w) as described by Steponkus et al. (1992). Prepare a small volume of this medium in water alone and filter sterilize (see chapter two). The medium can also prepared and autoclaved without BSA. Then, sterilize the BSA powder in a thin layer under an ultraviolet (UV) light for about an hour, add it to the cool sterile portion and swirl to dissolve. Wear safety glasses when working with the UV light.

#### Dilution medium
The dilution medium is used for removing the vitrification solution from shoot tips and is composed of MS salts and vitamins, 1.2 M sorbitol and 100 mg/L myo-inositol (pH 5.7). The same dilution medium is used for both vitrification methods.

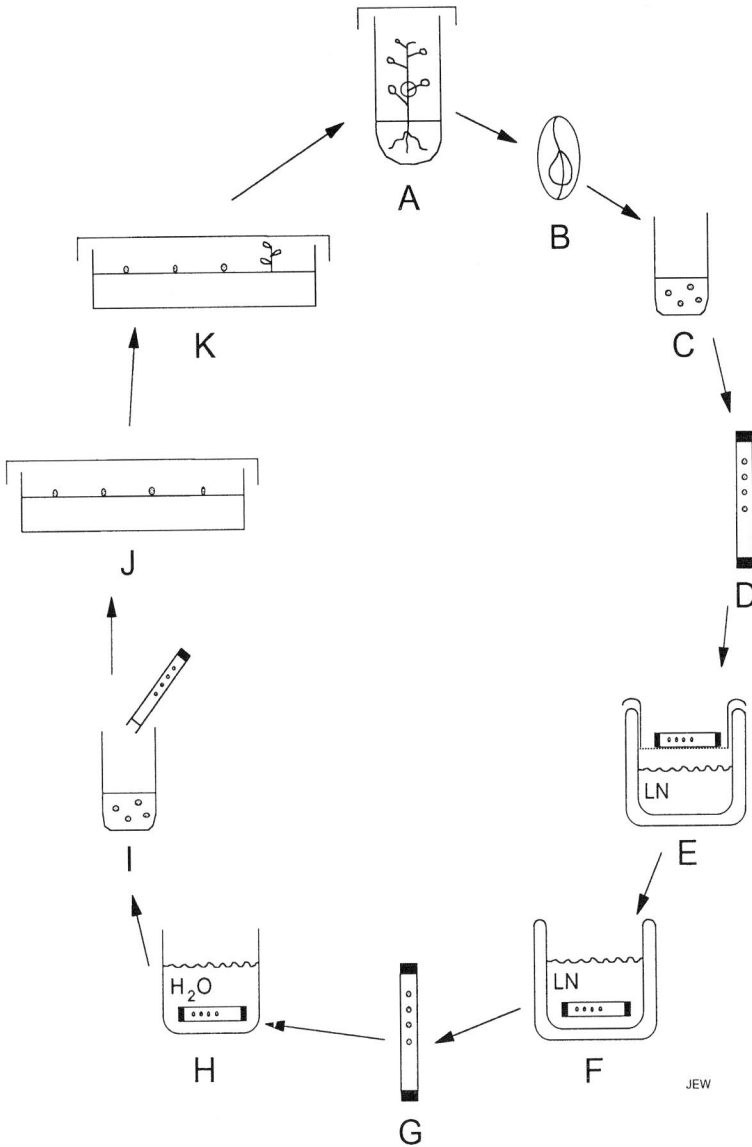

**Figure 34.1**  This schematic illustrates the steps involved in using a vitrification solution and rapid cooling for cryopreserving shoot tips. Stock plant (A) with well-developed axillary buds (circle). Excised buds (shoot tips) (B) are exposed to the vitrification solution (C) and then placed in semen straws (D). Semen straws are exposed to about −160 C (E) prior to immersion in LN (F). For warming, straws are held in the air for 10 sec (G) before immersion in water for 1 min (H). Shoot tips are diluted from the vitrification solution (I), cultured on agar (J) and, when sufficient growth occurs, transferred to normal growth medium (K).

### Experimental protocol

*Excision and culture of shoot tips.*    Select stock plants that are healthy and sufficiently old such that axillary shoot tips are well developed but not elongated. Using a dissecting microscope, excise tips with a razor blade, scalpel or edge of a hypodermic needle. The isolated shoot tip should be between 0.5 and 1 mm in length and should contain about two to four distinct leaf primordia. In some cases, the outer larger leaf primordia must be removed to trim the shoot tip to this size. Work quickly to minimize desiccation of the shoot tips. Transfer isolated shoot tips into a 12 × 75 mm test tube containing a small amount of growth medium without PGRs. Place shoot tips into a series of test tubes in random order to minimize the effect of original shoot tip position on the shoot axis. Four shoot tips are placed in each test tube with six replicate tubes per treatment.

Shoot tips should be used shortly after excision. Alternatively, they may be precultured for 1–2 days in liquid medium with PGRs and either low levels of cryoprotectants (i.e., 5% DMSO or 3% ethylene glycol) or elevated levels of slowly penetrating osmotica (i.e., 0.3–0.7 M sucrose).

Application of the vitrification solution to shoot tips differs depending on the method used. Method A: Remove liquid with a pasteur pipette from the test tubes containing shoot tips and add approximately 0.25 ml of 20% PVS2 at room temperature. Swirl the tubes to mix the solution with any residual liquid left in the test tube. After 60 min, drain this solution, add about 0.25 ml of ice cold 100% PVS2 solution and swirl. Place tubes in an ice bath. After about 10 min in this solution, transfer, using a long-tipped pasteur pipette, shoot tips and a portion of the PVS2 solution to a semen straw that was previously heat sealed at one end. Add the solution such that there is no air pocket at the sealed end. Heat seal the other end of the straw (usually a small air pocket remains). The total time in the 100% PVS2 solution is 20 min prior to exposure to LN (or to dilution in the case of controls exposed to just the vitrification solution). Practice transferring shoot tips with liquid to straws to assure that you can load all the straws within the 10-min time frame. Be careful not to draw shoot tips too far into the pasteur pipette during transfer or they may adhere to the glass and be difficult to expel. Straws can also be sealed with cotton or small polypropylene closures supplied by some manufacturers.

Plastic cryoampoules can be used in place of semen straws, although their use slows the cooling rate during immersion. Be sure that the tubes are specified for use with LN. Include about 1 ml of vitrification solution with the shoot tips in the cryoampoule. Since ability to form a glass within cells is cooling rate dependent, use of cryoampoules may result in less survival in some systems. Tests with mint lines have shown similar levels of survival using straws and cryoampoules.

An alternative method for vitrification of shoot tips is Method B. Shoot tips are cultured within 0.4–0.7 M sucrose for 1–2 days. No "loading" phase is used. Drain liquid from the tube with a pasteur pipette. Add the EG/sorbitol/BSA vitrification solution and incubate for 50 min at room temperature (e.g., potato) or for 15 min at 0 C (e.g., carnation). Load shoot tips into straws or cryoampoules as described in Method A above. Other species will require some modification of preculture duration, osmoticum concentration and exposure duration to vitrification solutions. Generally, addition of the vitrification solution at 0 C is more desirable.

*Cooling to low temperatures.*   The following safety precautions must be complied with during the exercise. You must wear safety glasses or a protective shield when using dewars and working with LN (filling containers, immersing materials, carrying, etc.) and when warming samples in case of an implosion of the dewar or explosion of a tube immersed in LN. Use insulated gloves when handling materials and wear a laboratory coat to protect arms. These safety concerns are extremely important!

Place semen straws or cryoampoules containing shoot tips and solution at about –160 C and hold for 30 min. This temperature is approximated by filling a wide mouth dewar flask about one third full with LN and suspending a screen slightly above the liquid level. A loose-fitting insulated lid is placed on the dewar. This step reduces the incidence of glass fracturing within the straw during subsequent transfer to LN. Keep the dewar loosely covered when not manipulating the straws. After 30 min in the vapor phase, immerse the straws directly into LN. Alternatively, if an LN refrigerator is available, store the straws directly in the vapor phase above the LN (ca. –180 C). Instead of a dewar flask, Styrofoam cups may be used to hold LN. Briefly surface disinfest the cup with a 70% ethanol solution, allow to dry and then add LN. The LN evaporates faster in a cup so be sure to keep the straws or cryoampoules within the LN. Samples should be stored in LN at least 30 min before warming. If necessary, samples can be stored until the next laboratory session, but be sure that they will not experience any warming.

*Warming.* Remove straws or cryoampoules from LN or vapor-phase storage using precooled insulated forceps. Hold in room temperature air for 10 sec for straws or 30 sec for cryoampoules prior to immersing it in sterile water at about 22 C. Keep in water for 1 min for straws or 2 min for cryoampoules. These times are long enough for the solution to become liquid and for the temperature to rise to about 0 C. Transfer of the vessel from LN directly into water often causes excessive glass cracking, which can reduce viability in some systems. Wipe excess water from the vial and briefly wipe with 70% ethanol. If cryoampoules are used, LN may seep into the vial; although infrequent, during warming the rapid expansion of the liquid to a gas may cause the tube to shatter or explode. Wear your safety glasses. Only thaw a few samples at one time; once they are warm the shoot tips are exposed again to vitrification solution and the potentially toxic conditions it imposes. Therefore, you must move quickly to the removal phase.

*Removal of vitrification solution and culture of shoot tips.* Cut the end of the straw with sterile scissors and expel the contents into 3 ml of dilution medium at room temperature. Flush the straw with diluent using a pasteur pipette if shoot tips adhere. For cryoampoules, transfer the shoot tips with about 0.25 ml of the vitrification solution to 3 ml of diluent. Remove shoot tips from diluent after 30 min and place on culture medium.

Dispense growth medium containing PGRs into $100 \times 15$ mm petri dishes and allow to solidify. Transfer shoot tips from dilution medium to the agar surface using a pasteur pipette. Remove all excess liquid from around the shoot tip. Be careful not to embed the shoot tip within the agar. Wrap petri dishes with Parafilm to retard moisture loss and store dishes under low light conditions. After 1 week, transfer dishes to normal light conditions. Transfer shoot tips at regular intervals (1–3 weeks) to fresh medium.

### Anticipated results

*Viability.* Observe whether or not the shoot tips are green immediately after plating. Make growth observations using a dissecting microscope periodically over 4–6 weeks. After LN treatment, the shoot tips should be green upon plating, but color may be lost within a day or so. Shoot tip expansion in controls often begins within 3 days, but vitrified shoot tips develop more slowly, often after 7 days or more. If they are brown immediately, they probably are not viable. Tips that slowly lose their color may still be alive and often redevelop color or show some signs of growth within 2 weeks. Transferring the shoot tips to fresh medium can enhance the growth rate.

*Experimental parameters to examine.* A number of variables can be identified in the experiment described above and could be examined to optimize or improve survival. When making comparisons, each treatment should comprise at least six replicate tubes of four buds per tube. A number of parameters are important. In order to test preculture of the isolated shoot tips, compare shoot tips that have been cultured in normal growth medium with those cultured in growth medium with 0.3 M sucrose for 1 or 2 days.

Examine a range of sucrose concentrations between 0.5 M and 0.7 M. To evaluate the effect of time in vitrification solution, which is toxic, examine times ranging from 15–60 min. Toxicity is a kinetic phenomenon, influenced by concentration, exposure time and temperature. Therefore, the temperature of application of the full-strength vitrification solution may be examined, for example using 0 C and 22 C applications. To test whether loading with the 20% PVS2 is necessary, compare samples incubated with 20% PVS with ones not exposed to it. Most other factors are somewhat minor, but may be important for some species or cultivars. These factors include time in the dilution medium and concentration of sorbitol in the dilution medium. Be sure to include controls to determine whether or not the dilution medium itself is injurious.

### Questions

* What advantages are there to storing germplasm in LN vs. tissue cultures or as plants?
* What is the purpose of the cryoprotectant? Could distilled water be substituted?

## Experiment 2. Cryopreservation using the encapsulation/dehydration method of vitrification.

### Materials

Encapsulation media are required. A liquid medium is prepared containing 3% sodium alginate (medium viscosity) in a modified MS minerals and vitamins solution (see chapter two for composition) in which calcium is omitted. A second medium contains a modified MS minerals and vitamins solution with a final concentration of 100 mM $CaCl_2$. Both media contain 102.6 g/L (0.3 M) sucrose, but could be prepared with higher concentrations.

### Procedure

This experiment is depicted diagrammatically in Figure 34.2. Excise shoot tips and briefly hold in growth medium until all shoot tips are harvested. If the shoot tips cannot be used the same day, they may be held overnight in growth medium with 0.3 M sucrose. Drain shoot tips well and resuspend in the alginate solution. Draw a single shoot tip with some liquid into the pasteur pipette. Form beads by dropping a small volume of the alginate medium containing a single shoot tip into the 100 mM calcium medium (within a beaker). The size of the drop can be adjusted by using pipettes of different internal diameter (bores); 3–5-mm-diam. beads are desired. Retain beads in the calcium medium for 30 min to allow them to gel completely.

Remove beads from the calcium medium and place them into a flask containing MS medium containing 171 g/L (0.5 M) sucrose and shake overnight. Use about 90 ml of the sucrose medium for each 10 ml of bead volume. The next day, transfer beads to MS medium with 239.4 g/L (0.7 M) sucrose and again shake overnight. The following day, blot the beads to remove excess moisture and place on a clean sterile surface (such as the top of a glass petri dish) within a laminar flow hood. Dry the beads in the air flow of the laminar hood for about 3–5 h. Beads may also be dried in a closed container with silica gel, but this method is somewhat slower.

Because drying rates vary with bead size and humidity, it is desirable in a preliminary experiment to measure changes in bead moisture content with time of drying in the laminar flow hood. You do not need shoot tips within these beads. You should also include some shoot tip–free beads during each experiment, such that when you remove a sample, you can determine percent moisture. Moisture content is determined gravimetrically; beads are heated overnight at 85–90 C and reweighed the next day.

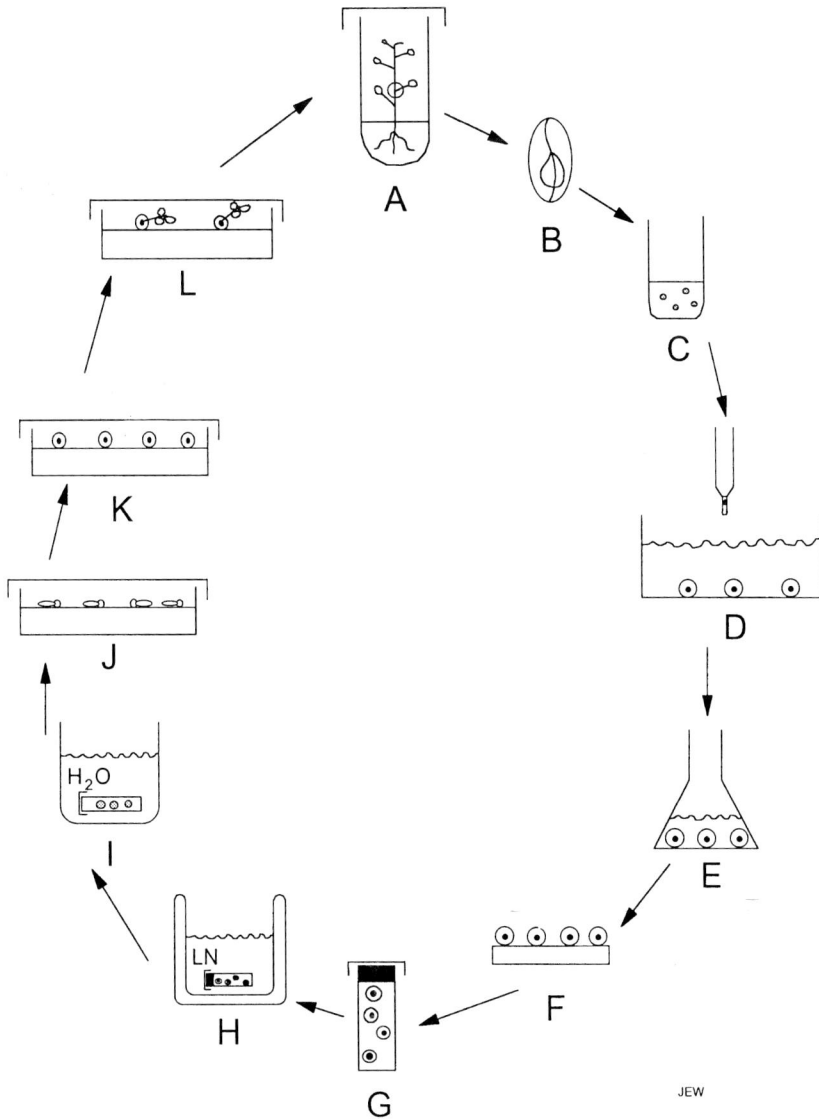

**Figure 34.2**   This schematic illustrates the steps performed in using encapsulation dehydration and cooling as a method for cryopreserving shoot tips. A stock plant (A) is selected and shoot tips (circled area in A) are excised (B) and placed in a calcium-free alginate solution (C). Shoot tips are dropped into a calcium solution (D) to form the bead. Beads are incubated in sugar solutions (E) and then partially desiccated (F). Beads are placed in cryoampoules (G) and the cryoampuoles directly immersed in LN (H). Warming is by immersion in water (I). Beads are directly cultured on agar (J) during which the bead rehydrates (K). When sufficient growth has occurred, shoots can be transferred to normal growth medium (L).

*Cooling and warming.*   Remember to follow the same safety precautions for working with LN listed previously. Use beads that have been dried to a moisture content of about 15–20% (wet weight basis). Place five beads in a sterile cryoampoule and immerse directly into LN. The ampoule will float in the LN. After a minimum of 1 h, retrieve vials and thaw in sterile 30 C water for 2–4 min. Wipe the excess water from the vial and briefly wipe with a 70% ethanol solution.

*Dilution and culture.*   Remove beads from the cryoampoule and directly place on a growth medium with a somewhat elevated sucrose level, approximately 68.4 g/L (0.2 M) sucrose. Culture the beads so that the shoot tips are close to the agar. Culture in dim light for 4–7 days. Then transfer the beads to a medium with the usual 2% sucrose. After about a week, remove the shoot tips from the beads and culture on fresh growth medium. For some species, the shoot tip does not need to be removed from the bead for normal development to occur and/or it may be beneficial to include 1% (10 g/L) activated charcoal in the solidified culture medium.

*Experimental parameters to examine.*   Several variables are important for optimizing survival. The major ones to examine are the preculture phase, including time in the sucrose solution and concentration of sucrose, and the time for dehydration of the bead in the air flow. Time for drying is empirically determined; plunge beads of about 25, 22, 18 and 15% moisture into LN to determine the best moisture level. In most cases rapid cooling is desired, but for some species slower cooling (0.25–1 C/min) is beneficial.

   *Questions*

   * Why is it necessary to rapidly cool the beads?
   * How does moisture level affect survivability of meristem tips stored in LN?

## Literature cited

Fabre, J. and J. Dereuddre. 1990. Encapsulation–dehydration: a new approach to cryopreservation of *Solanum* shoot tips. *Cryoletters.* 11:413–426.
Sakai, A., S. Kobayashi and I. Oiyama. 1991. Survival by vitrification of nucellar cells of navel orange (*Citrus sinensis* var. *brasiliensis* Tanaka) cooled to –196° C. *J. Plant Physiol.* 137:465–470
Steponkus, P.L., R. Langis and S. Fujikawa. 1992. Cryopreservation of plant tissues by vitrification. In: *Advances in Low-Temperature Biology*, Vol. 1. P.L. Steponkus, Ed., pp. 1–62. JAI Press Ltd., Greenwich, CT.

*chapter thirty-five*

# Secondary product expression in vitro

*Mary Ann L. Smith*

The accumulation of secondary products in plant cell cultures in response to deliberate treatments to the chemical or physical microenvironment provides students with striking and explicit experimental evidence to confirm basic physiological principles. Secondary products, such as vivid pigments, aromatic compounds, flavors and bioactive phytochemicals, can now be successfully accumulated in many undifferentiated plant cell cultures (callus and suspensions). In light of the current consumer demands for natural food products and plant-derived medications, in vitro production of valuable secondary products has become an industrially promising venture. Metabolite production in vitro also provides an excellent forum for in-depth investigation of plant metabolic pathways under highly controlled conditions. In the past, many tissue culture courses have concentrated on in vitro propagation strategies and breeding/selection, including biotechnology. However, this overlooks the entire realm of in vitro work concerned more with valuable plant products rather than with plant production.

Production of secondary products from plant tissue culture is based on the premise that the same valuable product produced in nature in an organ, fruit or other tissue from a plant can be stimulated to accumulate in undifferentiated cells. Researchers have succeeded in producing valuable secondary phytochemicals in unorganized callus or suspension cultures.

In contrast to conventional selection tactics, there may be no correlation between the level of product found in the whole plant and the level that can be induced to accumulate in tissue culture. A cranberry (*Vaccinium macrocarpon* Ait) cultivar selected in the field for intense color may not produce much pigment in callus, whereas a cranberry that is characterized by duller color in vivo may accumulate a greater concentration of complex anthocyanins in the callus stage.

In some cases where the metabolite is only produced in specialized plant tissues or glands, production has not been successfully stimulated in cell culture. A prime example is ginseng (*Panax ginseng* C.A. Meyer). Since saponin and other valuable metabolites are specifically produced in ginseng roots, root culture is required in vitro. As another example, biosynthesis of lysine to anabasine occurs in tobacco (*Nicotiana tabacum* L.) roots, followed by the conversion of anabasine to nicotine in leaves. Callus and shoot cultures of tobacco can produce only trace amounts of nicotine because they lack the organ-specific compound anabasine. In other cases, at least some degree of differentiation in a cell culture must occur before a product can be synthesized, e.g., vincristine or vinblastine from *Catharanthus roseus* (L.) G. Don, madagascar periwinkle. Reliance of a plant on a specialized structure for production of a secondary metabolite, in some cases, is a mechanism to keep a potentially toxic compound sequestered. One goal of genetic engineering research is to provide the gene encoding a limiting enzyme to all tissues in vitro in order to uncouple synthesis of secondary metabolites from differentiation.

## Current trends for in vitro secondary metabolite production

There is a series of arguments in favor of producing a valuable secondary product in plant cell culture, rather than in vivo in the whole crop plant. Plant cell culture eliminates potential political boundaries or geographic barriers to production of a crop, such as the restriction of natural rubber production to the tropics or anthocyanin pigment production to climates with high light intensity. When a valuable product is found in a wild, rare or scarce plant species, intensive cell culture is a practical alternative to wild collection of fruits or other plant materials. Secondary products in plant cell culture can be generated on a continuous year-round basis (no seasonal constraints), production is reliable and predictable (independent of ambient weather) and, in at least some cases, the yields per gram fresh weight may exceed those found in nature. Plant cell lines can be recurrently selected to amplify the productivity of the cell culture. Shikonin production in vitro at levels over 800 times that available from plant root is a case in point. Disagreeable odors or flavors associated with the crop plant can be modified or eliminated in vitro. Extraction from the in vitro tissues is much simpler than extraction from organized, complex tissues of a plant. Plant tissue culture techniques offer the rare opportunity to tailor the chemical profile of a phytochemical product by manipulation of the chemical or physical microenvironment to produce a compound of potentially more value for human use.

The three long-standing, classic examples of commercially viable production of a secondary metabolite in vitro — ginseng saponines, shikonin and berberine — each feature products that have some diversified uses including medicinal applications. Ginseng is produced in large-scale root cultures, whereas the other two products are produced in highly colored cell cultures. Tremendous research and development effort has advanced a number of other in vitro–derived secondary products to semicommercial status, including vanillin and taxol production in cell cultures. In a myriad of other cases, the in vitro processes for secondary metabolite production have fallen far short of expectations and have never approached commercializable status. Still, the arena of secondary product formation in cell cultures remains as an industrial pursuit. Engineers and biologists are currently joining forces on a global scale to develop new strategies for streamlining the critical bioprocesses (Smith et al., 1993). Research efforts on a broad range of plant cell culture–derived extracts can be cited in each of these major product categories: flavors (onion and garlic, peppermint and spearmint, fruit flavors, chocolate aroma, seaweed flavors, vanilla, celery, coffee, spice, sweeteners, etc.); edible colors for foods and medicines (mainly betalains and anthocyanins); non-food pigments for cosmetics and textiles (shikonin, berberine and various other products); several examples of fragrances, essential oils; and bioactive natural insecticides and phytoalexins useful in current integrated pest-management programs. Of course, intensive activity has centered on production of natural drugs or chemoprotective compounds from plant cell culture. Some of the most prominent pharmaceutical products in this latter category include ajmalicine (a drug for circulatory problems) from *C. roseus* and taxol (a phytochemical effective in treatment of ovarian cancer) from *Taxus* spp. (yews).

Because different levels of secondary metabolite production can be found within a cell line (e.g., sectoring of colored and noncolored callus in a culture), significant headway in the productivity of plant cell cultures has followed on the heels of intensive selection for high-producing cell lines. Single cell cloning and recurrent selection of the most-pigmented cells and sectors have significantly enriched shikonin- and anthocyanin-producing cultures, for example.

Another target for research has been the development of improved instrumentation for cultivation of plant cells en masse. Because intensive cultivation of plant cells differs from that of animal or microbial cells on a number of key fronts, the innovations have revolved around specific adaptations for the unique needs of plant cells in vitro. For example, improved bioreactors have been specifically designed to permit light to reach plant cells during division and product accumulation. In many cases, the infiltration of light at optimum intensities and spectral quality is prerequisite to successful synthesis and maximal expression of secondary metabolites. Modifications in aeration, agitation, and nutrient supply have similarly led to enhanced plant product yields in recent years. Elicitors (compounds that "stress" cells, leading to formation of a secondary product) from biotic and abiotic sources have been added to culture media to stimulate production, but this strategy depends on a strong understanding of the biochemical pathway leading to secondary product synthesis. Another approach is biotransformation — the deliberate feeding of metabolic precursors to a cell culture that already contains the necessary enzymes to convert them to product.

## Pigments from cell culture

Natural pigments, including anthocyanins, betalains, shikonin and other pigmented phytochemicals, are conspicuously accumulated in some cell cultures. Some of these products are of interest commercially as replacements for synthetic food dyes, which have fallen "under the gun" recently as adverse health and safety concerns have emerged. Negative reports and new medical studies in the popular press have made synthetic additives a target for contention and have created a strong market demand for safe, natural ingredients in foods. As an added bonus, significant chemoprotective benefits (including cardioprotective and anticarcinogenic properties) have been attributed to natural anthocyanin pigments and their precursors (Kinsella et al., 1993). These health-enhancing properties are unrelated to the nutritive value of the plants. Even natural non-food colorants are sought as replacements for environmentally damaging synthetic dyes fabricated with toxic solvents and heavy metals. Cell culture of pigment-rich crop selections presents a potentially more efficient alternative to extraction from the intact plant in nature (Stafford, 1991).

In many ways, natural plant pigments are ideal target compounds for laboratory or demonstration experiments on metabolite production for plant tissue culture classes because their accrual in vitro is quick, visible, easy to detect and fairly simple to quantify. In practice, and for validation of research results, pigment accumulation is typically measured by extraction, then quantification/separation with TLC (thin layer chromatography), HPLC (high performance liquid chromatography), NMR (nuclear magnetic resonance) and/or absorbance (spectroscopy). However, in a classroom, when traditional methods of pigment quantification are prohibited by the lack of sophisticated laboratory analytical equipment and/or insufficient student expertise, the production of a pigment in vitro can still be very easily detected and quantified by visual means alone or by coupling visual observation with very simple absorbance measurements using a spectrophotometer.

Many of the best-known in vitro pigment production systems are difficult to demonstrate because they either require cultures derived from specific, hard-to-find explants or must be developed through laborious repeated subculture and selection of only highly pigmented sectors before conspicuous concentrations of pigment are routinely produced. The experimental model presented in this chapter is not limited by these constraints and, therefore, is readily adaptable to classroom research.

*Ajuga reptans* N. Tayl. (Bugleweed) and the related species *A. pyramidalis* L. 'Metallica Crispa' are both widely distributed ground cover plants that grow rapidly and spread via stolons (Voigt et al., 1983). *Ajuga* tolerates adverse conditions so well that it is considered an intractable weed when it oversteps its boundaries in the landscape, and stock plants can be maintained long term in a greenhouse with minimal care. *Ajuga* cultivars are typically propagated by division of mother plants. A few select cultivars of *A. reptans*, which feature purple or bronze foliage, include 'Purpurea', 'Giant Bronze' and 'Burgundy Glow'. The latter cultivar has a characteristic chimeral green, creamy white and dark pink variegated foliage, but produces unicolor bronze, albino or green variants from adventitious buds in vitro (Lineberger and Wanstreet, 1983). *Ajuga pyramidalis* 'Metallica Crispa', which is frequently confused with purple forms of *A. reptans* in the trade, is characterized by savoy-type, glossy, bronze/purple foliage.

For a number of reasons, selections in the genus *Ajuga* are particularly well suited for use in plant tissue culture exercises. Like many members of the mint family (Labiatae or Lamiaceae), microplants flourish in culture and may be held in cold storage successfully for up to a year, which helps cut back on the need for repetitive subculture maintenance between semesters. *Ajuga reptans* proliferates readily in vitro from axillary or adventitious buds, exhibits classic, predictable responses to gradient increases of cytokinin concentration and is a good, rapidly growing model plant on which to illustrate stages in an in vitro production cycle through acclimatization within a single semester (Preece and Huetteman, 1994).

The purple or bronze leaves that distinguish some varieties of *Ajuga* are readily expressed also in microcultured plants and even undifferentiated callus can exhibit pigment expression. Callebaut et al. (1993) reported that primary callus from flowers, petioles, leaves and stems of *A. reptans* developed into a fairly homogeneous, blue-colored cell culture after over a year of selection and subculture. These suspensions produced anthocyanin pigments in the dark, and pigment expression responded to changes in the light regime and to chemical medium treatments. In our alternative approach, vegetative disks are directly explanted from purple foliage of *A. pyramidalis* 'Metallica Crispa' and bronze variants of *A. reptans* 'Burgundy Glow' and induced to produce callus, which rapidly acquires a vivid purple hue when exposed to illumination (Juthangkoon et al., 1994). Intense, uniform pigment expression is routinely achieved without selection, and friable callus colonies can be explanted into liquid suspension culture. Cell division and pigment accumulation are not mutually exclusive processes; both cell biomass and pigment accumulation increase steadily throughout each subculture cycle. The growth and pigment expression is so predictable and consistent that this system was used as a model to test the application of video image analysis to regulate bioprocesses by visual characterization of cell size, shape, aggregation and color intensity (Smith et al., 1993).

Current publicity surrounding "functional foods" and natural/medicinally active phytochemicals is gaining momentum in the popular press, and a renewed interest in natural product synthesis is emerging. Since production of these valuable plant metabolites in vitro is a viable alternative production route with notable commercial advantages, it is likely that industry and research opportunities will continue to expand. Students can be introduced to the natural products research arena through the laboratory exercises featuring *Ajuga* pigments in cell culture (see chapter thirty-six). These exercises have been repeatedly practiced by novice plant tissue culture students with satisfying and predictable outcomes. This experimental model is also amenable to individualized, student-directed experimentation with a vast range of physical or chemical microenvironmental parameters (treatments) which can be deliberately varied to determine the effects on metabolite accumulation. The experimental results also help students to appreciate the value of chemical analytical methods for measuring metabolite production as an adjunct to strict visual gauges for pigment production in a callus or suspension culture.

# Literature cited

Callebaut, A., M. Decleire and K. Vandermeiren. 1993. *Ajuga reptans* (Bugle): in vitro production of anthocyanins. In: *Medicinal and Aromatic Plants* V. Y.P.S. Bajaj, Ed., pp. 1–22. Springer-Verlag, Berlin.

Juthangkoon, S., D.L. Madhavi and M.A.L. Smith. 1994. Effect of chemical microenvironment on anthocyanin production in cell cultures of *Ajuga reptans*. *Eighth Intl. Tissue Culture Cong.*, Firenze, Italy, June 12-17, 1994. Abstract #M-34.

Kinsella, J.E., E. Frankel, B. German and J. Kanner. 1993. Possible mechanisms for the protective roles of antioxidants in wine and plant foods. *Food Technol.* 47:85.

Lineberger, R.D. and A. Wanstreet. 1983. Micropropagation of *Ajuga reptans* 'Burgundy Glow'. Ornamental plants — 1983: a summary of research. *Ohio State Univ. Wooster Ohio Res. Circ.* 274:19-22.

Preece, J.E. and C.A. Huetteman. 1994. An axillary shoot proliferation laboratory exercise — micropropagating *Ajuga reptans* L. *HortTechnology.* 4:312–314.

Smith, M.A.L., J.F. Reid, A. Hansen, Z.-W. Li and D.L. Madhavi. 1993. Vision characteristics for analysis of pigment-producing cell cultures. American Society of Agricultural Engineers, ASAE Meeting Paper #933513.

Stafford, A. 1991. Natural products and metabolites from plants. *Plant Cell and Tissue Culture*. A. Stafford and G. Warren, Eds., pp. 1–22. Milton Keynes, Open University Press, Buckingham, U.K.

Voigt, T.B., B.R. Hamilton and F.A. Giles. 1983. Ground covers for the Midwest. pp. 78–79. Univ. Ill. at Urbana-Champaign Coll. Agric. Spec. Pub. #65.

*chapter thirty-six*

# Pigment production in Ajuga cell cultures

*Mary Ann L. Smith*

The following laboratory exercises provide clear-cut demonstrations of secondary metabolite (anthocyanin pigment) production in unorganized (callus and suspension) cultures of *Ajuga pyramidalis* L. 'Metallica Crispa'; however, bronze and purple cultivars of the related species *A. reptans* N. Tayl. can be substituted and will also perform dependably. In each case, the *Ajuga* genotypes are readily available, and anticipated responses are predictable. Experiment 1 (the main exercise) explicitly illustrates that *Ajuga* callus (originating from leaf explants) is capable of expressing pigment in greater intensity/concentration than differentiated (leaf) tissues. The exercise provides a graphic demonstration of the influence of the chemical microenvironment on metabolite accumulation, specifically, the effects of the carbon source (sugars) on pigment expression.

Experiment 2 demonstrates the influence of explant size and source on productivity and pigment expression in amorphous masses of callus. A few variants of the main experiment are also suggested, including physical microenvironmental treatments to intensify results. Pigment production can be noted in direct response to stress imposed in physical or chemical microenvironments for *Ajuga* callus cultures. Pigment accumulation in nature and in vitro is often a response to environmental stress, and consequently pigment expression in cell cultures can conspicuously mark the outcome of experiments that impose a chemical or physical pressure on cultured cells.

## General considerations

### Plant materials

*Ajuga pyramidalis* 'Metallica Crispa' is readily available at most garden centers as flats of stoloniferous, spreading plants ready to plant in the landscape. To avoid confusion, identified stock plants should be obtained from a reputable source (e.g., Midwest Groundcovers, St. Charles, IL), because many retail shops sell flats labeled as "Purple Ajuga", without clear distinction between *A. pyramidalis* 'Metallica Crispa' and the phenotypically similar bronze/purple selections of *A. reptans* (Voigt et al., 1983). *Ajuga reptans* and *A. pyramidalis* freely hybridize, resulting in much confusion in the trade. Again, although any of these selections will yield a good experimental response in the exercises below, it is important to make sure that all stock plants are of a single genotype to avoid a source of uncontrolled variation unrelated to treatment effects and to be aware that adventitious shooting in a chimeral selection (e.g., *A. reptans* 'Burgundy Glow') may result in production of green or albino shoots not suitable for the exercises.

Stock plants are planted on 5-in. centers in a 1:1:1 mix (peat:perlite:soil) in a shallow flat and fertilized once per week with a 20:20:20 fertilizer. Only natural light (no supplemental lighting) is used in the greenhouse, which is maintained at 25–28 C. Both greenhouse plants and microcultured stock plants can be used as explant sources in these exercises. Approximately two to three healthy plants per student should be maintained to ensure sufficient new foliage for explanting trials. Greenhouse plants require low maintenance and may be divided and replanted approximately once per year to maintain stock. However, maximal success in explanting from the greenhouse is achieved when young, vigorously growing new leaves from new transplants are used as donor material.

Microcultured stock plants are initiated by explanting meristem tips (devoid of all leaves) or nodes from runners, each about 1–2 cm long. Any explant material taken from the greenhouse stock plants should not be in contact with the soil; runners or plants overhanging the sides of the flats are ideal for this purpose. Alternatively, *Ajuga* plants can be potted and placed in a growth chamber to avoid any overhead watering in the greenhouse until new growth is initiated; then explants can be taken from the new growth. Surface disinfest the explants by treatment in 15% commercial bleach with 0.1% Tween 20 (a few drops per liter). Agitate explants in the bleach solution for 15 min followed by three rinses in sterile, double-distilled water. After removing injured tissue, the explants can be placed in $150 \times 25$ mm culture tubes containing 10 ml of basal media composed of Woody Plant Medium (WPM) basal salts (Lloyd and McCown, 1981; see chapter two for composition) amended with 30 g (88 mM) sucrose, 1 g (0.55 mM) myo-inositol, 50 mg (0.28 mM) L-ascorbic acid, 0.5 mg (0.29 μM) thiamine HCl, 0.5 mg (2.44 μM pyridoxine HCl), 0.5 mg (4.06 μM) nicotinic acid, 2 mg (26.7 μM) glycine, 0.2 mg (1 μM) 2-iP, 150 mg PVP-T (polyvinylpyrrolidone-10) and 7 g of agar per liter (pH 5.7 prior to autoclaving). After cultures begin to produce new growth, four to five explants are subcultured into each GA-7 vessel (Magenta Corp, Chicago, IL) with 50 ml of basal medium. The microcultured plants are maintained indefinitely as shoot cultures by subculture of meristems every 4–6 weeks to fresh medium.

## Exercises

### Experiment 1. Pigment expression in Ajuga *callus and chemical microenvironmental effects*

The effect of the chemical microenvironment on secondary metabolite accumulation in a cell culture is well recognized, and medium composition is usually deliberately manipulated to enhance anthocyanin accumulation in vitro. While a number of parallels exist in different pigment production systems, determining the optimum chemical regime still requires empirical testing to some extent. Several research teams have realized analogous results when testing the effects of the culture medium composition on pigment expression, and these trends can be demonstrated readily in classroom experiments. For example, the culture medium that best supports cell growth may not be the same as the medium composition that induces the most intense pigment production (Zwayyed et al., 1991). Several in vitro pigment production systems for callus or suspension cultures are characterized by dual phases in the culture protocol; usually a "growth phase" followed by a "product accumulation phase" on a production medium that favors accumulation of pigment. For some of these systems, in the latter stage, the cell cultures have reached the status of little or no new growth during product accumulation. However, there is no question that the microenvironmental conditions imposed on cells during the growth phase exert substantial control over secondary metabolite production in the latter stage.

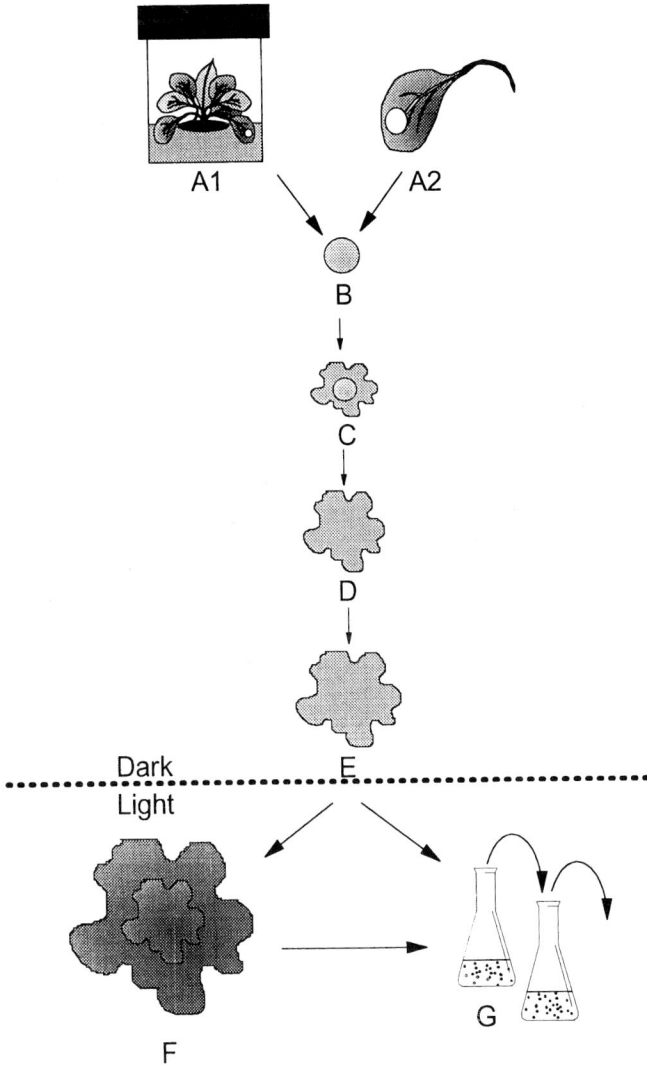

**Figure 36.1** Schematic representation of the pigment production protocol for *Ajuga*. (B) Explants obtained from either presterile microcultured plants (A1) or from greenhouse or growth chamber specimens (A2). C. Callus is excised from mother tissue and cultured independently for two additional rounds (D) and (E) of culture growth in the dark. E. Friable, ivory-colored callus is gently lifted from the callus induction medium, subcultured to pigment production medium, and placed on illuminated culture room shelves. F. Friable, colored callus may be maintained on solid pigment production medium for subsequent subcultures or introduced into liquid suspension cultures (G).

Anthocyanin production in vivo and in vitro can be provoked by a great number of environmental factors. Elevated levels of sucrose or other sugar (as a carbon source and/ or osmotic agent), a change in the nitrate to ammonium ratio, and changes in the plant growth regulator (PGR) composition are frequently needed in the production phase of a two-phase culture system to stimulate pigments in vitro.

This experiment was designed to illustrate the effects of carbon source and level on anthocyanin production in callus (Figure 36.1). The exercise can be efficiently performed by pairs of students and uses only in vitro microplants (shoot cultures) as the source of explants for callus induction.

This experiment is described from explanting (initiation of callus from vegetative leaf disks) through a final phase of pigment accumulation in callus colonies. Alternatively, students can begin with precultured callus already growing in the dark (Figure 36.1) to shorten the exercise. When starting with dark-grown callus, students can proceed immediately to the subculture of callus to different treatments of the production medium. The same experiment can be performed by subculturing the white, friable callus to liquid production media without agar and producing pigment in suspension culture.

### Materials
The following items are needed for each team:

* One to two shoot cultures of *Ajuga* in GA-7 vessels
* 15 GA-7 vessels containing a callus induction medium, which substitutes 2.3 μM 2,4-D and 1.0 mg/L 3.5 μM kinetin as the PGRs
* 10-mm cork borer
* Eight GA-7 vessels each containing experimental pigment production media, all with 2.26 μM IAA and 3.49 μM zeatin, but modified so that two vessels each contain either 10, 30 or 70 g/L glucose or 30 g/L galactose.

### Procedure
Leaf disks approximately 1 cm in diameter are cut from young, fully expanded leaves of the stock microculture plants with a sterile cork borer. Five identical leaf disks are placed immediately on the surface of the medium containing 2,4-D and kinetin. Cut surfaces of the disks should be in direct contact with the medium. Five replicate vessels are prepared. Cultures are maintained in the dark for 3 weeks, then fresh ivory white, friable callus is removed from mother tissue and cultured every 2 weeks to fresh medium in the dark for two cycles.

Finally, after a total of 7 weeks, ivory white, friable masses of callus about 2.5 cm in diameter are gently removed from the induction medium and transferred to pigment production medium. Five clumps of colorless callus are inoculated to the surface of the medium in each vessel, and two replicate vessels are prepared for each of the four treatments. Sufficient callus should be produced in the final five vessels of callus induction medium to supply material for the eight vessels containing experimental pigment production medium. These cultures are placed on lighted shelves under a photosynthetic photon flux (PPF) of 150 μmol·m$^{-2}$·sec$^{-1}$ provided by cool white fluorescent lamps.

At the end of 2 weeks under lighted conditions, each treatment is first evaluated visually to assess the production of pigment. In each of the culture vessels, two of the callus masses (subreplicates) are removed and evaluated for fresh weight (FW) and dry weight (DW) analyses. The remaining subreplicate fresh callus masses are ground with mortar and pestle, blended with 10 ml of ice cold methanol acidified with 1% HCl and stored overnight at 4 C. Absorbance of the solution is then measured at 535 nm with an appropriate spectrophotometer, which provides the relative anthocyanin content between treatments.

### Anticipated results
After as little as four days under lights, callus in each of the 4 carbon source treatments should exhibit marked differences in anthocyanin production. The information given in Figure 36.2 is typical of the effect of sucrose concentration in the production medium on yield of anthocyanins after 2 weeks under lights (Juthangkoon et al., 1994). The galactose treatment will support cell growth as well as sucrose at the same concentration, as illustrated by similar increases in FW and DW, but will stimulate significantly greater antho-

**Figure 36.2** Dry weight and anthocyanin concentrations of *Ajuga* callus at various sucrose concentrations in the pigment production medium.

cyanin expression. Typically, 15–20% greater anthocyanin concentrations will be produced in the galactose treatment as compared with sucrose at 30 g/L. It is important that anthocyanin concentrations be compared on the basis of yield per treatment, rather than on a milligram anthocyanin per gram FW or DW basis, because different carbon sources and levels will support different callus growth rates as well as different levels of anthocyanin accumulation. The treatment containing low sucrose concentrations (10 g/L) will exhibit significantly less growth during the 2 weeks on pigment production medium. Callus in this treatment will be visibly duller because of less pigment concentration.

*Questions*

* What function does the carbon source (sugar) play in the pigment production stage of culture for *Ajuga* callus? Is the sugar acting mostly as a metabolic agent or as a stress-evoking (osmotic) agent at the 30 g/L concentration? What about at the 70 g/L concentration?
* Why was callus initiated in darkness? How might initiation of callus from explants in the light change the outcome of these experiments?

## Experiment 2. Explant size and source effects on in vitro pigment production

The feasibility of stimulating in vitro secondary metabolites from unorganized cells can hinge on the ability to produce uniform, prolific, rapidly growing callus cells, of good quality, from tissues of the target species. Explant size can influence the initiation and growth rate of parenchymous callus cells, the friability of the callus masses, and the propensity for the tissue to revert to regeneration (organogenesis) instead of callus production. Characteristics of the callus depend on a complex relationship between the explant tissue used to induce callus, the composition of the medium and the microenvironmental conditions.

In this experiment, a range of explant sizes from two sources (greenhouse plants and microcultured plants) are compared in terms of callus induction rate, overall callus production and the ability of the produced callus to express anthocyanins.

### Materials
The following items are needed for each team:

* One or two shoot cultures of *Ajuga* in a GA-7 vessel
* One or two greenhouse plants of *Ajuga*
* 10% sodium hypochlorite solution and volumes of sterile double-distilled water
* 250-ml screw-top glass jars (Bellco Biotechnology, Vineland, NJ)
* Cork borers (2, 5, 7 and 10 mm)
* 16 GA-7 vessels containing callus initiation medium (2.3 µM 2,4-D and 3.5 µM kinetin)
* Eight GA-7 vessels containing pigment production medium (2.26 µM IAA and 3.49 µM zeatin with 30 g/L sucrose)

### Procedure
The procedures for this experiment are the same as those given for experiment 1, except that the callus initially induced on the leaf disks is not subcultured and grown for additional cycles on the induction medium in the dark. Instead, after 4 weeks of culture in the dark, callus is excised from the mother tissue and evaluated for FW and DW. Duplicate replicates from each treatment, if not subjected to fresh mass measurement at this time, can be subcultured directly to pigment production medium and evaluated after only an additional 2 weeks in the light in terms of both callus growth and pigment expression. This sequence of steps provides for a simple way to evaluate source tissue effects on callus production within a shorter (6-week total) time frame.

For each size of cork borer, ten leaf disks are cut from microcultured plants and placed flat on the surface of callus induction medium so that the cut surfaces are in direct contact with the medium surface. Five disks are explanted into each GA-7 vessel, so that two replicate vessels are prepared for each disk-size treatment. Young leaves are similarly excised from greenhouse plants, surface disinfested and rinsed prior to cutting five leaf disks for each cork borer size. All cultures are placed in darkness for 4–5 weeks. At the end of this period, callus is separated from leaf tissue from half of the subreplicates (explants) in each treatment, and FW and DW determined as before. Remaining callus masses are separated from explant tissue, transferred to pigment production medium and placed on lighted shelves. Callus appearance is observed daily for color and, after 2 additional weeks, total pigment production can be evaluated as described previously, or by visual rating alone.

### Anticipated results

Callus initiation rate, overall productivity and tissue quality are all affected by the size and source of leaf disk explant. Callus growth should be directly proportional to explant size for both greenhouse and micropropagated explants. Small explants produce substantially less biomass within a typical culture period. Although more callus is eventually produced from greenhouse explants, the rate of initiation is slightly faster from microcultured explants. Uniform callus tissue should be apparent after only 3 days on 7–10-mm microcultured leaf disks, but callus appears only sporadically around the perimeter of the same-sized greenhouse disks, and not until the fourth or even fifth day after subculture. This result may be because any tissues damaged during surface disinfestation may grow less actively.

Pigment production is usually markedly better from callus generated on medium-sized greenhouse explants (5- or 7-mm disks) as compared with the same diameter disks from microcultured explants. Residual influence of the elevated levels of stored carbohydrates in greenhouse explants may be responsible for this response. Pigment intensity is not significantly different for any of the microcultured explant sizes, although significantly more color-producing biomass is generated from larger explant sizes. Although callus initiates rapidly from 10-mm explants from both the microcultured and greenhouse explants, pigment expression is not favored.

### Questions

* Why do explant size and source (in vivo stock plant vs. in vitro microplant) affect the productivity in a pigment production system?
* Pigment produced in a callus often has a simpler profile (a less complex anthocyanin structure) than pigment in tissues of the plant in vivo. How do you think pigments from *Ajuga* callus would compare with pigments found in leaves? Flowers?

### Further experimental variations

*Ajuga* callus growth and pigment expression are also quite responsive to several other deliberate treatments to the physical or chemical microenvironment. Alternating concentrations of osmotica in the callus induction and pigment production media, for example, can elicit striking differences in coloration. The color is significantly influenced by the wavelength of light incident at the surface of the callus. When placed under narrowband wavelength lamps, red light stimulates little to no pigment production in the callus, whereas blue light stimulates intense coloration. Ultraviolet wavelengths provoke early, intense anthocyanin accumulation, but the callus eventually becomes necrotic while parallel callus cultures under white or blue light are just beginning to develop pigmentation. Optimal lighting conditions and requirements are underestimated factors in most plant in vitro pigment production schemes, and a series of simple, elucidative student experiments can be centered around the influence of light in pigment accumulation in vitro.

## Acknowledgment

Many thanks to Dr. L.A. Spomer for constructing the schematic diagram of the experiment.

# Literature cited

Juthangkoon, S., D.L. Madhavi and M.A.L. Smith. 1994. Effect of chemical microenvironment on anthocyanin production in cell cultures of *Ajuga reptans*. *Eighth Intl. Tissue Culture Cong.*, Firenze, Italy, June 12-17, 1994. Abstract #M-34.

Voigt, T.B., B.R. Hamilton and F.A. Giles. 1983. Ground covers for the Midwest. pp. 78-79. Univ. Illinois at Urbana-Champaign Coll. Agric. Spec. Pub. #65.

Zwayyed, S.K., G.C. Frazier and D.K. Dougall. 1991. Growth and anthocyanin accumulation in carrot cell suspension cultures growing on fructose, glucose, or their mixtures. *Biotechnol. Prog.* 7:288-290.

*chapter thirty-seven*

# *In vitro plant pathology*

## C. Jacyn Baker, Norton M. Mock and Elizabeth W. Orlandi

Tissue culture has been an essential tool in the field of plant pathology for many decades. One of the first uses of tissue culture by White and Braun (1941) cultured tumors induced by the pathogen *Agrobacterium tumefaciens* on sunflower that were devoid of viable bacteria. Peter Carlson (1973) was the first to demonstrate that tissue culture could be used to select for pathogen resistance in plants. Using a structural analog of the fireblight toxin, he selected resistant cells, developed callus and, eventually, regenerated whole plants that were resistant to fire blight. In 1986 Roger Beachy's laboratory successfully incorporated a coat protein of tobacco mosaic virus into tobacco cells, endowing the plants with resistance to subsequent viral infections and significantly delaying disease development (Abel et al., 1986). More recently, protoplast fusion techniques have facilitated the transfer of resistance between plant species (Garg et al., 1988). Many of these tissue culture techniques have been discussed in detail in previous chapters.

In addition to providing novel methods to develop disease-resistant plants, tissue culture has become an important tool in the study of the molecular aspects of plant/pathogen interactions. Plant cells react to environmental and pathogen-related stresses with a variety of responses (i.e., changes in ion fluxes across the plasmamembrane, increased synthesis or activation of specific enzyme systems and increased production of defense-related products). The use of plant cell suspension cultures or individual protoplasts has enabled easier and more-accurate monitoring of these cellular processes at the biochemical level. Many of these important biochemical processes cannot be monitored in whole plants. Unlike intact plants, plant suspension cultures provide relatively homogeneous populations of cells that are immediately accessible to treatment with pathogens or pathogen-related products.

This chapter will focus on the use of suspension cells to monitor plant cell responses to pathogenic bacteria. The ability of plants to recognize and respond to external stimuli, such as invading pathogens, is essential for their existence in nature. Successful recognition of a potential pathogen triggers various inherent defense mechanisms and in many cases results in a hypersensitive response (HR). The HR results in rapid plant cell death (9–24 h after inoculation) that essentially localizes the pathogen and prevents its spread throughout the plant. Pathogens that trigger the HR in a particular plant species or cultivar are considered to be "incompatible" on that plant. "Compatible" pathogens avoid recognition by the plant and, therefore, do not trigger the HR, resulting in more extensive invasion of plant tissues and, ultimately, disease symptoms.

Studies on the bacteria-induced HR have identified early plant responses that indicate recognition of the pathogen (Atkinson et al., 1985; Baker et al., 1991; 1993). One of these responses is the $K^+/H^+$ exchange response involving a net uptake of $H^+$ from the extracellular space and a corresponding efflux of $K^+$ from the plant cell. This is a reversal of the

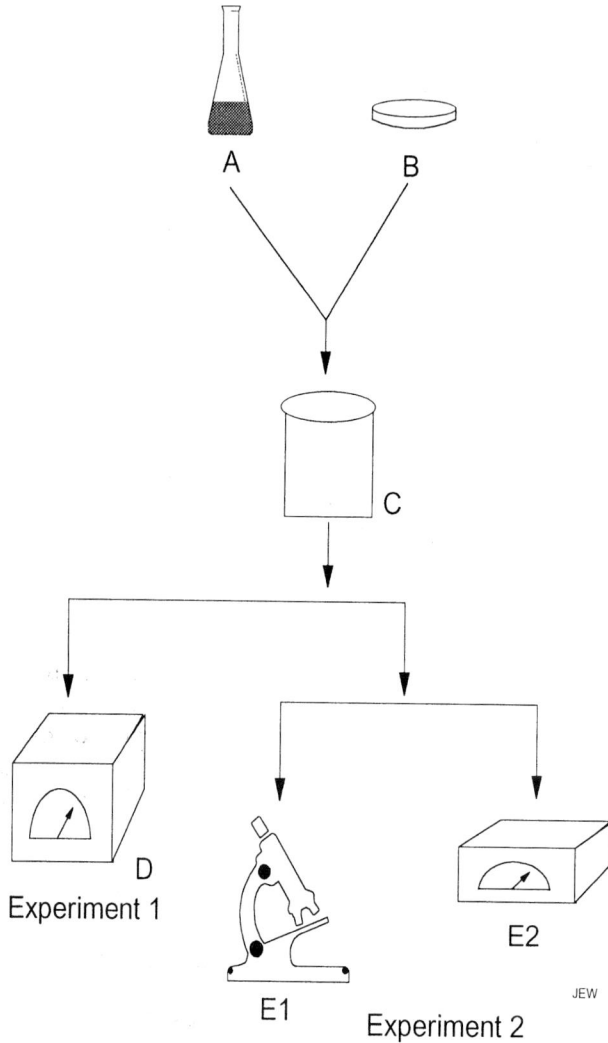

**Figure 37.1** Flowchart of host/pathogen interaction bioassay experiments. Cells from suspension cultures of soybean and tobacco (A) and (B) are treated with bacteria (C) and changes in [K+] and pH are monitored in D or experiment 1. In experiment 2, estimations of cell death by microscopic observation (E1) as well as spectrophotometric assay (E2) are made.

normal cell ion transport in which H+ is pumped out by ATPase activity and K+ is taken up, being the major inorganic ion within the cell. The K+/H+ response can be measured in disks cut from whole leaves, but it is more easily and rapidly monitored in plant cell suspensions. The K+/H+ exchange response precedes hypersensitive cell death by several hours in incompatible interactions.

The following laboratory exercises will focus on the measurement of the K+/H+ response and the subsequent hypersensitive cell death of suspension cells inoculated with bacteria (Figure 37.1). The experiments have been designed to demonstrate that cell suspensions can be used to monitor plant recognition responses, such as the K+/H+ response, which is specific for recognition of incompatible pathogens and whose activation is followed some hours later by hypersensitive cell death.

# General considerations

## Maintenance of plants

Tobacco and soybean plants are used for initiating suspension cell cultures and for testing the pathogenicity of bacteria used in subsequent suspension cell assays. Both tobacco and soybean plants should be grown under greenhouse conditions of approximately $28 \pm 5$ C and 12 h of light. Sow tobacco, *Nicotiana tabacum* L. 'Hicks', in moistened vermiculite, cover with a thin layer of dry vermiculite and moisten, being careful not to disturb the seeds. Seedlings will emerge within 1 week and should be thinned to approximately 1 in. apart. After 3–4 weeks the seedlings will be 1–2 in. tall and should be transplanted to 10-cm-diam. pots using Promix supplemented with osmocote 14-14-14 (both available through E.C. Geiger Co., Harleysville, PA). Once established (about 2–3 weeks later), plants are transplanted to 20-cm-diam. pots in Promix supplemented with osmocote. Tobacco plants should be watered only when the soil is dry since they are prone to root rot. It takes about 9 weeks from seeding to the time plants are ready to be used for most of these experiments.

Soybean (*Glycine max* L. Merr. 'Mandarin' or 'Harosoy 63') seeds are planted 5 cm deep in Promix. Cover these pots with dark plastic to facilitate uniformity in germination. Remove the cover after approximately 2–3 days or once seedlings have emerged. Soybean plants should be ready to use approximately 12 days after sowing.

## Preparation and maintenance of plant suspension cells

Tobacco suspension cells are initiated from callus derived from the pith of the plant. Remove a 10-cm section from the growing tip of a tobacco plant and cut off all of the leaves. Immerse the section into 95% ethanol for 15 sec. Using a sterile scalpel, cut off and discard the tip. Next, peel away and discard the outer vascular tissue to expose approximately 6 cm length of the pith. Place the exposed pith into a sterile petri dish, section into 5-mm portions and place sections on Linsmaier and Skoog (1965) medium (LS) solidified with purified agar, such as Agar Noble (Difco, Detroit, MI), and amended with 0.5 mg/L (2.2 μM) 2,4-dichlorophenoxyacetic acid (2,4-D). For composition of LS and other media mentioned, see chapter two. Incubate the pith cultures at 27 C in the dark. Callus will begin to grow within 1 week. Once callus is established, in about 1 month, transfer it to Schenk and Hildebrandt (1972) solidified medium (SH) amended with 0.5 mg/L 2,4-D and, thereafter, routinely transfer each month. The callus will proliferate and can be transferred in small chunks (8–10 mm diam.) into 40 ml of liquid SH medium contained in a 125-ml culture flask fitted with a sterile sponge closure (Bellco Biotechnology, Vineland, NJ). The flasks should be placed on a rotary shaker in the dark at 27 C and shaken at 150 rpm. Within a couple of weeks cells will have proliferated, cultures will have thickened and now can be transferred routinely. A typical transfer routine consists of aseptic transfer every 4 days of 15 ml of culture into 75 ml of fresh liquid SH medium contained in a 250-ml flask. It will take approximately 2 months to establish suspension cells beginning with the initiation of callus from pith tissue.

Soybean suspension cells are started from callus that is initiated from the hypocotyl. Surface disinfect soybean seeds with 95% ethanol for 1 min and then rinse with sterile distilled water. Soak the seeds in 25% commercial bleach (diluted with sterile distilled water) for 5 min. Rinse seeds well three times with sterile distilled water. Place seeds in a sterile petri dish that contains filter paper moistened with sterile distilled water. Do not seal the petri dishes as this will reduce gas exchange and negatively affect germination. Place the petri dish into a glass dish with moistened paper towels and cover with plastic wrap to hold in the moisture. Incubate in the dark at 27 C for 4–5 days. Once the seeds have

*Table 37.1* Recommended Pathavars of *Pseudomonas syringae* for HR Tobacco and Soybean Experiments

| Bacteria | Tobacco | Soybean |
|---|---|---|
| Compatible bacteria (Disease causing) | *P. syringae* pv. *tabaci* | *P. syringae* pv. *glycinea* race 4 |
| Incompatible bacteria (HR causing) | *P. syringae* pv. *tomato* *P. syringae* pv. *syringae* 61 *P. syringae* pv. *glycinea* race 4 | *P. syringae* pv. *tomato* *P. syringae* pv. *syringae* 61 *P. syringae* pv. *tabaci* |

germinated, aseptically excise the green hypocotyl, 1–2 cm below the cotyledon, and cut into 2–4-mm sections. Place these sections on Gamborg's B-5 solidified medium (Gamborg et al., 1968) supplemented with 1.0 mg/L (4.5 µM) 2,4-D (1B-5 medium) and incubate in a dark incubator at 27 C. Callus will begin to form within a few days. In about 2–3 weeks transfer callus to fresh 1B-5 solidified medium, then transfer routinely every month. Once callus is established, place small chunks (8–10 mm diam.) into 40 ml liquid 1B-5 medium contained in 125-ml flask with a sponge closure and incubate on shaker at 27 C and 150 rpm in the dark. Once cells have proliferated and the cultures have thickened, suspension cells can be routinely transferred aseptically every 4 days by diluting 1:1 with fresh medium for a total of 80 ml per 250-ml flask. Soybean suspension cells require approximately 2 months for establishment from excised hypocotyl sections.

## Maintenance of bacteria

The bacteria that are recommended for use in these experiments are listed in Table 37.1. Some of these isolates can be purchased from ATCC (Rockville, MD) or obtained from the authors with the proper permits. All bacteria isolates are aseptically loop inoculated onto appropriate agar medium in petri dishes, sealed with Parafilm, and allowed to grow 1–2 days in an incubator set at 30 C. Petri dishes are then stored at 4 C, and fresh cultures are prepared at monthly intervals. Bacteria listed in Table 37.1 are all grown on Kings B agar (KB) consisting of 20 g proteose peptone #3 (Difco) 1.5 g $K_2HPO_4$, 6.14 g $MgSO_4 \cdot 7 H_2O$, 15 ml glycerol, 20 g Bacto agar (Difco), brought to 1 L with distilled water.

## Initial testing of bacteria isolates by inoculation into whole plants

It is essential to test the bacterial isolates on plants prior to bioassays to ensure that their pathogenicity is properly characterized. This is best done by infiltrating healthy plants with a $2 \times 10^8$ colony-forming units (cfu)/ml bacterial suspension (bacteria are prepared as described below in Preparation of bacteria for bioassay). Prick the leaf with a tuberculin syringe needle. Place the barrel of a needleless syringe containing bacteria against the hole in the leaf and, with one finger on the other side of the leaf, gently infiltrate. Lightly outline and label the infiltrated water-soaked area with a permanent marker on the underside of the leaf. The leaf areas that were inoculated with incompatible bacteria should turn necrotic overnight (Figure 37.2a and c), whereas the areas inoculated with compatible bacteria should be symptomless or display yellowing over the next few days (Figure 37.2b and c). For tobacco, best results are achieved when the third fully expanded leaf from the top is inoculated; for soybean, inoculate the fully expanded primary leaves. It should be noted here that soybean leaves are more difficult to infiltrate than tobacco leaves and the infiltrated leaf area of soybeans will be considerably less than that of tobacco leaves.

**Figure 37.2**   Typical symptoms produced on tobacco (a) and (b) and soybean (c) by incompatible (I) and compatible (C) bacterial pathogens infiltrated into leaf tissue.

## Exercises

### Experiment 1. Correlation of the $K^+/H^+$ response with the incompatible plant/bacteria interaction

As discussed previously, the $K^+/H^+$ exchange is a recognition response indicating that the plant has detected the presence of an incompatible pathogen. In most cases this will result in the triggering of plant defense mechanisms resulting in a resistant reaction. In this experiment, the increase in external pH and $[K^+]$ will be monitored using pH and conductivity electrodes, respectively. The response should begin approximately 1.5 to 3 h after inoculation with incompatible bacteria. Previous studies have demonstrated that the increase in conductivity is due to $K^+$ efflux and not due to a general leakage of ions through the membrane (Atkinson et al., 1985). Membrane breakdown occurs at a later stage of the HR.

#### Preparation of suspension cells for bioassay

*Materials.*   The following materials and equipment will be needed for each student or team of students:

* 500 ml assay buffer (0.5 mM MES; 0.5 mM $CaCl_2$; 0.5 mM $K_2SO_4$; 175 mM mannitol) pH 6.0
* Six 50-ml disposable polystyrene beakers
* One 25-ml plastic cylinder (or 25-ml large tip opening serological glass pipette)
* One glass funnel (75 mm diam.)
* One 7-in.-square piece of grade 10 cheesecloth
* One 5-in.-square piece of Miracloth (obtained from Calbiochem Corp., LaJolla, CA)
* One rubber band
* One stainless steel microspatula

* Waterbath shaker fitted to hold the disposable beakers and calibrated to 27 C and 160 rpm, or flat bed rotary shaker in temperature controlled environment

*Procedure.* For this experiment, use suspension cells of either tobacco or soybean that are in log or exponential phase of growth, usually 2–3 days after the last transfer. Cover the mouth of the glass beaker with a single layer of cheesecloth and secure it with a rubber band. Then pour suspension cells on top of cheesecloth, collecting the finer aggregates of cells that filter through. Gently pour 100–200 ml of assay buffer over the cells that remain on top of the cheesecloth to wash more of the smaller aggregates into the beaker. Carefully remove the rubber band and discard the larger clumps of cells retained by the cheesecloth. Line the funnel with Miracloth and place on top of the 1000-ml flask. Collect cells on the Miracloth and then wash with 100–200 ml assay buffer. Drain well, but do not allow the cells to become dry. Weigh 5 g of cells into a 250-ml flask and slowly add 100 ml assay medium for a final concentration of 0.05 g/ml. If medium is added too rapidly, cells may clump and will need to be gently separated with a spatula. Gently swirl the cells, then pour 15 ml of the cell suspension into each of the 50-ml beakers, using a plastic 25-ml cylinder (the cells do not stick to the plastic). Place the beakers in the water bath shaker adjusted to 27 C and 160 rpm.

### Preparation of bacteria for bioassay

*Materials.* The following materials and equipment will be needed for each student or team of students:

* KB agar
* Sterile glass 5-ml pipettes (one for each bacterium)
* 1-ml pipettor and tips
* 200-µl pipettor and tips
* 50-ml centrifuge tubes (one for each bacterium)
* Disposable 3-µl cuvettes (one for each bacterium, plus one for blank)
* Vortex mixer
* Low-speed centrifuge
* Spectrophotometer

*Procedure.* Choose bacterial isolates that will give compatible and incompatible reactions on the suspension cells being used (Table 37.1). Transfer bacteria from the stock dishes to fresh medium by spreading a loop of the bacteria over the surface of the agar medium. Incubate the dishes at 30 C for 18–24 h. Gently scrape the bacterial lawns off the agar using 5 ml distilled water and a sterile glass 5-ml pipette. Remove bacteria from the dish using the pipette and transfer to a 50-ml centrifuge tube. If the bacteria are not well suspended, cover the tube with Parafilm and vortex gently. Add enough distilled water to fill the tube two thirds full, and centrifuge for 3 min at 8000 rpm at room temperature. Decant and discard supernatant, being careful not to disturb the bacteria pellet. Add 8 ml distilled water to the bacteria in the centrifuge tube, cover tube with Parafilm and vortex until pellet is suspended. Using the spectrophotometer, determine the optical density of the concentrated bacterial suspension. Prepare a 1:30 bacterial dilution by pipetting 2.9 ml distilled water into a cuvette and adding 100 µl of the concentrated bacteria (mix before removing aliquots). Cover the cuvette with Parafilm and invert a few times to mix well. The optical density units (ODU) of the diluted bacteria at 500 nm should be determined using a spectrophotometer that has been zeroed with water blanks. *Note:* Any dishes, plant materials, pipettes etc. that come into contact with pathogenic bacteria should be autoclaved before reuse or disposal.

The final bacterial concentration for inoculating plant leaves should be approximately $2 \times 10^8$ cfu/ml. This normally corresponds to about 0.2 ODU at 500 nm for most bacteria. This can easily be obtained by the following steps: (1) determine the ODU of the concentrated bacterial suspension, which is 30 times the ODU of the 1:30 dilution obtained above; (2) divide this number by 0.2 ODU (the desired final concentration of bacteria) to obtain the final volume to which 1 ml of the concentrated bacterial suspension should be diluted with distilled water. For example, if the ODU of the 1:30 dilution is 0.600, bring 1 ml of the concentrated bacterial suspension to 90 ml, which is $(0.600 \times 30)/0.2$.

For the suspension cell assays described in the following procedure, the final bacterial concentration should also be 0.2 ODU/ml in each beaker. For a 15-ml beaker, bring 1 ml of the concentrated bacteria to 1/15 of the total volume determined above for the plant inoculations; however, use assay buffer for dilutions instead of distilled water. Using the example above, if the ODU of the 1:30 dilution is 0.600 and the final volume of the solution prepared for inoculation into the plant is 90 ml, 1 ml of the concentrated bacteria should be diluted with assay buffer to 6 ml (90/15). One ml of this diluted bacterial solution will be added to beakers as described in the following section.

### Bioassay for the $K^+/H^+$ response

*Materials.* The following materials and equipment will be needed for each student or team of students:

* 1000-μl pipettor and tips
* pH meter (must be digital and capable of reporting pH to two decimal places)
* Conductivity meter

*Procedure.* Label the beakers containing the suspension cells (two replicates per treatment): control; name of compatible bacteria; name of incompatible bacteria. After the cells have incubated about 0.5 h in the water bath, add 1 ml of the appropriate bacterial suspension as prepared above. For controls, add 1 ml of assay buffer. Calibrate the pH and conductivity meters and store the electrodes in assay buffer when not in use. Take pH and conductivity readings every 0.5 h up to and including at least 5 h. To assure consistency of the pH readings, the contents of the beakers must be gently swirled by hand during readings. Alternatively, the pH electrode can be directly immersed into the beakers while they are rotating on the shaker.

### Anticipated results

Figure 37.3A and B depict typical changes in pH of soybean and tobacco suspension cells treated with compatible and incompatible bacteria. The pH of the tobacco controls usually decreases (perhaps as low as 5.1) over the 5-h period, whereas the soybean controls maintain a more consistent pH level during the experiment. Both tobacco and soybean suspension cells treated with compatible bacteria will undergo pH changes similar to their respective controls; suspension cells treated with incompatible bacteria will show an increase in pH that will begin within 3 h. Changes in conductivity should be similar to changes in pH. Changes in pH can be more accurately accessed if the results of the control are subtracted from the bacteria treatment (Figure 37.3C and D).

### Questions

* Define compatible vs. incompatible plant/pathogen interactions.
* *Pseudomonas solanacearum* is a pathogen of potato and causes an HR on tobacco. What kind of pH/$K^+$ response would you expect if (1) tobacco suspension cells or (2) potato suspension cells were treated with *P. solanacearum*?

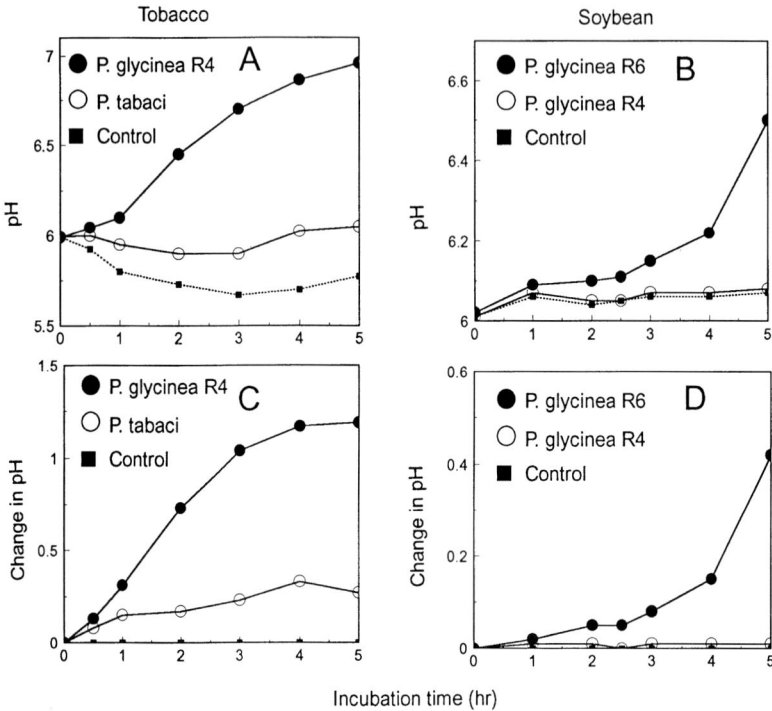

**Figure 37.3** Anticipated change in pH for tobacco and soybean suspension cells inoculated with compatible and incompatible pathogens. A and B show typical actual pH readings from a representative experiment; C and D show the change in pH for the various treatments with changes in the control subtracted from the data for the treatments.

## *Experiment 2. Measuring suspension cell death in response to compatible and incompatible bacteria*

The following exercise will demonstrate hypersensitive cell death due to the inoculation of suspension cells with incompatible bacteria. Many viability assays that are currently being used in the plant sciences are either relatively laborious and time-consuming or measure the temporary metabolic state of the cell, rather than irreversible cell death (Keppler et al., 1988). Evans blue has been a reliable stain for the microscopic determination of cell death (Turner and Novacky, 1974), and a spectrophotometric procedure using this stain has recently been developed that allows rapid, reproducible quantification of the stain retained by dead cells (Baker and Mock, 1994). Evans blue stain penetrates through ruptured membranes of dead cells, but not intact membranes of live cells; therefore, the amount of stain retained by the cell fraction reflects the amount of cell death. This will be verified by microscopic observation.

### *Materials*

The following materials and equipment will be needed for each student or team of students:

* 25 ml of 0.25% Evans blue in distilled water (Evans Blue, #E2129, Sigma Chemical Co., St. Louis, MO)
* 10 ml of 1% SDS in distilled water
* carborundum (approximately 180 mesh)

* Twelve $10 \times 35$ mm petri dishes
* Two 1-ml cuvettes
* Twenty-four 1.5-ml capped eppendorf tubes
* One plastic eppendorf tissue grinder/pestle
* Six cell strainers, 70-μm nylon mesh (Falcon #2350, Becton Dickinson Labware, Franklin Lakes, NJ)
* One stainless steel microspatula
* One 1000-μl pipettor and tips (tips should be cut off diagonally to allow passage of larger clumps of cells — use one per beaker)
* Pippette tip filters (optional)
* Fuchs-Rosenthal or other suitable counting chamber (Hausser Scientific Partnership, Horsham, PA)
* Table top centrifuge
* Spectrophotometer
* Compound microscope

*Caution:* Sigma Scientific classifies Evans blue as a possible carcinogen. Take appropriate precautions to limit exposure to this chemical — wear gloves, goggles, protective clothing and a particle mask.

### Procedure

Prepare suspension cells and bacteria as described earlier. Inoculate the plant suspension cells with bacteria as in experiment 1. At 0 h and 18–24 h after inoculation, cell death estimates should be made. Remove 2 ml from each treatment and put into a labeled cell strainer. Wash the suspension cells lightly with distilled water to remove the assay medium. Place individual cell strainers into petri dishes with 4 ml each of 0.25% Evans blue. Add additional Evans blue solution into strainers to cover cells. Incubate for 5 min. Remove one cell strainer at a time and gently rinse dye from the cells with distilled water, collecting waste in 1000-ml beaker. Place the strainers into petri dishes containing 4 ml distilled water for 5 min. Again, remove one cell strainer at a time and wash any remaining dye from the cells, collecting the waste as before. Using a spatula, gently transfer the cells into labeled eppendorf tubes containing a small amount of carborundum (<0.1 mg) and 0.5 ml 1% SDS. Grind cells with the plastic pestle, add 0.5 ml distilled water and centrifuge for 3 min. Remove 0.7 ml of the supernatant taking care not to disturb the pellet or remove any particulates that may be floating on the surface (use a pipette tip filter if necessary). Place into 1-ml cuvette and measure at 600 nm on spectrophotometer zeroed with water blank.

To observe estimated cell death microscopically, repeat the above procedure to treat cells with Evans blue solution. After the cells have been incubated with Evans blue and rinsed with distilled water, remove the cells from the strainers and put them into small test tubes with 0.5 ml distilled water. Place a drop of cells onto microscope slide and apply coverslip. Observe cells using 100–200× magnification and determine the percentage of cells that retained Evans blue.

### Anticipated results

At 0 h the amount of Evans blue extracted from the cells in all treatments should be relatively low. At 18–24 h, the readings from suspension cells treated with either buffer or compatible bacteria should still be relatively low signifying low levels of Evans blue retention. However, the suspension cells treated with incompatible bacteria should have higher levels of extractable dye from the cells (Figure 37.4). Beyond this time period, the amount of Evans blue retained in dead cells may decrease with time because of the complete disintegration of the cellular structures and membranes.

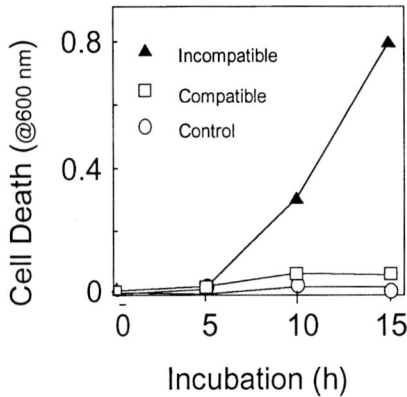

**Figure 37.4**   Anticipated results for cell death as monitored by extraction of retained Evans blue and spectrophotometric measurement at 600 nm. Compatible and buffer control will retain a background level of dye that will not increase substantially compared with the incompatible treatments.

## Questions

* Why does the spectrophotometer reading increase in cells treated with incompatible bacteria?
* Why is there more cell death in an incompatible (resistant) reaction?

## Literature cited

Abel, P.P., R.S. Nelson, B. De, N. Hoffman, S.G. Rogers, R.T. Fraley and R.N. Beachy. 1986. Delay of disease development in transgenic plants that express the tobacco mosaic virus coat protein gene. *Science.* 232:738–743.

Atkinson, M.M., J.S. Huang and J.A. Knopp. 1985. The hypersensitive reaction of tobacco to *Pseudomonas syringae* pv *pisi*: activation of plasmalemma $K^+/H^+$ exchange mechanism. *Plant Physiol.* 79:843–847.

Baker, C.J., N.R. O'Neill, L.D. Keppler and E.W. Orlandi. 1991. Early responses during plant–bacteria interactions in tobacco cell suspensions. *Phytopathology.* 81:1504–1507.

Baker, C.J., N.M. Mock, J.A. Glazener and E.W. Orlandi. 1993. Recognition responses in pathogen/nonhost and race/cultivar interactions involving soybean (*Glycine max*) and *Pseudomonas syringae* pathovars. *Physiol. Mol. Plant Pathol.* 43:81–94.

Baker, C.J. and N.M. Mock. 1994. An improved method for monitoring cell death in cell suspension and leaf disc assays using Evans blue. *Plant Cell Tissue Organ Cult.* 39:7–12.

Carlson, P.S. 1973. Methionine sulfoximine-resistant mutants of tobacco. *Science.* 180:1366–1368.

Gamborg, O.L., R.A. Miller and K. Ojima. 1968. Nutrient requirements of suspension cultures of soybean root cells. *Exp. Cell Res.* 50:151–158.

Garg, G.K., U.S. Singh, R.K. Khetrapal and J. Kumar. 1988. Application of tissue culture in plant pathology. In: *Experimental and Conceptual Plant Pathology.* Vol. 1. *Techniques.* R.S. Singh, U.S. Singh, W.M. Hess and D.J. Weber, Eds., pp. 83–119. Gordon and Breach Science Pub., New York.

Keppler, L.D., M.M. Atkinson and C.J. Baker. 1988. Plasma membrane alteration during bacteria-induced hypersensitive response in tobacco suspension cells as monitored by intracellular accumulation of fluorescein. *Physiol. Mol. Plant Pathol.* 32:209–219.

Linsmaier, E.M. and F. Skoog. 1965. Organic growth factor requirements of tobacco tissue cultures. *Physiol. Plant.* 18:100–127.

Schenk, R.U. and A.C. Hildebrandt. 1972. Medium and techniques for induction and growth of monocotyledonous and dicotyledonous plant cell cultures. *Can. J. Bot.* 50:199–204.

Turner, J.G. and A. Novacky. 1974. The quantitative relationship between plant and bacterial cells involved in the hypersensitive reaction. *Phytopathology.* 64:885–890.

White, P.R. and A.C. Braun. 1941. Crown gall production by bacteria-free tumor tissues. *Science.* 94:239–241.

*Part V*

---

*Special applications*

# chapter thirty-eight

# Commercial laboratory production

## Gayle R. L. Suttle

Over the past 15 years, the horticultural nursery trade has found ever-increasing ways to utilize micropropagation as a practical and cost-effective production tool. This chapter will describe some of the basic steps involved in plant micropropagation, discuss some of the practical aspects of the commercial micropropagation business and give some examples of how and why micropropagation is being used by growers today.

Many "scientific" papers have been published describing methods, formuli and techniques for the successful micropropagation of a multitude of plant species. In theory, once one has the right "recipe" all that is needed is a well-equipped kitchen, some trucks to haul the goodies to market and a good bank to hold all the money when it comes rolling back in. In reality, successful micropropagation on a commercial scale requires an understanding of two equally complex and dynamic factors — the plants and the marketplace. It is not enough to produce high-quality, healthy plant material (a major challenge in itself); one must also produce it in a timely fashion giving customers what they want, when they want it, for a price they can afford.

## The facility

Well-designed laboratories generally have a good work flow allowing for the actual movement of supplies, people, cultures and the finished product in an easy and logical pattern. Many small laboratories, unable to build a facility from scratch, use existing residential housing or mobile homes and remodel them to accommodate their needs. In order to keep airborne contaminants to a minimum, the ideal facility has very few entry sites (doors) and often has large areas designated as "clean rooms" with purified (HEPA-filtered) air flowing under positive pressure. The basic areas in any laboratory are listed below (Figure 38.1).

*Media preparation/dishes.* This area is set up similar to a kitchen with lots of counter space and shelves for chemicals and stock solutions, a pH meter, scales, autoclaves (usually more than one), media-dispensing equipment and a dishwashing area. Also important here is adequate storage space for empty culture vessels. Autoclaves create a tremendous heat load and usually are placed on an exterior wall with good, clean ventilation.

*Media storage.* Once media is "cooked" (sterilized) and dispensed into culture vessels, it must be cooled (again with good, clean ventilation) and stored until needed. Often this area is combined either with the media preparation room or in a section of the transfer room.

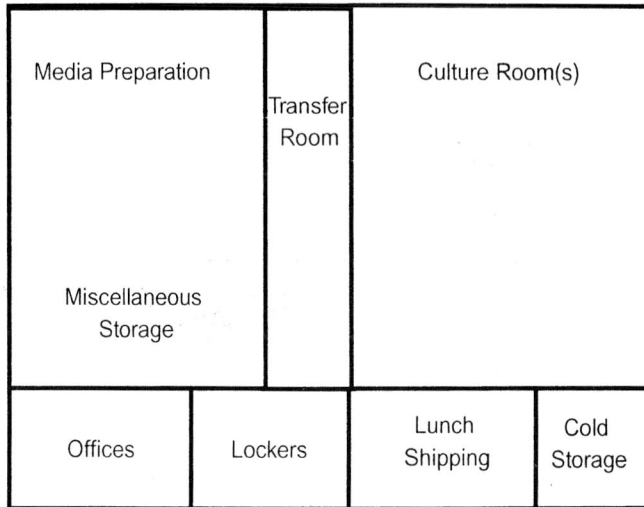

**Figure 38.1**   A design of a commercial tissue culture laboratory facility.

*Transfer room.*   This is the heart of any laboratory, filled with clean benches (laminar flow hoods) where all of the culture initiation and subsequent subculturing takes place. "Ripe" cultures (cultures with usable shoots) are harvested and cultures replanted (trans-ferred or reinitiated) to clean fresh media. As much as 80% of the labor force may be involved in this one room, so it must be a comfortable and pleasant working environment. Tools are sterilized at the hoods using glass bead sterilizers, bacticinerators or alcohol and flame. Often transfer rooms tend to be cramped, and the heat load generated by people, hoods and sterilizers is compensated for with a good air-conditioning system. People and plants suffer when it is hot and stuffy, and productivity declines.

*Culture room(s).*   This is usually the largest area of any laboratory; cultures are incubated on shelves under specific light and temperature regimes. Light quantity and quality can be key factors in culture success. Generally, regular or high output cool white fluorescent lighting is used, with higher intensities achieved by using more lights per square foot or by bringing cultures closer to the lights. Once again, heat buildup can be a major problem. Many laboratories remove the ballasts from the fluorescent light fixtures and mount them in an external area for easier cooling. Good air movement is critical in equalizing temperatures throughout the culture room. Temperature requirements are often crop specific, but generally range from 18–26 C. Different light or temperature regimes may require separate culture rooms for different crops. Usually air to the culture room is purified by HEPA filters.

*Cold storage.*   Cold storage is an absolutely essential component of any commercial micropropagation facility growing woody plants. Cold storage (2–4 C) is used for the following purposes: (1) to maintain stock blocks in culture for plants that are produced on a seasonal basis, (2) to allow production to continue during the cold winter months by "banking" finished plant material ahead of time and (3) to provide a chilling requirement for crops such as apple (*Malus*) and pear (*Pyrus*) species that respond in the greenhouse much more rapidly if given a "rest" or dormant period before out-planting.

*Shipping/lunch area.*   Ripe product (either microcuttings or rooted plantlets) must be harvested, boxed, packed and otherwise prepared for shipping. This activity requires sinks to rinse the plants, and the area also should be near the cold storage unit for holding until shipping. This area may serve double duty as a lunch or break area.

*Locker room.* When first entering any commercial facility, employees usually must change shoes and may be required to wear additional clean clothing such as laboratory coats, hair nets etc. into production areas. This locker area may also, as above, be incorporated with a lunch room and/or rest rooms.

*Offices.* Perhaps the least important place in a facility from an operations standpoint, but essential for running the minor details, such as sales, production planning, accounting and, of course, payroll, are the offices.

*Miscellaneous storage.* No laboratory ever has enough general storage space. Supplies, chemicals, shipping containers, extra equipment, spare parts and items, such as vessels used on a seasonal basis, all require space. Logical locations for storage are near an external door (for receiving) or near the media preparation and shipping areas where supplies are most often used.

## The process

### What to grow

There are two different approaches in determining product focus. The first is to grow what you think customers might want (perhaps a novel plant) based on what you hope is good market research and analysis and then to work hard to convince potential customers that they really do want it. While this approach is the very basis of our country's free enterprise system, it can be very risky if you happen to guess wrong. The most common reason for the failure of commercial laboratories over the past decade has been growing plants that the market either doesn't want or doesn't need. The second approach is to work with the customer on more of a contractual basis, asking what their problems are, what they need and when they need it, then going to work to try to solve the problem and fill the order. Arranging customers in advance is a much safer route to success considering the costs involved. Regardless of approach, every plant must be evaluated for suitability for micropropagation by asking the following questions:

1. What is the minimum number of plants needed? — Most laboratories have a minimum requirement (5–10,000 or more), although smaller facilities may be willing to have lower minimums.
2. How many customers are interested in a specific product? — The more the better, but sometimes exclusivity is important.
3. How long is the product needed? — Is it just for mother block establishment, or is it seasonal or will it be needed on a continuous basis?
4. How difficult is the plant to culture? — Is there existing literature or knowledge or will the technology need to be developed?
5. Who "owns" the plant? — Are there patent or customer restrictions on marketing the plants?
6. What is the estimated value of the plant once produced? — Can you charge enough to recover costs and make a profit?

All of these factors must be considered before starting any project. Often the price of a product cannot be defined until considerable research has been done on protocols for the given plant. Guarantees of success are impossible to make. The up-front cost of protocol development may or may not be shared with the customer. If exclusivity is needed, then the customer usually funds the work. Once a plant is deemed worthy of an attempt at micropropagation, the process of producing it may begin.

## Stock plant preparation

Healthy well-labeled stock plants are essential. The most important job the customer must perform is to make absolutely sure that the stock wood sent to the laboratory is exactly the named material that will be wanted in large numbers when finished. This step sounds simple, but in reality, mistakes on labeling can and do happen far too frequently. This point cannot be overemphasized.

Ideally, 10–15 young, juvenile stock plants are placed under an intense fertilization and pesticide regime to provide rapid, healthy shoot growth. The plants are watered from the base to avoid getting moisture on the foliage, a breeding ground for contaminants. Sometimes there may be a single very old tree located thousands of miles away as the only source or a customer may have only a small amount of stock and does not want to risk sending it away. In these cases, wood must be collected periodically throughout the growing season and sent by overnight express.

## Stage I: Culture initiation

The process begins by carefully harvesting soft, active shoot growth from the stock plant. Usually 5–15-cm cuttings are snipped early in the morning while the plants are fully turgid. The cuttings are immediately placed in a cooler until disinfestation takes place. The leaves and most, if not all, of the petiole are removed, and each stem is cut into one to two nodal pieces.

The disinfestation or clean-up process usually involves several cold soapy water washes, then soaking the shoots in a 10–20% commercial bleach (sodium hypochlorite) solution along with a surfactant such as Tween 20 for 10–20 min followed by a series of sterile water rinses. Damaged tissue is trimmed away using aseptic techniques. The treated cuttings are then placed individually in test tubes filled with a nutrient gel (medium) composed of a "best guess" formula or range of formuli that may allow the buds to grow. The tubes are usually incubated in a culture room under 16 h light at 25 C.

Visual screening takes place every 3–5 days, and the dead or dirty (visibly contaminated) cultures are discarded. Visibly clean, viable buds are transferred individually to fresh medium as new growth appears, usually 2–6 weeks after disinfestation. A portion of the stem piece is placed in a rich liquid or agar-based medium to detect bacterial contaminants that may be "invisible" or "unnoticeable" to the naked eye. This contaminant screening process (also called indexing — see chapters eight and thirty-nine) allows us to discard any additional and unseen dirty material from the pool of viable shoots. This procedure may be repeated periodically, however, and unfortunately, even this procedure does not catch everything, and growing "clean" stock is a continuous struggle.

Culture initiation can be more of an art than a science. Often the concentration of bleach and/or time of disinfestation must be adjusted based on the woodiness, hairiness and source of the stock tissue. Several variations may be tried if there is an abundance of source material, fine-tuning the process based on previous efforts and the time of year. If materials are limited, the best guess is tried and hopefully will succeed.

### Timing

The best success for getting plants into culture is typically realized by taking the first flushes of vegetative growth in the early spring. A very successful second option is to force dormant buds from sticks in the winter after the full chilling requirement has been satisfied. These sticks are placed in a "forcing solution" of 30 g/L Floralife Floral Preservative (Floralife, Inc., Burr Ridge, IL), incubated at approximately 20 C and 10–12-h photoperiod (typical office conditions). Floral preservatives generally have sugar, an antimicrobial agent and an acidifier to prolong bloom life. The dormant buds of many plant species will begin to sprout and grow under such conditions. Flower buds are removed and discarded when they appear, to allow for better vegetative growth. When these soft shoots reach the length of 2–4 cm, they are easily removed from the stem and processed. They tend to be extremely clean so that surface disinfestation and establishment in culture are usually easier.

One theory explaining why such early growth seems to go into culture so well is that endogenous plant growth regulators (PGRs) and growth factors may have primed the plants to "spring" into rapid growth phase (inhibitory factors are at a low ebb). Fortunately, many times this new growth is very "clean" — it hasn't had time to acquire a load of contaminating organisms. Whatever the reason, early spring growth is without question usually the best starting material. By starting early in the growing season, it also gives more time to generate numbers and bring more cuttings into culture if needed.

Late summer or outdoor field-grown stock poses particular problems from increased contamination and a lack of turgidity due to heat stress that causes plant tissues to be less tolerant of bleach disinfestation. Success rates for establishing material in Stage I can range from 0–95% because of all these factors and many more we do not yet comprehend. The ultimate goal of Stage I is to get the plant material clean and actively growing in culture. Once this is achieved, Stage II becomes the next challenge.

## Stage II: Shoot proliferation

The primary method of increase in woody plant micropropagation is by axillary shoot proliferation, although adventitious shoot proliferation may occur with some kinds of plants, such as species of the family Ericaceae. Axillary proliferation is preferred in order to avoid the potential of off-types (somaclonal variation) from occurring. Genetic mutation can occur by any vegetative propagation method, and if the characteristic is based on a relatively unstable chimera, such as some variegations are, micropropagation can enhance reversion. Micropropagators must continuously monitor the quality of plants they produce to ensure trueness. In our 15 years of experience of micropropagating trees, genetic mutation has not been even a minor problem. Useful strategies we employ to help reduce the potential of off-types are limiting the number of subcultures or length of time in culture, keeping PGR levels in the culture medium as low as possible and always going back to the original stock source to start new lines.

Once a plant is clean and actively growing in culture, the goal of Stage II is to induce all (or at least some) of the axillary buds along the new, young, aseptically growing shoot to sprout and grow. In other words, we want the shoot to branch. Every 2–8 weeks (culture cycles vary depending upon the type of plant) we can subculture or harvest the new branches, "replant" or transfer them to fresh medium and wait another 2–8 weeks until each of these new branches sprouts its own set of branches. There are rarely any roots involved during Stage II.

With woody plant micropropagation, there are two main methods of achieving multiplication of new axillary buds depending on the type of plant material involved. The first is by promoting many branches on a single stem piece resulting in a "clump" of new shoots. Species of apple and cherry (*Prunus*) would be two examples of plants best micropropagated by clump method. During subculturing, the nice tall shoots are removed and (1) replanted in fresh multiplication medium to produce their own clumps or (2) planted into a rooting medium or (3) shipped out as microcuttings to be rooted in the greenhouse. The basal clump mass is left intact and replanted to yield another crop of branches. The clumps get thicker and thicker with new branches and can be used as a mini-stock block. Eventually, the clumps decline in quality and quanity, usually after two to five culture cycles and must be discarded and replaced.

The second form of axillary shoot proliferation can be described as the single-node method. Plants exhibiting strong apical dominance, such as lilac (*Syringa*) and maple (*Acer*) species, are best micropropagated by single-node method. A single shoot grows up straight and tall, producing two to ten sets of leaves. One- or two-node sections are separated and replanted whereby each of these runs straight up again and is now ready to be chopped into nodal sections again. During any subculture, the nice terminals or tips can be harvested and planted into rooting medium or shipped out as microcuttings to be rooted in the greenhouse.

## Media formuli

The actual media formuli used during Stage I and II are often the same. There are three or four basic formuli that are adapted and modified as necessary, sometimes on a cultivar-by-cultivar basis. The proper ratios of the inorganic components in a formula actually are the main key to developing a successful protocol for a given plant. PGRs are also a key factor, but secondary to proper nutrition. It is not uncommon for a commercial laboratory to have over 50 modifications of basic media on file for multiplication alone and another 40 or so Stage III modifications. When a new plant is brought in, one takes an educated guess as to what might work based on past experience and literature searches, but usually modifications must be made by a systematic trial-and-error method of marching through the various components. In general, Stage I and II media usually have higher levels of salts, sugars and cytokinins. Stage III media generally have lower levels of salts, sometimes lower sugars and usually lack cytokinins, but include high levels of auxins.

## Stage III: Rooting

When sufficient numbers of healthy shoots have been generated by one of the proliferation processes described above, healthy active terminal buds are selected either for in vitro rooting or for shipping directly to customers as microcuttings for rooting and acclimatization in the greenhouse. The decision whether or not to root in vitro is quite often made by the customer, based on several factors. In vitro rooting adds about 30% to the final cost of the product, and some crops, such as birch (*Betula*) and lilac species, root so well in the greenhouse as microcuttings that it simply does not make sense to root them in vitro. Other crops, such as *Malus*, *Pyrus*, *Prunus* and *Acer*, are much easier to acclimatize as in vitro–rooted plantlets. Perhaps the most important factor of all for the grower, other than survival, is that in vitro–rooted plantlets require 2–5 weeks less time in the greenhouse to reach a size suitable for planting in the field. Many growers have limited greenhouse space and do not want to have to "baby-sit" tender microcuttings for such a long time. A grower may be able to process two, three or four crops through his greenhouse in the same time it takes to raise one crop from microcuttings. Often, it pencils out as more economical to buy rooted plantlets.

## Stage IV: Greenhouse acclimatization

Fresh Stage II microcuttings (without roots) or Stage III in vitro–rooted plantlets are both planted in any of several well-drained, soilless potting mixes in the greenhouse under high humidity. Various types of fogs, tents, domes and mist systems are used to wean the tender shoots into the real world by gradually reducing the humidity. The weaning process takes from 3 days to 6 weeks depending on what the plants are, the weather and if the plants were rooted in vitro. Fertilization is not recommended until the plantlets become established and begin to grow. Some crops such as *Betula* can become stunted if overfertilized.

## Production scheduling

When preparing an actual production schedule for any given plant, many factors play a role. How many plants are ordered and in what form (microcuttings or rooted plantlets)? When are they wanted? Can the order be prepared ahead of time (does it require or tolerate cold storage)? How does it multiply (nodal or clump)? What are the multiplication rates during buildup? What is the number of rootable or harvestable microcuttings (yield), and does the yield change over time? What rooting percentage can be expected? Can smaller batches be produced and stored or must it be produced in lump sum? How much time is required between cycles (during build-up and during harvest cycles)? What are the estimated labor requirements for each cycle? Once these factors are determined, one works backward from delivery date, calculating the number of cultures required at each of the various cycles along the way. Information on when each cycle must take place, what activities must be done at each step and estimated labor required are all plotted on a production-planning calendar. This planning process must be done for each cultivar on an individual basis since multiplication rates, yields, number ordered and dates ordered will vary from plant to plant. Critical labor peaks must also be smoothed out as well as possible ahead of time. Simply put, not everything can come out at once.

One of the most frustrating factors in planning a production schedule is the very dynamic and very unpredictable nature of the plant material itself. Plants are living systems, and we simply don't control or even understand all the factors involved in in vitro propagation. Multiplication rates and harvestable yields, for example, can only be best guesstimated. Any plan has to be continuously updated and revised.

## Seasonality

While a laboratory can pump out material on a year-around basis, anyone working on temperate deciduous crops knows that there is a season for everything. Plants can be fooled into thinking it is summertime in culture by continuously being subjected to long (16-h) days in the culture room, but as fall approaches, the rate of growth in the greenhouse slows considerably. To take maximum advantage of the growing season, customers want to get their product in the early spring through early summer. Use of cold storage to store crops ahead of time allows laboratory production of some plant material to take place in the winter, but much must come out fresh, making labor needs seasonal as well.

# Reasons micropropagation is used

While micropropagated material represents only a tiny percentage of the total number of plants produced in America today, there is no doubt that many of the largest and smallest nurseries in the United States view micropropagation as an essential tool that helps them maintain their competitive edge by growing better plants more efficiently. The remainder of this chapter will be devoted to a discussion of some of the advantages that in vitro micropropagation offers over conventional production systems.

## New introductions

Perhaps the most obvious use for micropropagation is to get a "jump start" on growing the newest and hottest items quickly. Micropropagation cuts 3–10 years off the time it takes to bulk up new selections and get them to market. For example, micropropagation was used by one grower to establish layer beds of a new apple understock. He sold over 1 million rootstocks in the same amount of time his competition had bulked up to only a few thousand using conventional propagation methods. The market is always looking for something new and exciting. Nurseries on the forefront of introducing new plant material build their reputation as leaders in the industry, causing customers to come back year after year to find out what else is new.

In some cases, growers use micropropagated plant material to establish mother blocks and to fill in production shortages while mother blocks are too young to be productive. Conventional propagation methods, such as hardwood or softwood cuttings, layer beds or budding and grafting, may take over long-term production needs once enough wood becomes available. In many other cases, micropropagation remains the method of choice for a variety of reasons.

## Rapid response to market demand

Large mother blocks or scion orchards are time-consuming and expensive to establish and maintain. It is often difficult for growers to adjust quickly to the rise and fall in popularity of a given plant. With micropropagation, the stock block is maintained in a $10 \times 10$ ft cold storage unit. If a customer gets a call for an additional 10,000 liners of a particular blueberry, for example, they simply call up and ask when is the earliest they can take delivery on the additional microcuttings. They then add the time they need for the greenhouse growing and call their customer back with a delivery date.

## Clean plants

Micropropagation is inherently a cleaner system for producing plants compared with traditional production methods. Since the plants are grown in culture, diseases are not transmitted from the field into the greenhouse and on to subsequent generations. A single disease-free mother plant can theoretically produce unlimited disease-free daughter plants without the possibility of reinfection. Conversely, individual mother plants in a traditional virus-free cutting block must be tested again every year in order to maintain and ensure virus-free status (see chapter thirty-nine). Testing fees add significantly to the expense of maintaining large mother blocks in the field.

## Ease of propagation

Bud incompatibility on budded or grafted stock and poor rooting percentages with softwood or hardwood cuttings make micropropagation the method of choice or the only option on many difficult-to-propagate plants such as *Syringa* and redbud (*Cercis*). Some red maple (*Acer rubrum* L.) cultivars, such as 'Karpick' and 'Bowhall', are absolutely impossible to root from cuttings. Unreliable seedling availability and poor or unpredictable bud stands on *Betula*, *Tilia* and *Morus* are problems growers are able to avoid by planting micropropagated material. Growing plants on their own roots offers major advantages for such plants as contorted filbert (*Corylus avellana* 'contorta'), where suckering of understocks can be a major problem.

Sometimes micropropagated material provides the grower with a nucleus of "juvenile" material from which additional cuttings can be more easily rooted. Each year or two, the customer starts over with a fresh batch of starter material from the laboratory.

## Speed

Research done at various universities indicates that it is possible to grow plants much more quickly to size than is traditionally seen in nurseries today, regardless of how the plants are propagated. Such rapid growth requires optimization of all growing conditions, including fertilizer, light and temperature. However, it is amazing the results that can be achieved with even modest adjustments of growing practices. Several field growers are now producing well-formed, small branched trees of *Prunus* 'Kwanzan', *Morus alba* 'Chaparral' (mulberry), and others in 1 year instead of 2. Blueberry plant production can be dramatically speeded up. Use of micropropagated cherry understock yields increased vigor and earlier fruitset (and earlier payback) for the orchardist.

## Better branching

Because the internode length is greatly reduced on micropropagated plants, there is generally more opportunity to develop a fuller head on the growing plant. Indeed, this is also one of the reasons why survival is often greater. If something happens to destroy the terminal bud (for example, damage caused by freezing, hungry rabbits or poor pruning), there are other buds below available to choose from. One grower accidentally sprayed a young block of *A. rubrum* liners twice with Surflan, a herbicide. The stems of the young micropropagated liners were girdled right at soil level. All was not lost, for the grower dug the soil out from around the base of the plants, removed the damaged tops, and the buds below the girdling all pushed out again. The grower lost some height on his crop (about 1–2 feet), but shorter plants are better than no plants at all. Having more buds to choose from also allows a grower to cut back closer to the ground, producing straighter trees.

The better branching is a real advantage when it comes to growing well-formed shrubs. *Hydrangea quercifolia* 'Snow Queen' (PP4458) produced from rooted cuttings tends not to throw many branches at an early age, whereas micropropagated plants are easily developed into a bushy habit with routine pinching. The increased branching of blueberry plants is seen as a great advantage by some growers; others prefer the more traditional vase shape formed from rooted cuttings.

## Greater survival and uniformity

While cultural practices play an important role in the ultimate performance of any block of plant material, the two key attributes most often given in describing micropropagated material are superior survival rates and much greater uniformity. Better survival is explained, in part, by heavier root systems and more buds to choose from on the top. The fact that the root-to-shoot ratio is more balanced from the very beginning and that the small plantlets are highly uniform right from the start helps to explain why subsequent growth is more reliable and consistent.

## Conclusion

As the use of micropropagation becomes increasingly varied and widespread, growers will find new and ingenious ways to take advantage of its power. Even so, growers must weigh the pros and cons of this relatively new technique against the old methods and determine for themselves, on a case by case basis, whether it is worth the trouble to change. It must help them solve problems, find new markets, or save time or money. The successful micropropagator will be one who fills these needs and many more.

*chapter thirty-nine*

# Indexing for plant pathogens

*Michael J. Klopmeyer*

Flowering plants are capable of reproducing either by seed or by vegetative propagules such as budwood, leaves, stems or cuttings. When plants are reproduced via vegetative propagules, there is a higher risk of transmitting plant pathogens from the original mother plant to its offspring compared with seed propagation. This problem can be circumvented by establishing methods to eliminate these pathogens and by maintaining pathogen-free stock through a production cycle.

This chapter will outline the various procedures used to eliminate fungal, bacterial and viral pathogens from vegetatively propagated plants. These basic methods can be applied toward plant species propagated in tissue culture (micropropagation), in greenhouses and in the field.

## What is indexing?

Indexing, simply stated, is testing various plant parts (i.e., cuttings) for the presence of any plant pathogen. If plant pathogenic fungi, bacteria, viruses or viroids are present in the plant part tested, that plant is either destroyed or subjected to various procedures designed to eliminate the pathogen. Only plant samples that test negative for any given plant pathogen are considered to be disease free, and these are increased in large numbers under strict sanitary conditions.

## Specific crops targeted

Indexing programs are being used in many university, government and commercial laboratories around the world. The major vegetatively propagated crops that are indexed include the following: fruit trees, such as stone fruits (i.e., apricot, cherry, peach and plum); pome fruits (apples and pears) and citrus (orange, lemon and lime); vegetables, such as potato and sweet potato; fruits, such as banana and papaya; and floricultural crops, such as foliage plants, bulbs (Easter lilies, Asiatic lilies, iris, gladiolus), chrysanthemum, carnation, New Guinea impatiens and geranium.

Each crop is unique in regard to the pathogens that infect it, what techniques are efficient in eliminating these pathogens and what are the most effective methods for verifying the pathogen-free status after pathogen elimination. Since many of these elimination techniques require passage of the plant through tissue culture, careful evaluation of an artificial growth medium is also required prior to initiation of the elimination procedures.

## Indexing techniques

There are various indexing techniques utilized to eliminate plant pathogens from vegetatively propagated plants and to verify that the pathogens have been eliminated. Indexing
strategies revolve around the host crop studied as well as the specific pathogens affecting
that crop. Indexing methods need to be developed that (1) exploit the biology of the
pathogen within the host, (2) utilize a detection and/or eradication technique that is
sensitive and economically feasible, (3) are repeated over months to reduce the probability
of pathogen infection and (4) are performed under strict sanitation in the field, greenhouse
and/or laboratory.

   The remainder of this chapter will focus on strategies employed in the floriculture
industry to eliminate the major plant pathogens from florists' geranium (*Pelargonium* ×
*hortorum*). The principles discussed are applicable to other crop/pathogen systems including plants maintained in tissue culture.

## Bacterial and fungal elimination

One method for the elimination of systemic fungal and bacterial plant pathogens from
geraniums is accomplished using a procedure called culture indexing. This procedure was
initially developed in the 1940s to free chrysanthemums of the fungal wilt pathogen
*Verticillium* (Raju and Olson, 1985). Because of the tremendous success with chrysanthemum, culture indexing was utilized to eliminate the major systemic plant pathogens of
geranium including *Xanthomonas campestris* pv. *pelargonii* (Xcp) and *Verticillium* spp.
(Oglevee-O'Donovan, 1993).

   Bacterial blight (caused by Xcp) is the limiting factor in the successful production and
finishing of zonal, ivy and regal geraniums. Under optimum environmental conditions,
this bacterial pathogen causes plant death in a relatively short time by physically blocking
the xylem vessels (water-conducting tissue) in the stem. Understanding the biology of the
pathogen allows for culture indexing to select plants free of the pathogen. Culture indexing
is simplistic in its method, yet very effective in the selection of plants free of systemic
fungal and bacterial pathogens. Briefly, culture indexing entails the following six steps.

1.  Establishment of mother plants in an incubation greenhouse at temperatures that
    are optimum for disease development. In this instance, bacterial blight caused by
    Xcp develops quickly at 25–30 C day, 20–25 C night.
2.  Top cuttings (5–8 cm long) are removed from the mother plant where they are
    individually numbered and brought to the laboratory for testing (without storage).
3.  The bottom 2-cm portion of the cutting is excised with a sterile knife and surface
    disinfected with a 10% bleach solution for approximately 10 min.
4.  The surface-disinfested stem section is aseptically sliced into thin (2-mm) sections,
    placed into a test tube containing sterile nutrient broth and incubated at room
    temperature (23–25 C) for 10–14 days.
5.  After incubation, if the nutrient solution appears cloudy, due to bacterial or fungal
    contamination emanating from the vascular system of the plant, the original num-
    bered cutting is discarded. Only those cuttings that yielded no growth of any
    microorganism in the nutrient broth tube are retained (Figure 39.1). The original
    mother plant is also destroyed.
6.  All cuttings that culture index free of pathogens are planted in the incubation block
    (at temperatures optimum for disease development) and retested after 3–4 months.
    If cuttings from these mother plants test clean, they are retained, whereas the
    original mother plant is again destroyed. This process is repeated at least one more

**Figure 39.1** Culture indexing test tubes containing nutrient broth plus surface-disinfested geranium stem sections. Tube containing cloudy medium (arrow) indicates growth of microorganisms and a positive finding.

time for a total of three times. At least 1 complete year of testing (three to four culture indexings) should be done to assure freedom from all systemic bacterial and fungal pathogens.

This method is easily adaptable to the monitoring of plants maintained in tissue culture on a year-round basis. It also provides a means of detecting both pathogenic and nonpathogenic bacterial and fungal pathogens and/or contaminants. Theoretically, a single bacterial cell or fungal spore could be detected. A combination of sequential mother plant incubation and culture indexing repeated over at least 1 year reduces the probability of fungal and bacterial contamination to near zero. Culture indexing is also a simple, inexpensive test that does not require sophisticated laboratory equipment and supplies.

## Virus indexing

Plant pathogenic viruses are submicroscopic particles consisting of nucleic acid (RNA or DNA) surrounded by a protein coat. Their effects on plants are varied, ranging from no symptoms to necrotic or chlorotic ring spots and mosaic patterns on the foliage to severe plant stunting and death (Agrios, 1988). These symptoms do not generally correlate with virus concentration or titer. For example, some viruses may be present at very high titer and not cause severe symptoms on susceptible hosts.

Virus indexing can accurately verify the cleanliness of a given variety and also determine if one or more viruses are present. Unfortunately, in many cases, one single testing method may not assure virus-free plants. Plant sampling methods, seasonality, testing method sensitivity and specificity all affect virus-indexing results.

The two major methods that are commonly used to detect plant pathogenic viruses in clean stock programs are serology and bioassays (biological indicator plant inoculation). Alternative indexing techniques include the use of nucleic acid probes, specific for the virus tested (Miller and Martin, 1988) or the visualization of the replicative form of the viral RNA isolated from infected plants (Valverde et al., 1990).

**Figure 39.2**   ELISA used for the detection of plant pathogenic viruses.

## Serology

The most common serological method used is enzyme-linked immunosorbent assay (ELISA) (Miller and Martin, 1988). ELISA has become the predominant technique used since it is reliable, sensitive and relatively inexpensive on a per-sample basis. It is also more sensitive than bioassays, but does require some specialized laboratory equipment.

An ELISA detection technique is developed by administering purified virus, i.e., tobacco mosaic virus (TMV), to rabbits, sheep or mice to induce the formation of antibodies against TMV. After 6–12 weeks, the antibodies are isolated from the animal's blood serum and purified and prepared for the ELISA test. ELISA is performed in commercially available, 96-well, polystyrene plastic plates (Figure 39.2). First, the viral antibodies are coated onto the surface of each well. Next, a macerated plant extract of the suspected virus-infected plant is incubated in the plate well. After incubation, the plant extract is removed and the well thoroughly rinsed of all remaining plant debris.

During the initial incubation period, any virus particles present in the extract bind exclusively to the antibodies coating the plate well. Next, additional antibodies (antibody conjugate), which are attached to a marker enzyme, such as alkaline phosphatase or horseradish peroxidase, are added to the well. During another incubation period, this antibody conjugate will bind to any virus particles present in the well that were previously bound by the coating antibody. This double binding, termed double antibody sandwich, is verified by adding an enzyme substrate that will indicate the presence of the antibody conjugate (attached to the enzyme). Many tests utilize colorless substrates that convert to a distinctive color after enzyme reaction. The presence and the amount of color development in each well signify the presence and relative concentration of the target virus.

## Bioassays

Bioassays are accomplished by mechanically inoculating an indicator plant with sap extracted from the test plant. The species of indicator plant used is dependent on the types of viruses that are capable of infecting the test plant. The most common indicator plants used are members of the Solanaceae or tobacco family including *Nicotiana tabacum* L., *N. benthamiana* Domin and *N. clevelandii* Gray. Another popular indicator plant is lamb's-quarters, *Chenopodium quinoa* Willd. and/or *C. amaranticolor* Coste and Reyn.

**Figure 39.3** Severe local lesion (white areas) response caused by an unknown plant pathogenic virus 16 days after inoculation onto the bioindicator plant, *Chenopodium quinoa*.

Infected plant sap is prepared by macerating 5–10 g of plant tissue in a specially prepared buffer solution at approximately a 1:10 dilution. This buffer usually contains anti-reducing agents to prevent the suspected virus(es) from being denatured during macera-tion of the plant tissue. The sap/buffer extract is gently rubbed onto the leaf surface of the indicator plant while trying to inflict as little damage as possible. The intent of the inoculation is to wound the leaf cell wall slightly, which allows for virus entry into the cell with subsequent cell wall healing. Since viruses are obligate parasites, the indicator plant cells cannot be destroyed during inoculation. Often an abrasive, such as carborundum or celite (diatomaceous earth), is added to the sap/buffer extract at inoculation to facilitate indicator plant cell wall wounding. Virus symptoms may develop on the indicator plant leaves anywhere from 3–21 days following inoculation. Typically, two major symptoms are observed. The first, a local lesion response, is typified by small, necrotic or chlorotic lesions (1–2 mm in diam.) on inoculated leaves (Figure 39.3). This local lesion response may occur soon after inoculation. The second, a systemic response, is typified by mosaic or vein clearing symptoms on newly emerging leaves. This systemic response usually appears 7–21 days after inoculation. As with all virus-indexing methods, repeated testing of indi-vidual plants at different times of the year is necessary for accurate results.

## Virus elimination

The virus-indexing procedures described above may reveal complete virus infection. If a positive reaction occurred in an ELISA test, the specific virus name, or class, is known and elimination procedures can be specifically designed. However, if virus symptoms appear on indicator plants, the identification of the specific virus may still be unknown.

If visual selection of virus-free clones is not possible, there are several techniques available to eliminate viruses, including the following: (1) exploiting erratic virus distribu-tion within plants, (2) chemotherapy, (3) heat or thermotherapy and (4) meristem tip culture (Matthews, 1991). Sometimes a combination of the above techniques may achieve greater success.

## Erratic virus distribution

Some viruses are found erratically located within plants, and shoot tips from rapidly growing stems on these plants may be propagated and found to be virus free. Virus titer may change depending on the time of the year. For example, during the summer months, high temperatures may slow virus replication and movement into the youngest growing points. Shoot tips from these rapidly growing shoots may be virus free.

## Chemotherapy

Some limited success has been reported utilizing chemicals to cure plants of pathogenic viruses. Ribavirin, a potent viricide, has been reported to eliminate major viral pathogens from potato and other crops (Griffiths and Slack, 1988; Hollings, 1965). In most instances, chemotherapy is more effective if incorporated into a tissue culture medium that allows for the plant to take up the chemical uniformly through the root system. Improved virus elimination can occur if chemotherapy is also combined with thermotherapy and subsequent meristem tip culture. The use of chemotherapy is difficult since virus metabolism is so closely integrated with plant metabolism.

## Thermotherapy

The most commonly employed method for elimination of viruses from infected plants is a combination of thermotherapy and meristem tip culture (Horst and Klopmeyer, 1993). The success of thermotherapy depends on the inability of the target virus(es) to multiply and move readily within plants being exposed to air temperatures of 35–40 C (Grondeau and Samson, 1994). Successful treatment times may vary from weeks to months depending on the plant cultivar and the specific virus.

Thermotherapy is very effective in reducing virus titer in geraniums. Geraniums, in general, are not very heat tolerant, since plants will die after 5–6 weeks at 37 C. One technique that can overcome this problem is to place plants in a temperature-controlled growth chamber at lower temperatures (24 C day, 21 C night) initially. Each day, the temperatures are raised approximately 3 C until a final setting of 38 C day and 33 C night is reached. This gradual acclimatization period allows for the plant to adjust to the higher temperatures. Plants can then be maintained under these conditions for as long as possible (3–4 weeks), after which meristem tips are removed for tissue culture.

## Meristem tip culture

The removal of meristem tips from plants grown under high temperatures is very effective in virus elimination. The excision of the meristem tip essentially excludes the virus from the "new" plant since this tip developed on the mother plant under the influence of virus-inhibiting high temperatures. The size of the meristem tip is critical for successful virus elimination. Obviously, the smaller the tip that is taken, the better chance for virus exclusion. However, as smaller tips are taken, meristem tip survival on tissue culture media is also reduced.

Meristem tip removal is accomplished by removing plant tissue surrounding the meristematic region and carefully excising the apical dome and one to two leaf primordia (Figure 39.4). Survival and growth are greatly improved if the tissue taken includes slightly expanded leaf primordia. After 1–2 weeks, the excised meristem tip expands and the leaf primordia develop and differentiate into mature leaves. At this point, the young plant can be transferred into a tissue culture medium containing IBA or IPA to initiate root development. After 8–12 weeks, the young plants are ready for transfer back into the

**Figure 39.4**  Chrysanthemum meristem tip (arrow) and two young leaf primordia (P).

greenhouse. This is accomplished by carefully washing the agar medium from the roots and placing the plant into a soilless greenhouse medium under intermittent mist for 5–7 days. Misting is important to help harden off the plant by allowing the formation of cuticle on the leaf surfaces.

After establishment in the greenhouse, the plants are ready for virus indexing by ELISA and bioassay to verify that all viruses have been eliminated. If virus elimination was not successful, thermotherapy and meristem tip culture will need to be repeated.

## Clean plant production

The development, maintenance and production of certified stock require strict adherence to the following five fundamental principles of clean plant production (Figure 39.5):

*Annual renewal.*   At the completion of each production year all old stock should be destroyed, the greenhouses disinfected and the stock base renewed with clean plants.

*Unidirectional flow.*   During annual renewal of the stock base, the new stock should always come from a cleaner source. Cuttings from old stock should never be used to reestablish a new stock base.

*Repeated testing.*   Indexing of certified stock needs to be repeated over a period of at least 1 year. This will allow for maximum opportunity to detect plant pathogens under different environmental conditions.

*Isolation.*   Certified stock, although pathogen-free, is still susceptible to infection by plant pathogens. Clean stock must be grown in isolated, protected areas under strict sanitation to prevent the reintroduction of plant pathogens.

*Clonal selection.*   After intensive indexing procedures, including tissue culture maintenance, the certified stock base should be flower trialed under controlled conditions. This will allow for selection of superior clones that flower earlier and more often on a plant with

New Varieties
↓
Isolation Block
↓
Bacteria/Fungi ←——— Incubation Block I
Eradication          Culture Indexing 3x
↓
Virus ←——— Heat Treatment
Eradication   Meristem Tip Removal
↓
Meristem Tip Culture
↓
Incubation Block II
Virus Indexing 3x
↓
Elite Nucleus Block
↓
Flower Trial/Selections
↓
Nucleus Block
↓
Build-Up
↓
Production

JEW

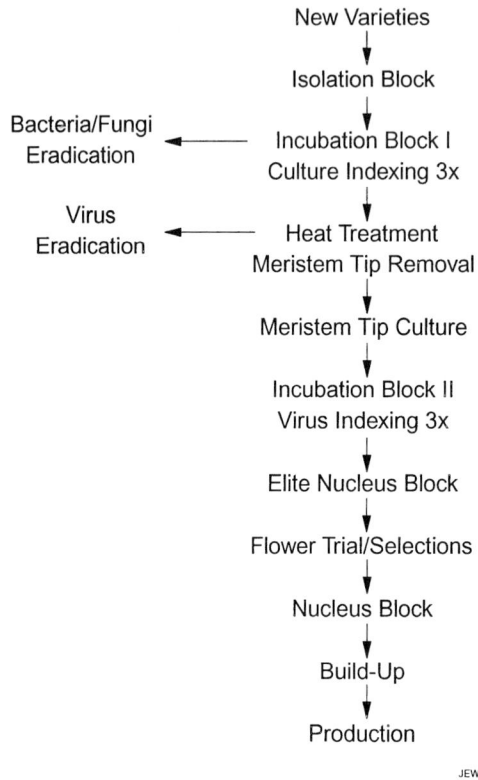

**Figure 39.5**   Flow chart for the production of bacterial, fungal and virus-free geraniums.

an excellent habit. Since most vegetatively propagated ornamental varieties are protected by a plant patent, clonal selection criteria should also closely approximate the variety description outlined in the patent application.

## Summary

Vegetatively propagated crops are constantly threatened by plant disease. Since their method of reproduction is amenable to spread of bacterial, fungal and viral diseases from generation to generation, the development of reliable clean plant production procedures is required. These procedures include effective elimination techniques and precise detection techniques to verify that all plants are clean. Finally, since all plants are just as susceptible to these pathogens after elimination, careful and controlled conditions must be maintained to produce successfully these commercially important crops.

## Literature cited

Agrios, G.N. 1988. Plant diseases caused by viruses. In: *Plant Pathology*. pp. 622–702. Academic Press, San Diego, CA.

Griffiths, H.M. and S.A. Slack. 1988. Potato virus elimination by heat and ribavirin. *Phytopathology*. 52:1230–1241.

Grondeau, C. and R. Samson. 1994. A review of thermotherapy to free plant materials from pathogens, especially seeds from bacteria. *Crit. Rev. Plant Sci.* 13:57–75.

Hollings, M. 1965. Disease control through virus-free stock. *Annu. Rev. Phytopathol.* 3:367–396.

Horst, R.K. and M.J. Klopmeyer. 1993. Controlling viral diseases. In: *Geraniums IV*, J. White, Ed., pp. 289–292. Ball Publishing, Geneva, IL.

Matthews, R.E.F. 1991. Control measures. In: *Plant Virology*. pp. 596–634. Academic Press, San Diego, CA.

Miller, S.A. and R.R. Martin. 1988. Molecular diagnosis of plant disease. *Annu. Rev. Phytopathol.* 26:409–432.

Oglevee-O'Donovan, W. 1993. Culture indexing for vascular wilts and viruses. In: *Geraniums IV*, J. White, Ed., pp. 277–286. Ball Publishing, Geneva, IL.

Raju, B.C. and C.J. Olson. 1985. Indexing systems for producing clean stock for disease control in commercial floriculture. *Plant Dis.* 69:189–192.

Valverde, R.A., S.T. Nameth and R.L. Jordan. 1990. Analysis of double stranded RNA for plant virus diagnosis. *Plant Dis.* 74:255–258.

*Part VI*

# *Appendix I.*
# *List of suppliers*

# Appendix I

## *List of some suppliers, addresses and toll-free telephone numbers*[*]

Amresco Inc.
30175 Solon Industrial Pkwy.
Solon, OH 44139-4300
800-829-2805

Electron Microscopy Sciences
321 Morris Road, P. O. Box 251
Fort Washington, PA 19034
800-523-5874

Fisher Scientific
711 Forbes Ave.
Pittsburgh, PA 15219-4785
800-766-7000

Gibco BRL
Gaithersburg, MD 20897-8406
800-828-6686

Lab Safety Supply
P. O. Box 1368
Janesville, WI 53547-1368
800-356-0783

Scientific Products
1750 Stoneridge Drive
Stone Mountain, GA 30083
800-234-5227

Sigma Chemical Company
P. O. Box 14508
St. Louis, MO 63178
800-828-2214

Thomas Scientific
P. O. Box 99
Swedesboro, NJ 08085-6099
800-345-2100

[*] Reference to a specific supplier or company does not constitute endorsement or criticism of those not listed.

*Part VII*

*Index*

# Index

## A

Abscisic acid
  laboratory requirements and, 18
  nonzygotic embryogenesis and, 143
  plant/cell history and, 5
AC. See Activated charcoal
Acclimatization, organogenesis, watermelon,
       115
Accumulation, secondary product expression,
       in vitro and, pigment, 307
Acetone, histological techniques and, 39
Activated charcoal
  cineraria, direct somatic embryogenesis, 180
  laboratory requirements and, 16
Active collections, germplasm, preservation and,
       293
Adenine, plant/cell history and, 5
Agarose
  histological techniques and, 34
  laboratory requirements and, 16
*Agrobacterium tumefaciens*
  Apiaceae, 271
  carrot, 263
    transformation protocol, 272
  chrysanthemum transformation, 275–281
  corn, 265
  electroporation, 263
  fluorimetry, 270
  GUS, 267, 274
    assay, 266
    reporter gene, 265
  kanamycin, 263, 267, 271, 273
  Klett-Summerson spectrocolorimeter, 273
  LBA4404, 263–264
  M9 medium, 262
  pBI121, 266
  plasmid, 261
  promoter-GUS fusion, 270
  reporter gene system, 261
  somatic embryogenesis, 271
  T-DNA borders, 261
  tobacco, 261–274
    regeneration medium, 266
    transformation, 265
  transgenic plants, 262
  triparental mating, 264

wounding, 268, 271, 274
X-gluc, 267, 273
*Ajuga* cell, pigment production, 311
  chemical microenvironment, 311
  growth phase, 312
  physical microenvironment, 311
Alcian blue, histological techniques and, 36
Alfalfa, nonzygotic embryogenesis and, 144
Alginic acid, orchardgrass and, 159
Aluminum
  laboratory requirements and, 15
  mounting stubs, histological techniques and,
       39
American Cancer Society, 5
Amino acids, orchardgrass and, 157
Ammonium nitrate, protoplast and, 206
Anatomy
  nonzygotic embryogenesis and, 139
  plant/cell history and, 5
Androgenesis, haploid
  culture, 241, 243
  tobacco, 249
Angiosperm
  embryology, plant/cell history and, 4
  life cycle, nonzygotic embryogenesis and, 134
  nonzygotic embryogenesis and, 133
Annual renewal, indexing for pathogens, 347
Anther culture
  haploid culture, 243, 245
  haploid tobacco, 244, 247, 250
Antioxidants, preexisting meristem propagation,
       66
Apiaceae, *Agrobacterium tumefaciens* and, 271
Apical dominance
  lilac micropropagation and, 91
  preexisting meristem propagation, 62
Apical meristems, preexisting meristem
       propagation, 61
Artificial seed, nonzygotic embryogenesis and,
       145
Asteraceae, cineraria, direct somatic embryogenesis,
       175
Asynchrony, nonzygotic embryogenesis and, 143
Autofluorescence, of chlorophyll, chrysanthemum,
       protoplast isolation and, 232
Auxin
  laboratory requirements and, 17

Sodium alginate, vitrification for cryopreservation, 302
Soilless medium, organogenesis, direct shoot and, 108
Solanaceae, petunia, organogenesis, 127
Somaclonal variation
  organogenesis, watermelon, 113
  protoplast and, 206
Somatic embryogenesis
  abnormal, nonzygotic embryogenesis and, 143
  *Agrobacterium tumefaciens* and, 271
  cineraria
    direct somatic embryogenesis, 176, 179
    receptacles, 175–181
  conifers, 191–198
  germination, peanut seed, somatic embryogenesis and, 173
  initiation, cineraria, direct somatic embryogenesis, 179
  maturation of, regeneration of plants, cineraria, direct somatic embryogenesis, 179
  melon seed, 161–166, 167–174
  orchardgrass, 149–160
  peanut seed, 167–174
  yellow poplar, 183–190
Somatic hybridizaiton, plant/cell history and, 6
Sorbitol, vitrification for cryopreservation, 298
Soybean, 321
  suspension cells, 321–322, 324
Specimen
  histological technique, for scanning electron microscopy, 38
  position, histological technique, 38
Speed, commercial laboratory production and, 339
Split-plot design, statistical analysis and, 51
Sports, organogenesis, direct shoot and, 105
Sputter coating, histological techniques and, 39
Stage
  of microspore development
    haploid culture, 242
    haploid tobacco, 249
  of pollen development, haploid tobacco, 250
Staining sections, histological techniques and, 36
Standard error of mean, statistical analysis and, 53
Statistical analysis
  benzyladenine, 52
  categorical data, 56
  Chi-square, 56
  cotyledon explants, 49
  embryogenic cultures, 54
  factorial experiments, 52
  linear regression, 54
  maize, 56
  multiple comparison, 53
  plant tissue culture, 47–57
    dry weight, tobacco callus, 53
    grape, 55
    maize, 56
    shoots per explant, 50

  sucrose effect, 55
  variance summary, 49, 50, 51
    tobacco callus, 52
  randomized design, 48, 50
  single factor CRD with subsampling, 49
  split-plot design, 51
  standard error of mean, 53
  tomato, 48
  variance, analysis of, 48
Stereomicroscope, plant cell/tissue culture, 42
Sterilization methods, 11–25
Stock
  plant preparation, commercial laboratory production and, 334
  solutions, laboratory requirements and, 21
Storage substances, intracellular, nonzygotic embryogenesis and, 141
Subculture cycles, lilac micropropagation and, 91
Sucrose
  laboratory requirements and, 15
  orchardgrass, 149, 150
  protoplast and, 206
Sugarcane, orchardgrass and, 149
Sugars, laboratory requirements and, 15
Sulfur, laboratory requirements and, 14
Sunflower tissues, plant/cell history and, 4
Suppliers, listing of, 353
Support systems, laboratory requirements and, 16
Surface disinfest
  cineraria, direct somatic embryogenesis, 176
  organogenesis, direct shoot and, 106
Suspension
  cell death assay, 326–328
  initiation, orchardgrass and, 157
  manipulation of, orchardgrass and, 157
  orchardgrass, 149–160
  soybean, 321–322, 324
  tobacco, 321–322, 324
  yellow poplar, protoplast isolation and, 237
*Syngonium* micropropagation, 73–80
  bacteria, 78
  foliage plants, 73
  indexing
    media, 79
    procedure, 78
  mixed cultures, 78
  shoot culture, 74, 75
  syngonium podophyllum, 73
Synthetic seed, orchardgrass and, 149, 150, 158
*Syringa*, lilac micropropagation and, 87

## T

T-test, for paired variates, cineraria, direct somatic embryogenesis, 177–178
Tall fescue, orchardgrass and, 149
Thiamine
  laboratory requirements and, 15
  plant/cell history and, 4